Remanufacturing and Waste Management

Remanufacturing and Waste Management

Edited by **Jude Stinton**

SYRAWOOD
PUBLISHING HOUSE

New York

Published by Syrawood Publishing House,
750 Third Avenue, 9th Floor,
New York, NY 10017, USA
www.syrawoodpublishinghouse.com

Remanufacturing and Waste Management
Edited by Jude Stinton

© 2016 Syrawood Publishing House

International Standard Book Number: 978-1-68286-049-6 (Hardback)

Printed in the United States of America.

Contents

Preface

Remanufacturing is constantly contributing towards waste management. It helps manufacturers by bringing down production cost along with reducing their carbon footprint. The aim of this book is to present researches related to life cycle analysis, product development and management, resource recovery, waste reduction, etc. It will provide new insights to readers that will transform this discipline and aid its advancement. This book, with its detailed analyses and data, will prove immensely beneficial to professionals and students involved in this area at various levels.

All of the data presented henceforth, was collaborated in the wake of recent advancements in the field. The aim of this book is to present the diversified developments from across the globe in a comprehensible manner. The opinions expressed in each chapter belong solely to the contributing authors. Their interpretations of the topics are the integral part of this book, which I have carefully compiled for a better understanding of the readers.

At the end, I would like to thank all those who dedicated their time and efforts for the successful completion of this book. I also wish to convey my gratitude towards my friends and family who supported me at every step.

Editor

Analysis of the quantities of the remanufacturing plan of perfect cost

Pedro Piñeyro[*†] and Omar Viera[†]

Abstract

Background: The remanufacturing plan of perfect cost makes reference to the remanufacturing plan of an optimal solution of the economic lot-sizing problem with remanufacturing (ELSR). In this paper, we address the problem of determining the quantities of the remanufacturing plan of perfect cost in an independent way.

Results: Assuming that the periods where remanufacturing is carried out are known in advance and certain other assumptions on the costs, we can show that the total remanufacturing quantity of a remanufacturing plan of perfect cost can be determined separately and in a time-effective way.

Conclusions: We consider that the theoretical results obtained in this paper contribute to a deeper knowledge of the characteristics of the ELSR optimal solutions. Thus, the results obtained can be used to develop an effective algorithm for solving the ELSR.

Keywords: Remanufacturing, Economic lot-sizing problem, Inventory control, Reverse logistics

Background

We consider a single item economic lot-sizing problem where the demand can also be satisfied by remanufacturing used items backed to the origin. This problem is commonly known in the literature as the economic lot-sizing problem with remanufacturing (ELSR) and refers to the problem of determining the quantities to produce, remanufacture, and dispose in each period over a finite planning horizon in order to meet the demand requirements of a single item on time, minimizing the involved costs. Used products returned by the customers are available at each period for remanufacturing. In addition, we consider that the returns can be disposed off, e.g., when there is an overstock of used products. This kind of problem has been receiving an increasing academic attention in recent years as the industry has been involved with the recovery of used products. This has been the result of governmental and social pressures as well as economic opportunities. Remanufacturing can be defined as the recovery process of returned products after which it is warranted that the remanufactured products offer the same quality and functionality than those newly manufactured [1]. Remanufacturing tasks often involve disassembly, cleaning, testing, part replacement, and reassembling operations. Products that are remanufactured include automotive parts, engines, tires, aviation equipment, cameras, medical instruments, furniture, toner cartridges, copiers, computers, and telecommunication equipment. Among the recovery options, the remanufacturing offers benefits for all of the parties involved. We refer the readers to de Brito and Dekker [2], Guide [3], Gungor and Gupta [4], and Hormozi [5] for the detailed descriptions about the remanufacturing benefits.

This paper is focused on the analysis of the quantities of the remanufacturing plan of an optimal solution of the ELSR that we refer as the remanufacturing plan of perfect cost. The remanufacturing plan plays a key role in the ELSR resolution since both the optimal production plan and the optimal final disposal plan can be determined separately and in an effective time way if the remanufacturing plan is known [6]. Thus, we can say that solving the ELSR reduces to the problem of finding the remanufacturing plan of perfect cost, i.e., the remanufacturing plan of an optimal solution of the ELSR. We note that the problem of finding a remanufacturing plan of perfect cost is NP-hard, since it is equivalent to the

* Correspondence: ppineyro@fing.edu.uy
†Equal contributors
Departamento de Investigación Operativa, Instituto de Computación, Facultad de Ingeniería, Universidad de la República, Julio Herrera y Reissig 565, Montevideo, CP 11300, Uruguay

ELSR which is a known NP-hard problem even under stationary cost structures [7-9]. Considering this difficulty, we tackle the problem of determining the quantities of a remanufacturing plan of perfect cost under the assumption that the periods where the remanufacturing is carried out are known in advance. This can occur in practice if cores, parts, machinery, or workers are only available in certain periods within the planning horizon. We also assume certain constrains on the costs that can be fulfilled in the real life, such as non-speculative motives or that the costs related to used items are at most equal to those related to new items. In addition, we provide a constraint on the costs which makes it more profitable to maximize the remanufacturing quantity in those periods where remanufacturing is allowed. This can be fulfilled in practice if the unit cost of producing is much greater than other unit costs of the problem, or in those cases for which the inventory holding costs can be neglected.

The rest of the paper is organized as follows: the 'Background' section finishes with a short literature review, followed by the problem formulation. The 'Results' section is devoted to the analysis of the quantities of the remanufacturing plan of perfect cost with fixed periods for remanufacturing. In the 'Discussion' section, we present a numerical example along with a discussion about the effectiveness of the theoretical results obtained. The 'Conclusions' section concludes the paper with possible directions for future research.

Literature review

According to our best knowledge, Richter and Sombrutzki [10] and Richter and Weber [11] are the first to analyze the ELSR. They consider the particular case that the returns in the first period are sufficient to satisfy the total demand over the planning horizon. Golany et al. [7] suggest a network flow formulation for the ELSR and provide an exact algorithm of $O(T^3)$ time for the case of linear cost functions. They also show that the ELSR is a NP-hard problem for the case of general concave cost functions. Yang et al. [8] show the same result of complexity for the case of stationary concave cost functions and provide a heuristic procedure of $O(T^4)$ time for the ELSR. van den Heuvel [9] shows that ELSR is NP-hard for the case of the setup and unit costs for the activities and unit costs for holding inventory, even in the case that they are stationary. Teunter et al. [12] consider the ELSR with joint setup costs for the production and remanufacturing activities and suggest an $O(T^4)$ time algorithm based on a dynamic programming approach. Also, several heuristic for the problem are provided. Piñeyro and Viera [6] suggest and compare several inventory policies for the ELSR using a divide-and-conquer approach and a tabu search based on

procedure. They also show the key role that the remanufacturing plan plays in the ELSR resolution and introduce the concept of the remanufacturing plan of perfect cost. Piñeyro and Viera [13] consider the ELSR with different demand streams for new and remanufactured items where, in addition, substitution is allowed for remanufactured items, but not vice versa. Recently, Nenes et al. [14] provide an analysis of the ELSR taking into account the quality of the returns, and Helmrich et al. [15] provide and compare different mathematical formulations for the ELSR with separate and joint setup costs for the activities.

Problem formulation

Figure 1 below shows a sketch of the flow of items for the inventory system that represents the lot-sizing problem that we are facing.

We consider a lot-sizing problem for which the demand and return values are known in advance for each period over the finite planning horizon. The demand for serviceable items must be satisfied on time, i.e., backlogging demand is not allowed. Infinite capacity for producing, remanufacturing, and disposing is assumed with zero lead times. The inventory level of both used and serviceable items is determined after all activities were carried out. Setup and unit costs are incurred for producing, remanufacturing, or disposing, and unit costs for carrying ending positive inventory from one period to the next. Finally, we assume that the initial inventory levels of both used and serviceable items are zero, and the demand is positive for each period in the planning horizon. The objective is to determine the amounts to produce, remanufacture, and dispose for each one of the periods in the planning horizon such that all demand requirements are satisfied on time, and the sum of all the involved costs is minimized. We refer to this problem as the ELSR, and it can be modeled as the following mixed integer linear programming (MILP) problem:

$$\min \sum_{t=1}^{T} \left\{ K_t^p \delta_t^p + c_t^p p_t + K_t^r \delta_t^r + c_t^r r_t + K_t^d \delta_t^d + c_t^d d_t + h_t^s y_t^s + h_t^u y_t^u \right\}$$

$$(1)$$

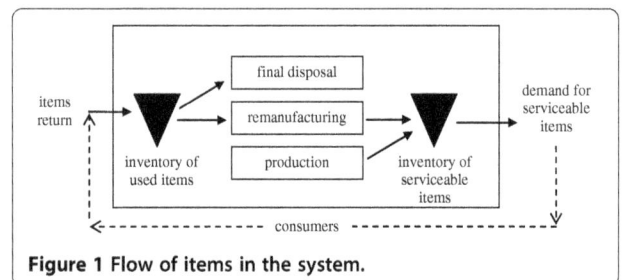

Figure 1 Flow of items in the system.

subject to

$$y_t^s = y_{t-1}^s + p_t + r_t - D_t \qquad \forall t = 1, 2, \ldots, T \qquad (2)$$

$$y_t^u = y_{t-1}^u - r_t + R_t \qquad \forall t = 1, 2, \ldots, T \qquad (3)$$

$$M\delta_t^p \geq p_t \qquad \forall t = 1, 2, \ldots, T \qquad (4)$$

$$M\delta_t^r \geq r_t \qquad \forall t = 1, 2, \ldots, T \qquad (5)$$

$$M\delta_t^d \geq d_t \qquad \forall t = 1, 2, \ldots, T \qquad (6)$$

$$y_0^s = y_0^u = 0 \qquad (7)$$

$$\delta_t^p, \delta_t^r, \delta_t^d \in \{0, 1\} \quad p_t, r_t, d_t, y_t^s, y_t^u \geq 0 \quad \forall t = 1, 2, \ldots, T. \qquad (8)$$

In models (1) to (8), the parameters T, D_t, and R_t denote the length of the planning horizon, demand, and return values in periods $t = 1, \ldots, T$ respectively: K_t^p, K_t^r, K_t^d, c_t^p, c_t^r and c_t^d the setup and unit costs for production, remanufacturing, and final disposing in periods $t = 1, \ldots, T$, respectively; h_t^s and h_t^u, the unit cost of holding inventory for serviceable and used products in periods $t = 1, \ldots, T$, respectively; and M, a number at least as large as $\max\{D_{1T}, R_{1T}\}$, where D_{ij} and R_{ij} are the accumulative demand and returns between periods i and j, with $1 \leq i \leq j \leq T$. The decision variables p_t, r_t, and d_t denote the number of units produced, remanufactured, and disposed in periods $t = 1, \ldots, T$, respectively; δ_t^p, δ_t^r and δ_t^d, binary variables equal to 1 if production, remanufacturing, or disposing is carried out in periods $t = 1, \ldots, T$, or 0 otherwise, respectively; and y_t^s and y_t^u, the inventory level during periods $t = 1, \ldots, T$, for serviceable and used items, respectively.

Constraints (2) and (3) are the inventory equilibrium equations for serviceable and used items, respectively. Constraints (4) to (6) indicate that a setup is made whenever an activity is carried out in a period for a positive quantity. Constraint (7) states that the initial inventory level for both serviceable and used items is zero. Finally, the set of possible values for each decision variable is specified by constraint (8).

The ELSR as modeled above is a NP-hard problem [9]. As we mentioned earlier, solving the ELSR is equivalent to find a remanufacturing plan of perfect cost, i.e., the remanufacturing plan of an optimal solution of the ELSR. In the following section, we analyze this last problem assuming that the periods for which the quantity of remanufacturing is positive are known in advance.

Results
Fixed periods for remanufacturing
In this section, we tackle the problem of determining the quantities of the remanufacturing plan of perfect cost

under the assumption that the periods with positive remanufacturing (periods for which the quantity of remanufacturing is greater than zero) are known in advance. We begin considering the particular case of only one positive-remanufacturing period and then we consider the case of more than one period. To conduct the analysis, we resort to certain assumptions on the costs as well as on the number of the available returns in the periods fixed. The first assumption that we introduce below is about the costs related to the used items.

Definition 1. We say that the costs of the returns are at most equal to the costs of the new items when the expressions below are fulfilled by the cost components:

$$K_i^r \leq K_j^p, \qquad (9.1)$$

$$c_i^r \leq c_j^p, \qquad (9.2)$$

$$h_i^u \leq h_j^s, \qquad (9.3)$$

for any couple of periods i and j in $1, \ldots, T$.

Expressions (9.1) and (9.2) state the fact that the remanufacturing of used items is economically preferred to the production of new items. This can happen in practice due to the saving of energy and raw material of the remanufacturing activity. Expression (9.3) is fulfilled as it is assumed that the value is added to the used items in order to make them serviceable. In addition, we assume that the setup costs are at least equal to the unit costs for each activity, i.e, $K_t^p \geq c_t^p \geq 0$, $K_t^r \geq c_t^r \geq 0$, and $K_t^d \geq c_t^d \geq 0$, for each period $t = 1, \ldots, T$.

The single-period case
Consider an ELSR instance of T periods with only one period i fixed as positive-remanufacturing period, i.e., $r_i > 0$, with $1 \leq i \leq T$, and $r_t = 0$ for all t with $1 \leq i \leq T$ and $t \neq i$. The objective is to determine the optimal remanufacturing quantity Q_i^r of the period i, with $0 < Q_i^r \leq y_{i-1}^u + R_i$ and $y_{i-1}^u + R_i > 0$.

First, consider the case that the number of available returns in period i are at most equal to the demand of the period, i.e., $y_{i-1}^u + R_i \leq D_i$. Then, by (9), the optimal remanufacturing quantity must be equal to all of the available returns, i.e., $Q_i^r = y_{i-1}^u + R_i > 0$. On the other hand, for the case that $y_{i-1}^u + R_i > D_i$, we must determine the last period j within the planning horizon for which it is more profitable to meet at least one unit of its demand by remanufacturing in period i. Assume first that the number of available returns is sufficient to exactly meet the accumulative demand from the current period i to certain future period k, i.e., $y_{i-1}^u + R_i = D_{ik}$, with $1 \leq i \leq k \leq T$. Then, the optimal remanufacturing quantity of period i is $Q_i^r = D_{ij}$, with j as the last period for which

$c_i^r D_j + \sum_{t=i}^{j-1} h_t^s D_j \leq K_j^p + c_j^p D_j + \sum_{t=i}^{T} h_t^u D_j$ is fulfilled, with

$1 \leq i \leq j \leq k \leq T$ and $D_j > 0$. However, we note that in general the number of available returns in period i is sufficient to meet only a portion of the demand of a certain future period k, i.e., $y_{i-1}^u + R_i = D_{i(k-1)} + \alpha < D_{ik}$, with $1 \leq i \leq k \leq T$ and $\alpha \geq 1$. Without loss of generality, let us assume that it is profitable to remanufacture in period i at least the needed quantity to cover the demand requirements from i to $(k-1)$, i.e., $D_{i(k-1)} \leq Q_i^r < D_{ik}$. This means that at least one unit of the demand requirement of period k is satisfied by means of the production of new items in a certain period t with $1 \leq t \leq k$. If $1 \leq t \leq i$, then there must be that $Q_i^r = D_{i(k-1)} + \alpha$, since $c_i^r \leq c_t^p + \sum_{\tau=t}^{i-1} h_\tau^s$ is true by (9). In the case of $i < t \leq k$, we have that $Q_i^r = D_{i(k-1)} + \alpha$ only if the condition $c_i^r + \sum_{t=i}^{t-1} h_t^s \leq c_t^p$ is fulfilled, otherwise $Q_i^r = D_{i(k-1)}$. This last condition can be relaxed by $c_i^r + \sum_{t=i}^{j-1} h_t^s \leq c_j^p + \sum_{t=i}^{T} h_t^u$ in the case that final disposal of used items is not considered, which is supported by economic as well as ecological reasons. Teunter et al. [12] point out that disposing option 'does not lead to a considerable cost reduction unless the remanufacturable return rate as a percentage of the demand rate is unrealistically high (above 90%) and the demand rate is very small (less than 10 per year)'. We resume the reasoning above by means of the following assumption about the profitability of maximizing the remanufacturing quantity in a certain period.

Definition 2. Given two periods i and k of an ELSR instance of T periods, with $1 \leq i \leq k \leq T$, such that $r_i > 0$, we say that it is profitable to maximize the remanufacturing quantity of period i if the expression

$$c_i^r + \sum_{t=i}^{j-1} h_t^s \leq c_j^p \qquad (10.1)$$

is fulfilled for each period j, with $1 \leq i \leq j \leq k \leq T$, or

$$c_i^r + \sum_{t=i}^{j-1} h_t^s \leq c_j^p + \sum_{t=i}^{T} h_t^u \qquad (10.2)$$

in the case that the final disposal of used items is not considered.

Thus, if Definition 2 is fulfilled for any couple of periods i and j in $1,\ldots,T$, with $i \leq j$, we can assure that the optimal remanufacturing quantity Q_i^r f a single period i fixed as positive-remanufacturing period is the minimum between the amount of available returns and the accumulative demand from the current period and the end of the planning horizon, i.e., to remanufacture as much as possible. We note that for a given instance, it is sufficient that Definition 2 is fulfilled between the period fixed as positive-remanufacturing period and the last one for which at least a portion of its demand is attainable by remanufacturing in the period fixed. On the other hand, if Definition 2 is not fulfilled, it is unlikely that we can determine the optimal remanufacturing quantity of a certain period without knowing the periods where production is carried out since, in the case that the available returns in period i are only sufficient to partially meet the accumulative demand to certain future period k, we need to know if the rest of the demand of period k is produced either in the same period or in a previous one.

Real situations where Definition 2 is fulfilled include cases where holding costs of both used and serviceable items are similar or negligible, very low remanufacturing costs, as well as instances with few periods. We also note that the problem of finding the optimal positive-remanufacturing period for an ELSR instance for which it is profitable to remanufacture as much as possible at any period can be solved in $O(T^3)$ time since we must consider T different periods, and the corresponding optimal production and final dispose plans can be obtained in $O(T^2)$ by means of a Wagner-Whitin algorithm type [16].

The multi-period case

We now consider the problem of finding the remanufacturing quantities of a remanufacturing plan of perfect cost with at least two periods fixed as positive-remanufacturing periods. We first note that the amount to be remanufactured in a certain period depends in part of the remanufactured quantity in previous periods as well as affects the amount to be remanufactured in future periods. Then, it may not be possible to determine efficiently the optimal remanufacturing quantity for each period, even under the assumptions introduced in the previous section. In view of this difficulty, we focus on the problem of determining the total quantity of a remanufacturing plan of perfect cost. Before we tackle this problem, we provide a result about the form of the remanufacturing plan of perfect cost for a particular case.

Proposition 1. Consider an ELSR instance for which the number of available returns in a certain period i fixed as a positive-remanufacturing period is sufficient to fully cover the demand until the end of the planning horizon, i.e., $R_i + y_{i-1}^u \geq D_{iT}$, $r_i > 0$, with $1 \leq i \leq T$. If the optimal solution set is not empty, there is at least one optimal solution for which the total remaining demand from period i is satisfied only by remanufacturing from period i onwards, i.e., $r_{iT} = D_{iT}$, with $r_{ij} = \sum_{t=i}^{j} r_t$, $1 \leq i \leq j \leq T$.

Proof. Let us consider an optimal solution of the ELSR with $r_i > 0$, $R_i + y_{i-1}^u \geq D_{iT}$, and $r_{iT} < D_{iT}$. Then, the quantity $(D_{iT} - r_{iT}) > 0$ is satisfied by means of the production of new items. We can determine a new solution with $r_{iT} = D_{iT}$ from the current solution as follows: First, for each period t with $i \leq t \leq T$ and $p_t > 0$, we replace the entire production in t by remanufacturing, i.e., $r_t \leftarrow p_t$, $p_t \leftarrow 0$. Note that the replacement operation is possible as we are assuming the returns are sufficient. Second, while $r_{iT} < D_{iT}$ take the last period t with $p_t > 0$ and $1 \leq t < i$, and transfer units of the production of period t to the remanufacturing of period i, until $r_{iT} = D_{iT}$ or $p_t = 0$. By (9), the cost of the new solution is at most equal to the cost of the original. Therefore, there must be an optimal solution of the ELSR for which $r_{iT} = D_{iT}$ if $r_i > 0$ and $R_i + y_{i-1}^u \geq D_{iT}$ is are complied.

Proposition 1 helps us to identify the form of a remanufacturing plan of perfect cost for the ELSR in the particular case that the number of available returns in a period fixed as positive-remanufacturing period is sufficient to meet all the remaining demand until the end of the planning horizon. We must note that if the amount of available returns in a certain period is sufficient to meet all the remaining demand but the period is not fixed as a positive-remanufacturing period, we cannot ensure the result above unless the period under consideration is the first one (see [10]).

We consider now the problem in general sense, i.e., no kind of relationship is assumed between the returns and the demand values. First, we provide the following definitions about the costs and the quantities of remanufacturing.

Definition 3. We say that the remanufacturing costs are non-speculative with respect to the transfer when they satisfy the following expressions:

$$K_i^r + c_i^r + \sum_{t=i}^{j-1} h_t^s - \sum_{t=i}^{j-1} h_t^u \geq K_j^r + c_j^r + \sum_{t=i}^{j-1} h_t^u \tag{11.1}$$

$$c_i^r + \sum_{t=i}^{j-1} h_t^s - \sum_{t=i}^{j-1} h_t^u \geq c_j^r + \sum_{t=i}^{j-1} h_t^u \tag{11.2}$$

for any couple of period i and j in $1, \ldots, T$.

Expression (11.1) states that it is profitable to transfer the entire remanufacturing quantity from a certain period to other future period that was inactive, while (11.2) states that it is profitable to transfer forward at least one unit between two periods with positive remanufacturing. We note that the expressions given in (11) are fulfilled in different settings of practical interest, e.g., when all the costs involved are stationary or they do not increase over time.

Definition 4. Given an ELSR instance with a set of periods fixed as positive remanufacturing periods and a feasible remanufacturing plan r, we define the *upper bound of remanufacturing* of a certain period i to the quantity $u_i = 0$ if $r_i = 0$ and $u_i = \min(R_i + y_i^u, D_{i(j-1)})$ if $r_i > 0$, where j is either the next positive-remanufacturing period within the planning horizon, or $(T + 1)$ if i is the last positive-remanufacturing period, i.e., $r_t = 0$ for all periods t in $(i + 1), \ldots, T$.

Proposition 2. Given an ELSR instance, there is at least one optimal solution for which the remanufacturing quantity of each period is at most equal to its upper bound of remanufacturing, i.e., $0 \leq r_t \leq u_t$, for all periods $t = 1, \ldots, T$.

Proof. Without loss of generality, consider an optimal solution of an ELSR instance with only one period i for which $r_i > u_i = \min(R_i + y_i^u, D_{i(j-1)})$ and $r_j > 0$ with $1 \leq i \leq j \leq T$. First, we note that the case $u_i = R_i + y_i^u$ is not feasible since the remanufacturing quantity is greater than the amount of available returns. Now, consider the case that $u_i = D_{i(j-1)}$. Then, by (11), we can obtain a new solution with at most the same cost than the original by transferring remanufactured units from period i to the consecutive period j with $r_j > 0$, until $r_i = D_{i(j-1)}$ in the new solution. Therefore, an optimal solution for the same ELSR instance for which $r_t \leq u_t$ can be obtained, for all periods $t = 1, \ldots, T$.

Proposition 2 states that the remanufacturing quantity of a certain period is upper-bounded by the minimum between the number of available returns and the accumulative demand until the period preceding the next period with positive remanufacturing. We note that the upper bound value of certain period depends on the remanufacturing quantities of the previous periods. In addition, it may not be possible to determine how close or how far to its upper bound is the remanufacturing quantity of a certain period in an optimal solution of the ELSR. Despite these facts, the upper bound of remanufacturing allows us to determine the total remanufacturing quantity of a remanufacturing plan of perfect cost, as we show in the following proposition:

Proposition 3. Consider an ELSR instance with a set of periods F fixed as positive-remanufacturing periods such as for any pair of consecutive periods i and j of F, the Definition 2 is fulfilled for any pair of meaningful periods, i.e., pairs (i, t) with $i \in F$ and t the last period before j for which at least a portion of its demand is attainable by remanufacturing in i with $i \leq t < j$. Then, consider the remanufacturing plan $\bar{r}$ obtained by remanufacturing in each period the amount given by the upper bound of remanufacturing applied in ascending order, i.e., $\bar{r}_t = u_b$ assuming that $\bar{r}_1 = u_1, \bar{r}_2 = u_2, \ldots, r_{(t-1)} = u_{(t-1)}$, for all periods $t = 1, \ldots, T$. Then, there is an optimal solution with a remanufacturing plan r^* for which $r^*_{1T} = \bar{r}_{1T}$

where $r^*_{ij} = \sum_{t=i}^{j} r^*_t$ and $\bar{r}_{ij} = \sum_{t=i}^{j} \bar{r}_t$, with $1 \leq i \leq j \leq T$.

Proof. We note that by Proposition 2 and Definition 4, there must be that $r^*_{1T} \leq \bar{r}_{1T}$. Without loss of generality, let us assume that $r^*_{1T} = \bar{r}_{1T} - 1$. Then, there exists a period i, with $1 \leq i \leq T$, for which $0 < r^*_i = \bar{r}_i - 1$, $r^*_{1(i-1)} = r_{1(i-1)}$, and $y^{*u}_{(i-1)} = y^{u}_{(i-1)}$. This means that the upper bound of remanufacturing of period i is the same for both remanufacturing plans under consideration, with $0 < r^*_i < u_i = \bar{r}_i$. We also note that $y^{*u}_t \geq 1$ is fulfilled for all periods $t = i, \ldots, T$. Therefore, we can obtain a new feasible solution for the same ELSR instance with at most the same cost by increasing the remanufacturing in period i in one unit, i.e., $r^*_i \leftarrow r^*_i + 1 = \bar{r}_i$, without affecting the remanufacturing of the future periods and in the meantime by reducing the production of a certain period j in $1, \ldots, T$. This new solution fulfills that $r^*_{1T} = \bar{r}_{1T}$, and this cost is at most the same than the cost of the original optimal solution as we are assuming that maximizing the remanufacturing quantity of the periods with positive remanufacturing is profitable according to Definition 2.

Proposition 3 states that in order to determine a remanufacturing plan of perfect cost for an ELSR instance with certain periods fixed as positive-remanufacturing periods, we only need to explore those remanufacturing plans for which the total remanufacturing quantity is equal to the sum of the upper bounds of remanufacturing. These values can be determined efficiently (linear time) by applying Definition 4 period by period, beginning with the first period fixed as positive-remanufacturing period. We show the usefulness of Proposition 3 through the following numeric example.

Discussion
A numerical example
Consider an ELSR instance with $T = 5$, a demand vector $D = (5,3,6,4,5)$ and a return vector $R =$

Table 1 Candidate remanufacturing plans

t	D	R	r				
1	5	3	0	0	0	0	*0*
2	3	2	5	5	5	4	*3*
3	6	2	0	0	0	0	*0*
4	4	2	4	3	2	3	*4*
5	5	3	3	4	5	5	*5*

$(3,2,2,2,3)$, where the periods 2, 4, and 5 are fixed as positive remanufacturing periods. The cost values are as follows: $K^p_t = 200$, $c^p_t = 20$, $K^r_t = 150$, $c^r_t = 15$, $K^d_t = 100$, $c^d_t = 10$, $h^s_t = 5$, and $h^u_t = 2$, with $1 \leq t \leq 5$. Note that the remanufacturing is profitable according to Definition 2 for all the meaningful pair of periods, i.e., $(2,3)$, $(4,4)$, and $(5,5)$. Applying Definition 4, we have that the total remanufacturing quantity is 12 since the upper bounds of remanufacturing obtained sequentially are $u = (0, 5, 0, 4, 3)$. Table 1 below provides the candidate remanufacturing plans that we must consider in order to determine the remanufacturing plan of perfect cost for the ELSR instance.

These candidate plans were obtained by assigning to each period the maximum quantity according to its upper bound and then transferring unit by unit from period 4 to period 5, and from period 2 to period 4. The last column of Table 1 in italics corresponds to the remanufacturing plan of perfect cost. The corresponding production and final dispose plans of the optimal solution are $p = (11, 0, 0, 0)$ and $d = (0, 0, 0, 0)$, respectively.

Effectiveness of the upper bound of remanufacturing
In [6], a basic tabu search (BTS) based on procedure was suggested and evaluated for the ELSR. The procedure receives among other parameters, an initial $(0,1)$ T-tuple, where a value of 1 in position t indicates that remanufacturing is allowed to be positive in period t; otherwise, it must be zero. The procedure explores different remanufacturing plans by means of swapping the periods where remanufacturing can be positive. The remanufacturing quantity of each period i fixed as positive-remanufacturing period is equal to the minimum between the number of available returns in i and the accumulative demand from i to the period preceding the next period j with positive remanufacturing, i.e., the upper bound of remanufacturing of Definition 4. The BTS procedure was tested for a wide range of return-demand relationships, cost settings, and planning horizon lengths of 5, 10, and 15 periods. For all of the tested cases, the BTS showed a very good behavior (less than 2% of average gap between the cost of the solution obtained from BTS and the cost of the optimal solution), finding in many instances the optimal solution.

The good performance observed for the BTS procedure can be explained in part by the theoretical results provided in this paper about the quantities of the remanufacturing plan of perfect cost, at least for those cases where the conditions of Definitions 1 to 3 are fulfilled. In this sense, we note that for the numeric experiments of the BTS procedure of Piñeyro and Viera [6], it is assumed that the costs of the returns are at most equal to the costs of the new items according to expression (9) of Definition 1. In addition, horizon planning lengths of 5, 10, and 15 are used; thus, it can be assumed that the conditions of Definitions 2 and 3 are fulfilled in many of the tested instances. On the other hand, for those cases where the conditions are not fulfilled, it may be that the upper bound of remanufacturing is not a good option which in turn explains why the BTS procedure is not able to achieve high-quality solutions for some of the tested instances, e.g., when the positive-remanufacturing periods are widely separated or the holding costs of serviceable items are relatively greater.

Conclusions

In this paper, we have addressed the problem of determining the quantities of the remanufacturing plan of perfect cost for the ELSR assuming that the periods where remanufacturing is carried out are known in advance and that it is profitable to remanufacture as much as possible in a period fixed as positive remanufacturing period. Thus, we are able to determine the optimal remanufacturing quantity for the particular case of only one period fixed as positive-remanufacturing period. We also note that the problem of finding the optimal period for remanufacturing can be solved in $O(T^3)$ time. For the general case of more than one period fixed as positive-remanufacturing period, we note that it may not be possible to determine the optimal remanufacturing quantity for each one of them in an effective time way. Nevertheless, we show that the total remanufacturing quantity of an optimal solution can be determined as the sum of the upper bounds of remanufacturing, assuming also that the remanufacturing costs are non-speculative respect to the transfer, i.e., remanufacturing occurs as late as possible. The upper bounds of remanufacturing can be computed period by period in a linear time way as the minimum between the number of available returns and the accumulative demand from the current period to the period preceding the next period with positive remanufacturing. The theoretical results obtained about the quantities of a remanufacturing plan of perfect cost serve to explain the effectiveness of the tabu search based on procedure suggested in [6] for the ELSR.

Future research

More attention should be placed in the future on the problem of determining the quantities of the plan of perfect cost in an independent way by relaxing some of the assumptions imposed in this paper. More specifically, in identifying situations in which it is desirable to maximize the remanufacturing, even the condition of Definition 2 is not fulfilled. In addition, the problem of determining the periods with positive remanufacturing should be tackled. In this sense, we can resort to the useful remanufacturing problem (URP) introduced in [6]. The URP refers to the problem of determining the useful remanufacturing plan that minimizes the involved costs and maximizes the use of the returns. Then, we can assume that the positive periods of a useful remanufacturing plan are close to the positive periods of a remanufacturing plan of perfect cost. We may include also different demand streams for new and remanufactured items, as in [13].

Methods

This research provides theoretical contributions for solving the ELSR which are based on the study of the mathematical properties of the problem under certain particular assumptions.

Competing interests

The authors declare that they have no competing interests.

Authors' contributions

PP carried out the research presented in this paper as part of his Ph.D. OV supervised the research and corrected the draft versions of the paper. All authors read and approved the final manuscript.

Authors' information

PP is a Ph.D. student and an assistant professor at Universidad de la República, Uruguay. OV is a titular professor at Universidad de la República, Uruguay.

Acknowledgements

This work was supported by PEDECIBA, Uruguay. The authors thank the anonymous referees for their suggestions.

References

1. Ijomah, W: A model-based definition of the generic remanufacturing business process. (2002)
2. de Brito, MP, Dekker, R: Reverse logistics – a framework. Econometric Institute Report EI 2002-38. Erasmus University Rotterdam, Netherlands (2002)
3. Guide Jr, VDR: Production planning and control for remanufacturing: industry practice and research needs. J. Oper. Manag. 18, 467–483 (2000)
4. Gungor, A, Gupta, SM: Issues in environmentally conscious manufacturing and product recovery: a survey. Comput. Ind. Eng. 36, 811–853 (1999)
5. Hormozi, AM: The art and science of remanufacturing: an in-depth study. In: (ed.) 34th Annual Meeting of the Decision Sciences Institute., Washington D. C (2003). 22–25 November 2003
6. Piñeyro, P, Viera, O: Inventory policies for the economic lot-sizing problem with remanufacturing and final disposal options. Journal of Industrial and Management Optimization 5, 217–238 (2009)

7.   Golany, B, Yang, J, Yu, G: Economic lot-sizing with remanufacturing options.
     IIE Trans. **33**, 995–1003 (2001)
8.   Yang, J, Golany, B, Yu, G: A concave-cost production planning problem with
     remanufacturing options. Nav. Res. Logist. **52**, 443–458 (2005)
9.   van den Heuvel, W: On the complexity of the economic lot-sizing problem
     with remanufacturing options. Econometric Institute Report EI 2004–46.
     Erasmus University Rotterdam, Netherlands (2004)
10. Richter, K, Sombrutzki, M: Remanufacturing planning for the reverse
     Wagner/Whitin models. Eur. J. Oper. Res. **121**, 304–315 (2000)
11. Richter, K, Weber, J: The reverse Wagner/Whitin model with variable
     manufacturing and remanufacturing cost. Int. J. Prod. Econ. **71**, 447–456
     (2001)
12. Teunter, R, Bayındır, Z, van den Heuvel, W: Dynamic lot sizing with product
     returns and remanufacturing. Int. J. Prod. Res. **44**, 4377–4400 (2006)
13. Piñeyro, P, Viera, O: The economic lot-sizing problem with remanufacturing
     and one-way substitution. Int. J. Prod. Econ. **124**, 482–488 (2010)
14. Nenes, G, Panagiotidou, S, Dekker, R: Inventory control policies for
     inspection and remanufacturing of returns: a case study. Int. J. Prod. Econ.
     125, 300–312 (2010)
15. Helmrich, M, Jans, R, van den Heuvel, W, Wagelmans, APM: Economic
     lot-sizing with remanufacturing: complexity and efficient formulations.
     Econometric Institute Report EI 2010–71. Erasmus University Rotterdam,
     Netherlands (2010)
16. Wagner, HM, Whitin, TM: Dynamic version of the economic lot size model.
     Manag. Sci. **5**, 89–96 (1958)

2

Life cycle approach to sustainability assessment: a case study of remanufactured alternators

Erwin M. Schau[*], Marzia Traverso and Matthias Finkbeiner

Abstract

Sustainability is an international issue with increasing concern and becomes a crucial driver for the industry in international competition. Sustainability encompasses the three dimensions: environment, society and economy. This paper presents the results from a sustainability assessment of a product. To prevent burden shifting, the whole life cycle of the products is necessary to be taken into account. For the environmental dimension, life cycle assessment (LCA) has been practiced for nearly 40 years and is the only one standardised by the International Organization for Standardization (ISO) (14040 and 14044). Life cycle approaches for the social and economic dimensions are currently under development. Life cycle sustainability assessment (LCSA) is a complementary implementation of the three techniques: LCA (environmental), life cycle costing (LCC - economic) and social LCA (SLCA - social). This contribution applies the state-of-the-art LCSA on remanufacturing of alternators aiming at supporting managers and product developers in their decision-making to design product and plant. The alternator is the electricity generator in the automobile vehicle which produces the needed electricity. LCA and LCC are used to assess three different alternator design scenarios (namely conventional, lightweight and ultra-lightweight). The LCA and LCC results show that the conventional alternator is the most promising one. LCSA of three different locations (Germany, India and Sierra Leone) for setting the remanufacturing mini-factory, a worldwide applicable container, are investigated on all three different sustainability dimensions: LCA, LCC and SLCA. The location choice is determined by the SLCA and the design alternatives by the LCA and LCC. The case study results show that remanufacturing potentially causes about 12% of the emissions and costs compared to producing new parts. The conventional alternator with housing of iron cast performs better in LCA and LCC than the lightweight alternatives with aluminium housing. The optimal location of remanufacturing is dependent on where the used alternators are sourced and where the remanufactured alternators are going to be used. Important measures to improve the sustainability of the remanufacturing process in life cycle perspective are to confirm if the energy efficiency of the remanufactured part is better than the new part, as the use phase dominates from an environmental and economical point of view. The SLCA should be developed further, focusing on the suitable indicators and conducting further case studies including the whole life cycle.

Keywords: Life cycle sustainability assessment, Life cycle assessment, Life cycle costing, Social LCA, Remanufacturing, Alternator, Automotive parts, Germany, India, Sierra Leone

* Correspondence: erwin.m.schau@gmail.com
Department of Environmental Technology, Chair of Sustainable Engineering, Technische Universitaet Berlin, Office Z1, Strasse des 17. Juni 135, Berlin D-10623, Germany

Background

The alternator is the automotive part with the highest remanufacturing rate [1]. The function of the alternator is to deliver electrical energy to charge the battery and to the on board equipments like light [2].

Remanufacturing can play an important role as a way to close the material cycles and thereby contribute to less material and energy use [1,3-7], which are the important steps to realise a sustainable development.

However, in a life-cycle perspective, not only the production or remanufacturing phase but also the use stage is needed to be taken into account. In the use stage of engines and generators, energy use, associated emissions and costs are of high concern [8]. Up to now, few studies have looked at the whole life cycle of automotive parts that requires energy in the use phase including remanufacturing of the used parts [9].

Sustainability encompasses not only the environmental dimension, but also social and economic ones, as it is defined by the Brundtland Commission [10]. Consequently, a methodology to measure sustainability is getting extremely important. The measurement of the environmental dimension of sustainability is the most mature method of the three.

LCA is a standardised method [11,12] widely used to investigate the potential environmental impacts of products and services through the whole life cycle from cradle to grave [13,14]. The life cycle approach helps to avoid shifting of burden from one phase to another.

Life cycle costing (LCC) is proposed for the assessment of the economic dimension of sustainability. LCCs have been used since the 1930s [15]; however, it is a relatively new tool within sustainability assessment. The Society of Environmental Toxicology and Chemistry (SETAC) working group on LCC [16] classifies three types of LCC - conventional, environmental and societal LCC and considers the method of environmental LCC [16] currently as the most suitable for combining with LCA [15,17,18].

Social life cycle assessment (SLCA) is the life cycle tool to assess the potential social and socio-economic impacts of the products and their consumption throughout their life cycles [19].

To combine LCA, environmental life cycle costing (LCC) [16,20] and SLCA [19], a methodology called Life Cycle Sustainability Assessment (LCSA) [21] has been suggested and can be formally expressed in the symbolic equation [22-25]:

$$LCSA = LCA + LCC + SLCA, \qquad (1)$$

where

1. LCSA = Life cycle sustainability assessment,
2. LCA = Environmental life cycle assessment,
3. LCC = Environmental life cycle costing and
4. SLCA = Social life cycle assessment

Based on the well-known depiction of sustainability, where the three dimensions of environment, economy and society intersect, as depicted in Figure 1a, the LCSA can be illustrated synchronously as previously described (Figure 1b).

Similar to the LCA method, environmental LCC and SLCA are life cycle approaches which have been proven useful to prevent shifting of burden from one process to another in the product life cycle [23]. Despite the long history of conventional life cycle costing, the environmental LCC is a relatively new method in a sustainability context [20]. SLCA is still in its infancy, where one of the current focuses is developing the indicators to be used [22].

This paper presents the results from a multidisciplinary research project applying LCSA on different scenarios for remanufactured alternators - three different countries and three different alternator designs are investigated - and thereby lead contribution to the development of the LCSA methodology. The whole life cycle is considered (for the LCA and the environmental LCC), but the main focus in this paper is on the remanufacturing process. Thereby, the measurements to improve the sustainability of the

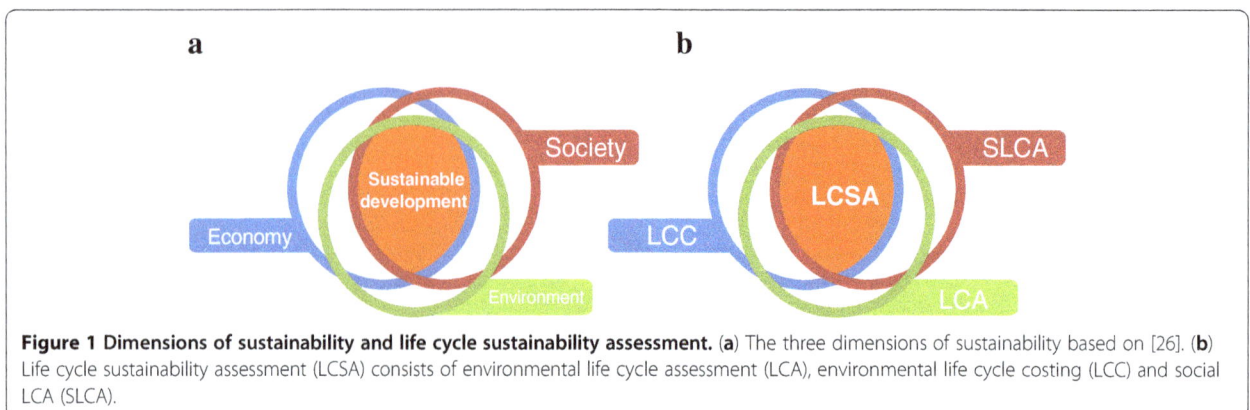

Figure 1 Dimensions of sustainability and life cycle sustainability assessment. (**a**) The three dimensions of sustainability based on [26]. (**b**) Life cycle sustainability assessment (LCSA) consists of environmental life cycle assessment (LCA), environmental life cycle costing (LCC) and social LCA (SLCA).

remanufacturing have been identified. Where data permits, the comparison between the new alternators and the remanufactured ones is performed.

Methods

The life cycle assessment of the three dimensions: environment, economy and society should ideally use the same system boundary and the same reference unit [27] - called the functional unit (FU) - a unit which all the results are related to and which quantify the performance (valuable main output) of the system [11]. The life cycle inventory for a LCA includes all exchanges or flows e.g., materials and energy between the technosphere (economy) and the nature (environment) crossing the system boundary [13,28]. However, due to the different flows to consider, method maturity and data access of the LCA, LCC and SLCA techniques, the use of the same life cycle inventory is difficult to realise.

LCA is a well-known tool standardised in the ISO14040/ 14044 and used to investigate the potential environmental impacts of products and services. LCA is divided into four phases: goal and scope definition, life cycle inventory analysis, life cycle impact assessment and interpretation in an iterative process [11,12].

Hunkeler et al. [16] defined environmental LCC as *'An assessment of all costs associated with the life cycle of a product that are directly covered by any one or more of the actors in the product life cycle (e.g., supplier, manufacturer, user or consumer, or EoL actor) with complementary inclusion of externalities that are anticipated to be internalized in the decision-relevant future. (...) Environmental LCC has to be accompanied by a life cycle assessment and is a consistent pillar of sustainability.'*

This definition is our starting point for the LCC. Since the flows investigated in LCC is of monetary art, all costs have their counterpart in an income. Therefore, the system boundaries and the stakeholder perspective taken, e.g. that of the user, are of importance for the results of the LCC [8].

SLCA assesses the social impacts on workers, the local communities, the consumers, the society and all other value chain actors affected by the production and consumption of products under consideration [19]. According to the guideline of social LCA [19] (published by United Nations Environment Programme UNEP/SETAC Life Cycle Initiative), a generic SLCA can be implemented as a first step to identify the social hotspots.

The life cycle of the alternator is modelled as shown in Figure 2 and used for the LCA. Starting with the left part of the figure, the production phase consists of raw material extraction, material processing and manufacturing. In the use phase, the alternator generates the necessary electricity for the automobile during its 200,000 km or about 13 years lifetime. Subsequently, the alternator is remanufactured in a mini-factory and placed in a container for worldwide use. Arrows indicates transport; however, at this stage, a detailed logistic system is not in place, such that it is assumed that the transport is the same for all design alternatives.

The remanufactured alternator can be used again as an electrical generator in the vehicle. The use phase is modelled once for Germany only. The remanufactured alternators are applied in the used vehicle already driven some distance. Therefore, the 200,000 km FU may be restricted by the (rest) of the vehicle - as this may be scrapped before driving at 200,000 km with the

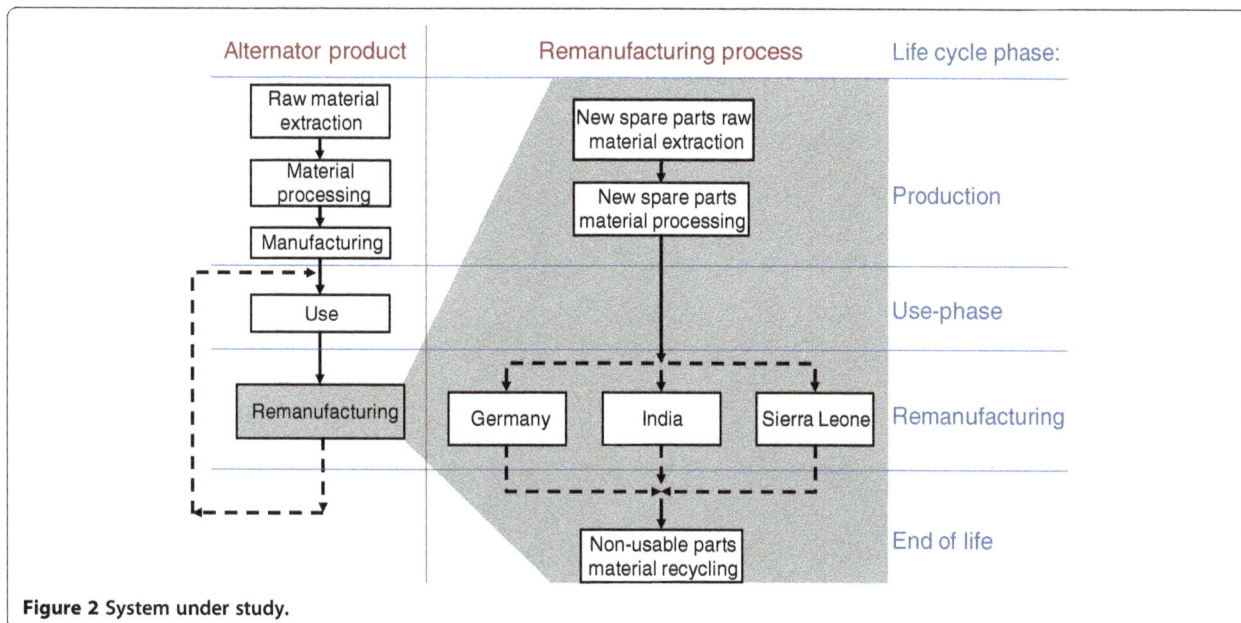

Figure 2 System under study.

remanufactured alternator. However, to facilitate the comparison of the new alternators to the remanufactured alternator, the FU of 200,000 km is used also on the remanufactured alternator.

The middle part of Figure 2 focuses on the remanufacturing process. In addition to the used alternator, the remanufacturing process needs some new alternator spare parts which are sourced globally. Similar to the new alternator production, raw material extraction and material processing are needed for the new spare parts. The remanufacturing scenario will take place in the container mini-factory equipped with all necessary tools and model to be set in Germany, India or Sierra Leone. The final stage, which is the end of life, is modelled as a part of the remanufacturing phase and includes also those fractions of the used alternators that cannot be used anymore (10% to 100% *cf.* Table 1). The right hand side of Figure 2 names the different life cycle phases.

The perspectives of the remanufacturer and the user (of the remanufactured alternator) are presented for the environmental LCC. Due to the case study's prospective nature and connected limited data access, only the three different potential remanufacturing sites in Germany, India and Sierra Leone are investigated for the SLCA.

Data for the LCA is mainly taken from and modelled in the GaBi 4.0 database [31], and the characterization factors used was CML2001 [13] (with update in [32]); whereas the environmental LCC is estimated using literature and invited quotations [8]. The data for the SLCA are from the social hotspot database [33,34] and other international database available online in addition to scientific literature.

Three different design alternatives are investigated by LCA and environmental LCC. The design alternative 1 is a conventional alternator (weight; 6.069 kg) with belt fitting, fan and steel bearings and cast iron housing. Design alternative 2 is a lightweight alternator (4.378 kg) with a plastic fan and aluminium housing. Design alternative 3 is an ultra-lightweight alternator (3.952 kg), where also the belt fitting and bearings are replaced by lightweight parts (aluminium and plastic respectively). Table 1 shows the material, weight, and replacement probability (the likelihood of a part being replaced within the alternator by the remanufacturer) of the different parts of the alternator for each design alternatives. The new materials, weight and replacement probabilities (in alternatives 2 and 3) are best estimates made by the designers. These are highlighted in italics in Table 1.

Results

First in this section, the environmental dimension is presented; second, the economic dimension; third, the social dimension. The results of the LCSA are summarised at the end of this section by applying the Life Cycle Sustainability Dashboard [35].

Environmental dimension: LCA of the remanufactured alternator

In this section, the LCA results for all steps of the product life cycle: production, use and remanufacturing are presented. Afterward, the comparison of the three different design alternatives and the different localization options are expressed.

Table 1 Alternator parts, materials, weights and replacement probabilities [29] for each of the design alternatives 1–3

	Design alternative 1			Design alternative 2			Design alternative 3		
	Conventional generator [29]			Lightweight generator [29,30]			Ultra-lightweight generator [29,30]		
Part	Material	Weight (kg)	Replacement probability (%)	Material	Weight (kg)	Replacement probability (%)	Material	Weight (kg)	Replacement probability (%)
Stator	Steel	0.773	20	Steel	0.773	20	Steel	0.773	20
Rotor coil	Copper	0.550	22	Copper	0.550	22	Copper	0.550	22
Rotor	Iron cast	1.094	19	Iron cast	1.094	19	Iron cast	1.094	19
Drive shaft	Steel	0.262	10	Steel	0.262	10	Steel	0.262	10
Belt fitting	Steel	0.519	10	Steel	0.519	10	*Aluminium*	*0.180*	*75*
Fan	Steel	0.138	10	*Plastic/PP*	*0.016*	*100*	*Plastic/PP*	*0.016*	*100*
Spacer	Aluminium	0.003	50	Aluminium	0.003	50	Aluminium	0.003	50
Bearings	Rolled steel	0.099	50	Rolled steel	0.099	50	*Plastic/PP*	*0.011*	*100*
Slip ring N	Copper	0.033	100	Copper	0.033	100	Copper	0.033	100
Slip ring S	Copper	0.071	100	Copper	0.071	100	Copper	0.071	100
Housing	Iron cast	2.527	15	*Aluminium*	*0.958*	*40*	*Aluminium*	*0.958*	*40*
Sum		6.069	-	-	4.378	-	-	3.952	-

Entities in *italics* are best estimates made by the designers.

The life cycle impact assessment results are pictured in Figure 3, which show that the use phase plays a dominating role. The exception is that in abiotic depletion potential (ADP), marine aquatic ecotoxicity potential and radioactive radiation (RAD) where the production phase is dominating. Figure 3 also indicates that the remanufacturing causes about 1/8 (12%) of the emissions compared to the production of the new part. The ADP indicator is a measurement of the resources and energy needed and is displayed in two components: ADP elements and ADP fossil fuels. The ADP elements are dominated by the production phase (71%), followed by the remanufacturing phase (24%). The use phase (5.5%) is relatively unimportant in the ADP elements. However, in the ADP fossil fuels, the use phase is totally dominating the overall life cycle result (99%) as expected as the alternator needs energy (taken from the internal combustion motor running on fossil fuel) to work. The ADP elements, which describe the use of mineral resources (e.g. copper) excluding fuels, can be explained further as the results are somehow counter-intuitive. First, the low share of the ADP elements in the use phase is explained by that the consumption during this phase is mainly fuel from abiotic resources (Petrol) and thereby part of the ADP fossil fuel. Second, the remanufacturing requires roughly 1/3 of the ADP element compared to the production (including upstream processes). This relatively high share (compared to the roughly 1/8 of the emissions) can be explained by the copper needed in the remanufacturing to replace the rotor coil (in 22% of the cases) and the slip rings each time (in 100% of the cases). The production of pig iron and primary aluminium contributes to the radioactive radiation. The primary aluminium production is the main cause of

marine aquatic ecotoxicity potential, mainly lead, by the emission of hydrogen fluoride to the air [31]. In the use phase, ADP fossil fuels and the global warming potential are dominated by the direct combustion of fuel. For the other impact categories, the use phase dominated due to the upstream processes of the fuel production.

Figure 4 shows the LCA results for a complete life cycle (from raw material extraction to use and finally remanufacturing) for design alternative 2 (lightweight) and design alternative 3 (ultra-lightweight) compared to design alternative 1 (conventional alternator). The remanufacturing site shown in Figure 4 is Germany. The conventional alternator has the best performance for all impact categories investigated. This is caused by (a) the conventional parts (e.g., made of cast iron) which have a low replacement probability in contrast to the lightweight parts (e.g., aluminium and plastics) and (b) the upstream environmental impacts of the conventional materials is smaller compared to the lightweight materials. If we observe the remanufacturing process (cf. Figure 2), these effects are very clear, as Figure 5 discloses.

Figure 5 represents the LCA results for design alternatives 2 (lightweight) and 3 (ultra-lightweight) relative to design alternative 1 (conventional) of remanufactured alternators. Only the remanufacturing process is showed for only one site (Germany). The range of the difference between the lightweight alternatives 2 and 3 and the conventional alternator is from two to eight times, except in the ADP elements where the differences are much smaller (11.3% and 2.5% in favour of the conventional alternator).

For the abiotic depletion impact category, the differences between the three design alternatives are small compared to the other impact categories investigated

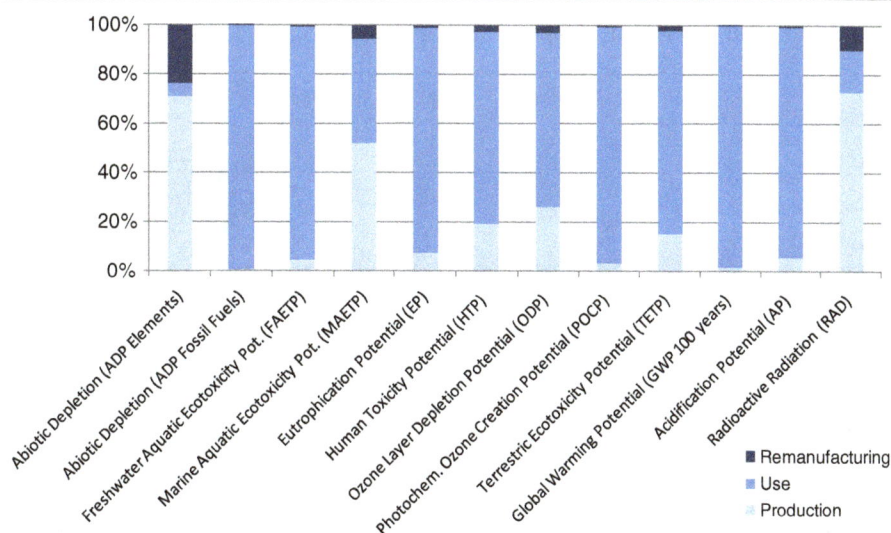

Figure 3 Results of the environmental LCA of remanufactured alternators (alternative 1, conventional - location, Germany).

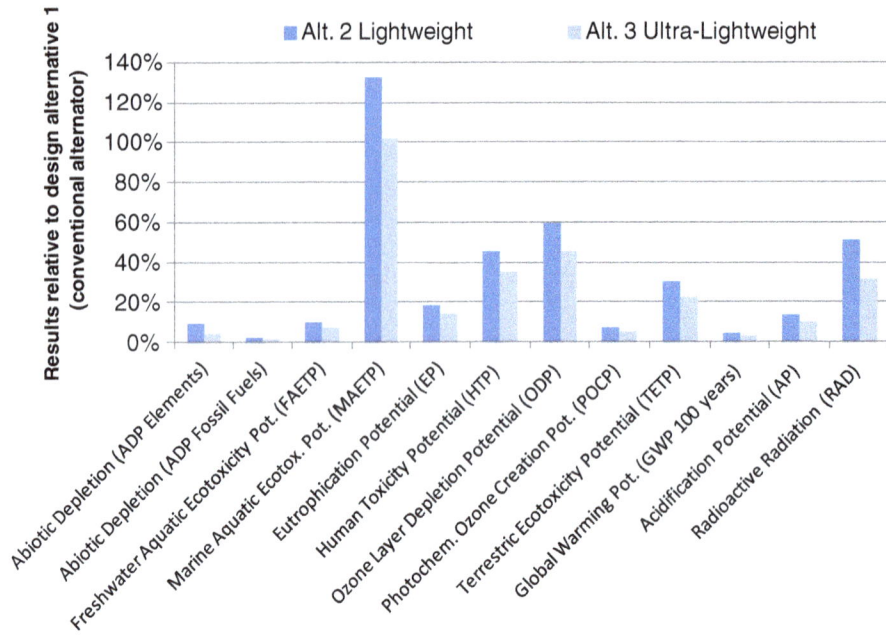

Figure 4 Results from the environmental LCA of remanufactured alternators. Complete life cycle (Germany only), design alternative 2 (lightweight) and 3 (ultra-lightweight) compared to design alternative 1 (conventional alternator).

(factors 2–8 *cf.* Figure 5). This can be explained by the ADP elements that roughly reflect the material use in the different design alternatives. This effect is scaled up as the fossil fuels (ADP fossil fuels) and emissions in the upstream processes needed are taken into account.

Comparison of different localization option (alternator remanufacturing)

Figure 6 shows the LCA results for the alternator 1, the conventional one, for the remanufacturing at the three different localizations investigated. The LCA results for the different localization options do not show very much variation compared to the different alternator designs

(*cf.* preceding sections). New spare parts are assumed to be sourced globally. Their upstream processes contribute the largest portion to the LCA results. The electricity use, which is drawn from the national grid in Germany and India (with their respective grid mix of power sources) and produced locally at the remanufacturer by diesel aggregates in Sierra Leone, explains some of the differences.

Economic dimension: LCC

In this section, firstly, the economic dimension from the remanufacturer perspective is presented and then the user perspective.

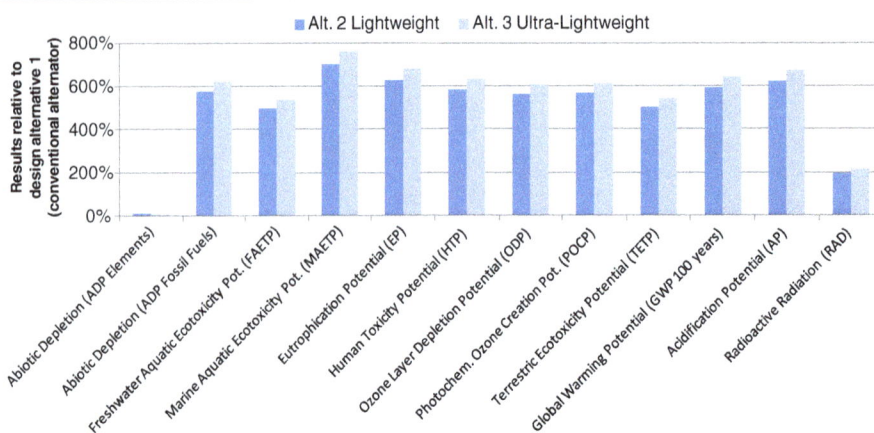

Figure 5 LCA results for the remanufacturing only (in Germany). Design alternatives 2 (lightweight) and 3 (ultra-lightweight) relative to design alternative 1 (conventional).

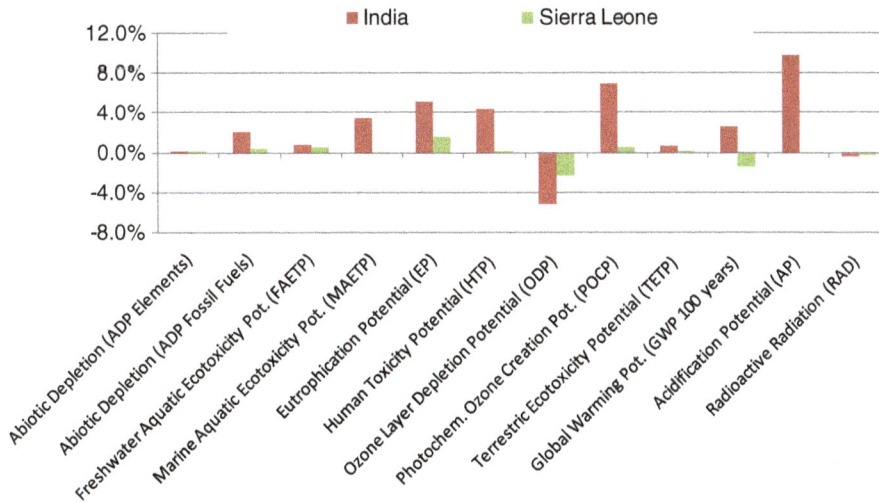

Figure 6 LCA results for different remanufacturing localizations - India and Sierra Leone relative to Germany (alternative 1, conventional-remanufacturing only).

Figure 7 shows the results from the environmental LCC from the remanufacturer perspective. Potential costs for the used alternators and new spare parts are assumed to be the same in India and Sierra Leone as in Germany. Transport cost includes oversea shipping to Germany from the India and Sierra Leone locations.

Cost for warranties is the only cost from the remanufacturer perspective that occurs in the use phase. The costs of (new) spare parts and used alternators acquisition are dominating. The differences between the different alternator design alternatives are significant, where alternator 1, the conventional one, has the lowest potential costs. The differences between the location choices are, however, mainly decided by the transport cost, and thereby depended on where the remanufactured alternator is used. Labour costs play a minor role in Germany, and are almost negligible in India and Sierra Leone, as the same as the cost of energy for cleaning used parts and cost for warranties. It

was surprising that the labour cost was so less important even in Germany, as the remanufacturing industry is normally considered as being labour intensive [36].

Figure 8 displays the LCC results from the user perspective. The fuel use cost for power production is dominating and is the same for all three alternatives. The largest difference between the design alternatives is found in the cost of repair and maintenance, where the lightweight alternatives 2 and 3 are estimated to need some repair and maintenance during their lifetime. The lightweight material used here are less durable than in the conventional design (alternative 1). The corresponding slightly lower weight-induced fuel use cost for the lightweight alternatives 2 and 3 cannot balance out the higher acquisition cost of these designs compared to the conventional one. This can be explained by the higher cost of the new lightweight spare parts for the remanufacturer (cf. Figure 7), which is passed on to the user

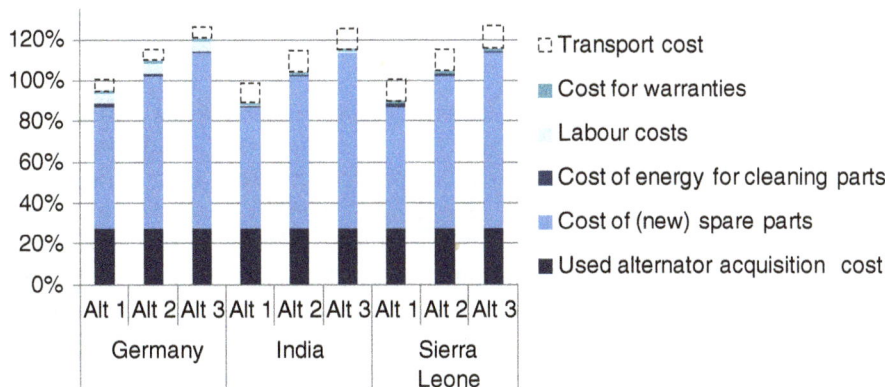

Figure 7 LCC results from the remanufacturer perspective.

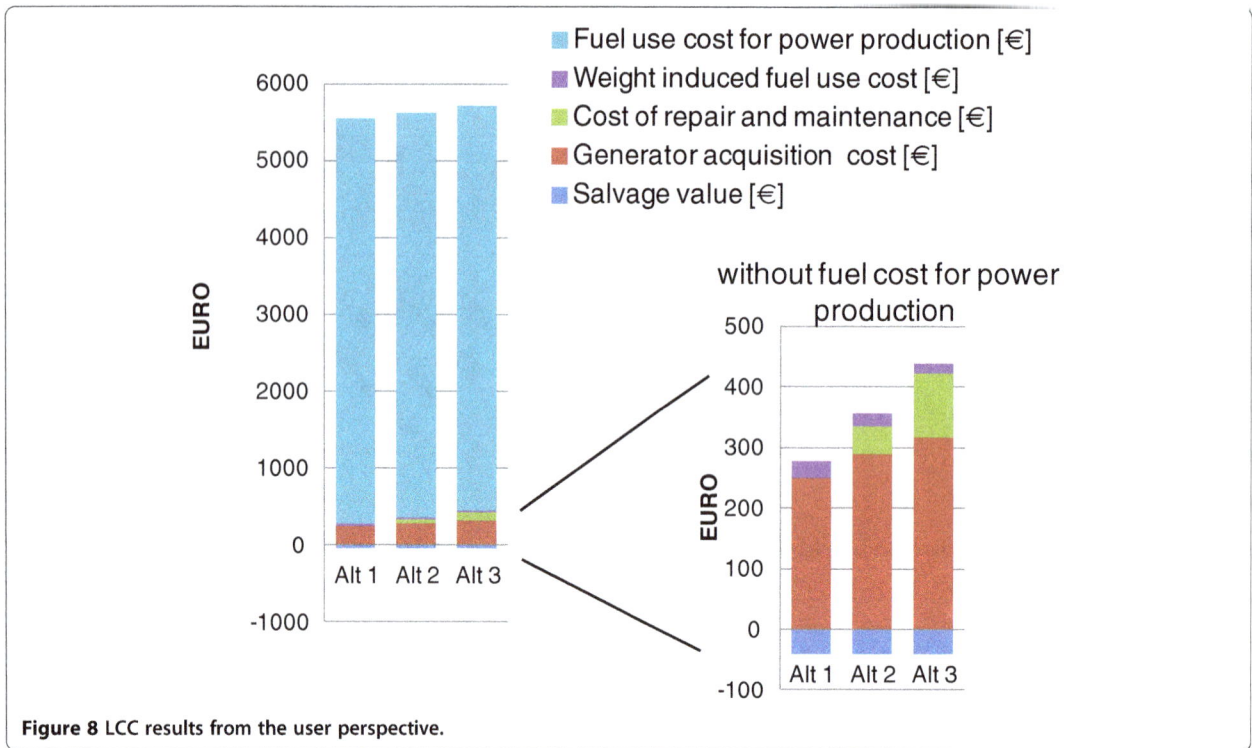

Figure 8 LCC results from the user perspective.

(plus tax and profit margins). The salvage value is depicted below the *x*-axis in Figure 8 as this is an income (negative cost) for the user at the end of the lifetime.

Social dimension: SLCA

The main social factors of Sierra Leone and India can be identified by using the Social Hotspots database available [34,37]. The Social Hotspots database reveals very high risks in Sierra Leone for the following topics [34]:

1. Potential of country not adopting Labour Conventions
2. Percentage of population living on less than US$2/day
3. Risk of child labour
4. Overall fragility of legal system
5. Overall fragility of gender equity
6. Risk of not having access to improved sanitation (total, rural and urban) and
7. Risk of not having access to improved drinking water (total).

A better scenario is shown about India, where the social hotspots shows very high risks for topics such as [34]:

1. Potential of country not adopting Labour Conventions

2. Percentage of population living on less than US$2/day
3. Potential for high conflict
4. Overall fragility of gender equity and
5. Risk of not having access to improved sanitation (total, rural and urban)

For Germany, the Social Hotspots database does not show very high risks for any of the topics covered [34].

Among these social hotspots, the ones that can be directly affected by a remanufacturing plant are child labour (children aged less than 15 years), salary and gender equity. According to the Social LCA guidelines [38], we should consider the main affected stakeholder group, which in this case is the workers. The Social LCA guidelines report the main impact subcategories of the workers' group, such as salary, child labour, health and safety and freedom of association, just to mention a few subcategories. Data collection on the working conditions should be carried out, and primary data about the previous indicators should be collected from the three remanufacturing mini-factories. Because the three remanufacturing mini-factories are not in place yet and the focus of this work is the assessment of the three scenarios, we mainly identify their potential positive and negative impacts.

Child labour is about 48% (2005) [39] in Sierra Leone and about 12% (2005, 2006) in India [40]. Child labour is not an issue in Germany, which has ratified the ILO 'Minimum Age Convention' [41] that legally prohibits child labour.

In relation to women discrimination, some data are reported in Table 2. The newest data for Sierra Leone existing is from 2001 [42]. To facilitate a better comparison between the countries, 2001 data from Germany and India were also shown in the table. An increase in the values for India can be recognised; for example, the ratio of female to male secondary enrolment increased to 88% in 2007. The values for Germany are stable on a high level [42].

The population living under US$2/day is about 76.1% in Sierra Leone [43] and 75.6% in India [43] (also in this case, it is not an issue for Germany). Therefore, it is important to consider the worker salary or in this case, the minimum wage. In Germany, there is no general statutory law on the minimum wage. However, labour organisations are relatively strong in the German remanufacturing sector and wages are according to the general tariff negotiated between the employer and employee organisations. In the LCC estimations, 24.50 €/h as cost for employers during the remanufacturing has been used [44], but this number includes all social costs which are relatively high in Germany.

In India, wages in the remanufacturing sector have been estimated based on some available databases [45,46] to be about 1.54 €/h in 2006. However, the trend in Indian wages shows a relatively rapid growing trend [45]. The national minimum wage floor has been risen from 80 INR (Indian rupees, about 1.42 €) in 2007 to 100 INR (about 1.48 €) in 2009, but the actual minimum wages are set regionally [47]. For example, in the automobile repair section of the Maharashtra region, the total minimum wage for a semi-skilled worker was 5,813.60 INR per month [48] or about 2.30 €/h (March 2011).

Sierra Leone is one of the poorest countries in the world, and the wages are generally low. There is a statutory minimum wage in Sierra Leone of 25,000 leones per month (less than 1 Euro), too low to secure a decent living [49]. In the remanufacturing sector, the average wage is about 4 million Le [50] (about 1,000 Euro) per year and equals an hourly rate of about 0.50 €/h. This wage includes 13.65% social contribution [50]. Table 3 summarises the social factors for Germany, India and Sierra Leone.

An implementation of Life Cycle Sustainability Dashboard

To summarise and present the previous results from the LCSA in a consistent way to the decision makers, we use the Life Cycle Sustainability Dashboard (LCSD). The LCSD is a spreadsheet tool and a specific application of the Dashboard of Sustainability and is used to compare different products or scenarios [22]. The LCSD uses a colour scale ranging from red (bad) via yellow (intermediate) to green (good) to present the comparison of the results of the three dimensions of sustainability [51]. Figure 9 shows a screenshot of the sustainability dashboard for the three different alternator designs remanufactured in Germany.

As indicated from Figure 9, alternator 1 is the best design alternative, showing the best performance in the LCA and also showed good results for the LCC. The SLCA in Figure 9 for all alternatives are the same, as we only assess the social dimension for the different location choices, but not for different design alternatives.

Discussion

The results are dependent on the underlying assumptions and data. In this study the replacement probabilities for the different parts of the alternator are of major importance. These are collected from the different remanufacturer in the greater Berlin region (Berlin-Brandenburg) by another member of the larger research project in which this paper is a part. As the underlying replacement probabilities data are not collected by the authors of this paper, it has not been possible to investigate the range of this data. Furthermore, to which degree the data from the Berlin region represent Germany without having the data from the rest of Germany is difficult to judge. But as remanufactured alternators and cores are frequently shipped throughout Germany, the replacement probabilities used are assumed not to differ very much from the German average. This may also be the case where used alternators from Germany are shipped to India and Sierra Leone for remanufacturing there. However, for the Indian and Sierra Leone location, the replacement probability may be very well different if the used cores are locally or globally sourced. It has not been part of the scope of this research to investigate the replacement probabilities for alternator parts in these countries. Therefore, the results presented on the different design alternatives in this study are focused on the German location only. Furthermore, the

Table 2 Social indicators on the gender equity [42]

Indicator name (data from year 2001) (%)	Germany	India	Sierra Leone
Ratio of female to male primary enrolment	99.4	85.1	68.1
Ratio of girls to boys in primary and secondary education	98.8	79.6	68
Ratio of female to male secondary enrolment	98.5	71.7	68
Proportion of seats held by women in national parliaments	31	9	9

Table 3 Social factors relevant for the remanufacturing plant

	Germany	India	Sierra Leone
Ratio of population living under US$2/day (%)	na	75.6	76.1
Estimated hourly rate in the remanufacture industry (€/h)	24.50	1.54	0.50
Minimum wage (€/month)	na	92	<1

na, not applicable.

replacement probability for the lightweight parts is an expert estimate from the design team involved in the research reported here. As these are new designs, actual statistics from the remanufacturing of such parts is very limited. Further research should focus on the better numbers for the replacement probabilities, and especially for the lightweight parts, as this will increase the reliability of the results.

The main environmental (and economic) impacts are from the use phase, when the alternator delivers electricity for the battery and on board electronic devices. This result heavily relies upon the efficiency of the motor to transform chemical energy in the diesel/petrol to rotational energy and the alternator itself to transform rotation energy from the motor to electric energy. While the motor efficiency is well-documented, the average efficiency is heavily dependent on different factors like speed, load and operating conditions [52]. To find reliable data on the alternator efficiency was, however, not that easy. The assumed average used in this study is 55% alternator efficiency based on [52]. This is the same number and source used in a master thesis of modelling remanufactured automotive alternators - however, it was not used in cars but in small wind turbines [53]. Bosch,

a large manufacturer (and remanufacturer) of alternators, publishes an edition of know-how for automobiles. In this book, Meyer [2] lists a maximum alternator efficiency of 65%; he noted, however, that the mean efficiency is between 55% and 60%. A minimum efficiency is not given [2]. In other older literature, the alternator efficiency is reported to be between 40% and 64% [54] such that our choice of 55% alternator efficiency seems reasonable. Note also in this respect that our basic assumption is that the remanufactured alternators are as good as new ones (which is close to the definition of remanufacturing - see e.g. [55]) such that the energy transformation efficiency is the same.

A finding in this study is that the much favoured lightweight strategy is not necessarily fruitful when the life cycle of the automotive part like the alternator is taken into account. In our LCA and LCC studies of three different designs for remanufactured alternators, the conventional (heavy) alternator scores are better than both the lightweight (alternative 2) and the ultra lightweight ones (alternative 3). The use phase is dominating the life cycle environmental impact of the remanufactured alternator. This is not surprising for a product requiring energy in the use phase. When comparing the different

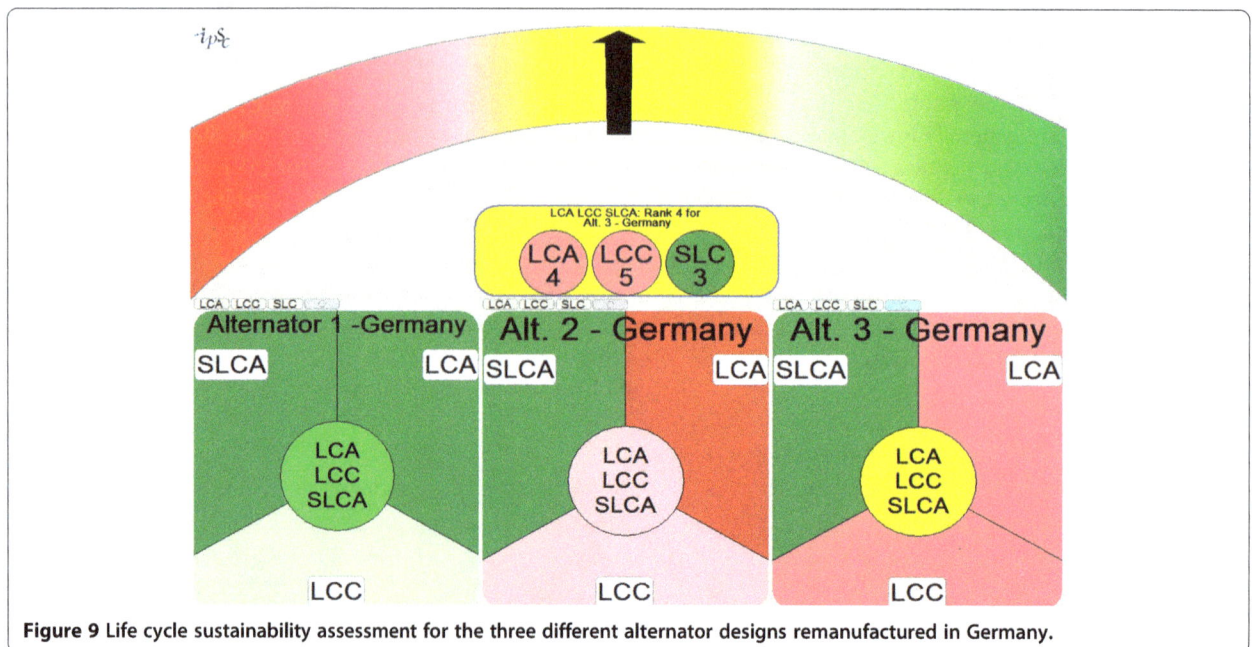

Figure 9 Life cycle sustainability assessment for the three different alternator designs remanufactured in Germany.

design alternatives, however, the heaviest alternative, the conventional alternator, is the best option from an environmental and economic life cycle perspective. Our starting point for the analysis is that the three different design alternatives have the same performance (e.g. energy conversion efficiency) in the use phase. Therefore, the major share (i.e. for the electricity production) of the environmental and economic cost in the use phase is the same for all design alternatives. On the other side, the lightweight alternatives save some fuel in the use phase compared to the conventional heavier alternator due to the lighter weight. However, in a life cycle perspective, the saved energy in the use phase due to the lightweight materials is not enough to compensate for the increased life cycle environmental impact of the increased probability for changing this part (and replace with new spare parts) in the remanufacturing phase. For example, the iron cast housing in the conventional alternator needs to be replaced in 15% of the cores, while a lightweight aluminium housing needs to be replaced in 40% of the cores (cf. Table 1).

As discussed above, the weight-induced fuel use (cf. the costs in Figure 8) is relatively small in our study compared to the overall picture. Thus, the weight advantage during the use phase of the lightweight alternatives is not able to offset the higher impact of these alternators and parts (lightweight spare parts which are more frequently need to be replaced during the remanufacturing). In this regard, a curiosity occurs; if the alternators are used in Germany and shipped to India and Sierra Leone for remanufacturing and back again for further use in Germany, the economic and environmental costs for the oversea transport may be lower for the lightweight alternatives. However, in the present study, we have assumed that the transportation costs are the same for all design alternatives. A more precise estimate of the oversea shipping costs requires a more detailed logistic planning which we leave for further research.

The LCC results show that the costs for new spare parts are higher than a complete used alternator. Hence, an option to reduce the potential cost is to buy a used alternator. However, one of the main challenges in the remanufacturing industry is the difficulty of getting access to used alternators [56], so we have assumed that the future supply of used alternators will be limited. Also typically, the parts that need to be exchanged are the same for all used alternators. Thus, used alternators for serving spare parts are not considered as a valid option. This assumption also influences the LCA results.

The SLCA shows that exporting the working conditions standard from Germany to India and Sierra Leone could improve the situation for child labour, gender equity and poverty in these two countries. Child labour

should not be tolerated in any considered remanufacturing scenario. This means a potential positive impact by the reduction of child labour in Sierra Leone and India, but without impact in Germany. While a ban of child labour for a new remanufacturing plant sounds reasonable, this may not be the case for existing child workers, where the alternatives for child performing labour has to be taken into account. A fruitful option to impede the child labour throughout the value chain can be to cooperate with fair trade non-governmental organisations. They often have experience from less-developed countries and can cooperate with local authorities and organisations to promote educational activities.

For the better balance between the genders, it is clear from Table 2 that further efforts have to be paid in Sierra Leone to reduce the difference between men and women. In this direction, the remanufacturing mini-factory should be designed in a way such that women can also be employed. The poverty (here, measured on the ratio of the population living under US$2 a day) is extremely high not only in Sierra Leone but also in India. To place the remanufacturing plant there and pay the workers a decent salary (the minimum wage in Sierra Leone is not enough for living) could make a small contribution to the right direction.

The emphasis on the life cycle is important also for the remanufacturer. The use phase and the upstream processes of the production of (new) spare parts contribute an essential part to the LCA and LCC results. This may also be vital to the SLCA, but here further data collection and research are still necessarily required.

Ideally, the same life cycle inventory (LCI) should be used for LCA, LCC and SLCA: 'The best solution would be the use of one identical LCI for all three components' (p. 90 in [27]). However, in this respect, it is worth to note that the inventory consists of the different types of data: for LCA typically, physical inputs and outputs; for LCC monetary flows and for SLCA, e.g., the social factors for the stakeholders. In addition, the life cycle phases of interest do not completely overlap for the three dimensions. For example, the remanufacturer perspective also includes the cost of warranties which belongs to the use phase. Therefore, some adjustment of the life cycle inventory is needed to fit the different life cycle techniques.

As the LCA and LCC results show, the most preferred environmental-friendly alternative also equipped the best cost efficiency, both from the remanufacturer and user perspectives. This case study shows that involved stakeholders can protect the environment and reduce the cost at the same time. Due to the predictive art of this study, limited data access and immature methodology, it has not been possible to clearly distinguish the three different designs alternatives regarding the SLCA.

Even though the primary aim of the study is to compare the different alternator designs and remanufacturing sites, some findings for the LCA and LCC regarding the difference between new alternators and remanufactured alternators are possible to be drawn. By comparing the life cycle stages raw material extraction, material processing and manufacturing of the alternator production (*cf.* upper left part in Figure 2) with the remanufacturing process (middle part of Figure 2), some findings are possible to reveal. We have shown in Figure 3 that the remanufacturing only need a fraction of the inputs (ADP elements and ADP fossil fuels) and only cause minimal emissions compared to the production of a new alternator. However, except in the ADP elements and RAD impact categories, the use phase is dominating under our assumption that the performance of the remanufactured alternator in the use phase is the same or better than a new alternator. This finding also holds for the LCC. Thus, our findings are in line with the study of Liu et al. [57] who found that the use of energy for remanufacturing of an engine motor is 10.22% compared to producing a new engine [57]. Therefore, the investigation of the performance of remanufactured alternators compared to new alternators in the use phase is extremely important. However, such an investigation requires more data [58,59] on the performance and the driving distance of new and remanufactured alternators. We leave this question for further research.

With the increasing use of start/stop motor and hybrid vehicle, in addition to fully electrified drive train in modern automotives, the use of conventional alternators (in addition to the starters) are increasingly replaced by electro-motors. This calls for attention especially for remanufacturing companies specialised in remanufacturing alternators and starters. Where used automobiles are increasingly equipped without alternators and starters, changes both from the supply and the demand side are required. A shift towards remanufacturing of electro-motors instead of alternators and starters could be an option.

As the energy efficiency of products (in the use phase) is improving, the reuse of materials is becoming relatively more important than the energy use and associated emissions. In the LCA, this means a shift from impact categories dominated by energy use to which categories that material use dominates. At the same time, when the energy use is decreasing (with its environmental and economic costs), the other dimension of sustainability, the social aspect which is measured with SLCA, is becoming more important for the life cycle sustainability impacts.

Conclusions

By applying LCA, LCC and SLCA, we have been able to quantify some indicators for the life cycle sustainability

of the three different remanufactured alternators and remanufacturing localization options. While environmental LCA is widely used and the result for the environmental dimension of sustainability is relatively easily to interpret, the (environmental) LCC and SLCA still need further development.

The advanced development of the (environmental) LCC should focus on its relationship to sustainable development such as the links to wage: All monetary flows from one perspective (a cost) have a corresponding receiver (an income) from the other perspective. The further development of the SLCA should focus on case studies, where the complete life cycle and the suitable indicators are investigated. In this study, we have chosen to investigate child work, fair wages and gender equity out of more than 200 indicators available [60] for the remanufacturing phase only. Base line indicators suitable for comparison between product alternatives and (on a later stage) different SLCA studies should be agreed upon. This could bring better transparency to the results of SLCA studies. Important measures to improve the sustainability of the remanufacturing process seen from a life cycle perspective include the following: (a) to make sure the energy efficiency of the remanufactured part are like new or even better than the new part (this is important as the use phase dominates from an environmental and economic point of view), (b) to place the remanufacturing mini-plant in India or Sierra Leone while exporting German labour standards (this could improve the results for the social dimension of LCSA), and (c) to consider that lightweight parts are not necessarily better. In our case study, the conventional alternator is performing the best for the LCA and LCC. The conventional (heavy) parts are not replaced so often such that their life cycle environmental and economic performances are better. This implies that life cycle thinking is necessary in the decision process.

Competing interest
The authors declare that they have no competing interests.

Authors' contributions
EMS reviewed the literature on LCSA and remanufacturing and had the leading role in performing the LCSA. He wrote the paper and prepared the figures and tables. MT had the leading role in conducting the SLCA. MF defined the case object (alternator) and directed the research from the start to the end. MT and MF provided important advices throughout the study and helped in editing the manuscript. All authors read and approved the final manuscript.

Authors' information
EMS led the research project 'Methodological sustainability assessment of machine components in the development process' which was part of a larger university-wide multidisciplinary research project 'Sustainable value creation exemplified by a mini-factory for remanufacturing' at the Technische Universitaet Berlin (TU-Berlin). He is now involved in the Collaborative Research Centre (CRC) 1026 'Sustainable Manufacturing - Shaping Global Value Creation.' He holds a Master of Science and Ph.D. in Industrial Economics and Technology Management from the Norwegian

University of Science and Technology (NTNU) in Trondheim, Norway. EMS has experiences from life cycle assessment of automotive products, transport and food products. He wrote several papers for international scientific journals and conferences. He is also a reviewer for a scientific research organisation and international journals such as *Environmental Science & Technology*.

MT is an expert on LCSA and SLCA. She was working as a post-doctoral scientific assistant in the Chair of Sustainable Engineering of TU-Berlin. She is an environmental engineer and holds a Ph.D. of 'Environmental Applied Physics' from the University of Palermo, Italy. She was working in the applications of LCA on building sector and strategic environmental assessment to the urban transportation systems. She was also a reviewer of a European Ph.D thesis 'Sustainability assessment within the residential building sector: a practical life cycle method applied in a developed and a developing country'. She joined the working group of UNEP/SETAC Life Cycle Initiative for developing the first guideline of LCSA. She currently involves in a UNEP Life Cycle Initiative of developing methodological sheets of Social Life Cycle Assessment guidelines. She wrote several papers for national and international scientific magazines and congresses, and she reviewed papers for international scientific magazine such as *Sustainability Journal*.

MF is currently Chair of Sustainable Engineering and Vice-Director of the Department of Environmental Technology at TU-Berlin. In addition, he is Advisory Professor for Sustainability Assessment at Aalto University Helsinki. He spent several years in industry including Mercedes-Benz Car Group of Daimler AG, Stuttgart, Germany as manager for Life Cycle Assessment and Life Cycle Engineering at the Design-for-Environment Department. Furthermore, he was leading vehicle projects with alternative powertrains (fuel cell, hybrid, etc.). He is a member of the Chairman's Advisory Committee and CHAIR of the Portfolio Task Force in TC207 for the ISO 14000 series of standards, and chairman of the ISO TC207/SC5 Life Cycle Assessment. He serves as a member of the International Life Cycle Board of the UNEP/SETAC Life Cycle Initiative and leader of their carbon footprint project. Moreover, he is a member of the Steering Committee of the Greenhouse Gas Protocol Product/Supply Chain Initiative of WBCSD/WRI and was appointed as member of the governing body of the 'Blue Angel' in Germany. MF is involved in several projects of UNEP/SETAC life cycle initiative e.g., guidelines of SLCA and guidelines of LCSA; he is also the vice speaker and a project leader of the CRC 1026 Sustainable Manufacturing - Shaping Global Value Creation. MF published more than 80 papers and serves on several editorial boards (*International Journal of Life Cycle Assessment*, and *ReSource, Waste Management & Research*).

Acknowledgements
We acknowledge Sara J. Ridley from Caterpillar Reman Service who gave us the idea to cooperate with fair trade non-governmental organisations to impede child labour on the conference dinner at the International Conference on Remanufacturing, Glasgow 27–29 July 2011.
We thank Barbra Ruben and Jan Fiebig for help collecting data and Ya-Ju Chang for help on English proofreading. We thank Kai Lindow, Robert Woll and Rainer Stark at Department of Industrial Information Technology, Technische Universitaet Berlin for their helpful cooperation. The position of EMS is supported by the German Research Foundation DFG (Fi 1622/1-1 and SFB 1026/1 2012). The article processing charge is kindly waived by the organiser of the International Conference of Remanufacturing (ICoR), Dr Winifred Ijomah and Dr James Windmill in cooperation with BioMed Central.

References
1. Kim, H-J, Skerlos, S, Severengiz, S, Seliger, G: Characteristics of the automotive remanufacturing enterprise with an economic and environmental evaluation of alternator products. INT. J. Sustain. Man. **1**, 437–449 (2009)
2. Meyer, R: Generatoren und Starter [Energieerzeugung und Bordnetz, Physikalische Grundlagen, Geräteausführungen für PKW und Nkw, Qualitätsmanagement, Werkstatt-Technik]. Bosch, Stuttgart (2002)
3. Lindahl, M, Sundin, E, Östlin, J: Environmental issues within the remanufacturing industry. In: 13th CIRP International Conference on Life Cycle Engineering, pp. 447–452. K.U. Leuven, Leuven (May 31–June 2 2006)
4. Boustani, A, Sahni, S, Graves, SC, Gutowski, TG: Appliance remanufacturing and life cycle energy and economic savings. In: Sustainable Systems and Technology (ISSST), pp. 1–6. 2010 IEEE International Symposium, Arlington, VA (17–19 May 2010)
5. Mayer, HW, Pester, W: Die Aufarbeitung von Kfz-Komponenten birgt Potenzial für Kunden und Umwelt. VDI Nachrichten **64**, 24 (2010)
6. Warsen, J, Laumer, M, Momberg, W: Comparative life cycle assessment of remanufacturing and new manufacturing of a manual transmission. In: Hesselbach, J, Herrmann, C (eds.) Glocalized Solutions for Sustainability in Manufacturing, pp. 67–72. Springer Berlin, Heidelberg (2011)
7. Sundin, E, Lee, HM: In what way is remanufacturing good for the environment? In: Design for Innovative Value Towards a Sustainable Society: Proceedings of EcoDesign 2011: 7th International Symposium on Environmentally Conscious Design and Inverse Manufacturing, pp. 552–557. Springer, New York (2012)
8. Schau, EM, Traverso, M, Lehmann, A, Finkbeiner, M: Life cycle costing in sustainability assessment - a case study of remanufactured alternators. Sustainability **3**, 2268–2288 (2011)
9. Ke, Q, Zhang, H-c, Liu, G, Li, B: Remanufacturing engineering literature overview and future research needs. In: Hesselbach, J, Herrmann, C (eds.) Glocalized Solutions for Sustainability in Manufacturing, pp. 437–442. Springer, Berlin (2011)
10. Brundtland, GH: World Commission on Environment and Development: Our Common Future. Oxford University Press, Oxford (1987)
11. ISO 14040: Environmental management - life cycle assessment - principles and framework (ISO 14040:2006), 2nd edn. ISO, Geneva (2006)
12. ISO 14044: Environmental management - life cycle assessment - requirements and guidelines (ISO 14044:2006). ISO, Geneva (2006)
13. Guinée, JB, Gorrée, M, Heijungs, R, Huppes, G, Kleijn, R, de Koning, A, van Oers, L, Sleeswijk, AW. Suh, S, de Haes, HA: U, de Bruijn, H, van Duin, R, Huijbregts, MAJ: Handbook on Life Cycle Assessment - Operational Guide to the ISO Standards. Kluwer, Dordrecht (2002)
14. Klöpffer, W, Grahl, B: Ökobilanz (LCA): Ein Leitfaden für Ausbildung und Beruf. Wiley-VCH, Weinheim (2009)
15. Lichtenvort, K, Rebitzer, G, Huppes, G, Ciroth, A, Seuring, S, Schmidt, W-P, Günther, E, Hoppe, H, Swarr, T, Hunkeler, D: Introduction - history of life cycle costing, its categorization, and its basic framework. In: Hunkeler, D, Lichtenvort, K, Rebitzer, G (eds.) Environmental Life Cycle Costing, pp. 1–6. CRC Press & SETAC, Boca Raton (2008)
16. Hunkeler, D, Lichtenvort, K, Rebitzer, G: Environmental Life Cycle Costing. CRC Press, Boca Raton (2008)
17. Rebitzer, G, Nakamura, S: Environmental life cycle costing. In: Hunkeler, D, Lichtenvort, K, Rebitzer, G (eds.) Environmental Life Cycle Costing, pp. 35–57. CRC Press, Boca Raton (2008)
18. Klöpffer, W: Outlook - role of environmental life cycle costing in sustainability assessment. In: Environmental Life Cycle Costing. CRC Press, Boca Raton (2008)
19. Benoît, C, Mazijn, B: Guidelines for Social Life Cycle Assessment of Products. United Nations Environment Programme, Paris (2009)
20. Swarr, TE, Hunkeler, D, Klopffer, W, Pesonen, H-L, Ciroth, A, Brent, AC, Pagan, R: Environmental Life-Cycle Costing: A SETAC Code of Practice. SETAC Press, Pensacola (2011)
21. Valdivia, S, Sonnemann, G, Hildenbrand, J: Towards a Life Cycle Sustainability Assessment - Making informed choices on products. UNEP/SETAC Life Cycle Initiative, Paris (2011)
22. Finkbeiner, M, Schau, EM, Lehmann, A, Traverso, M: Towards life cycle sustainability assessment. Sustainability **2**, 3309–3322 (2010)
23. Klöpffer, W, Renner, I: Life-cycle based sustainability assessment of products. In: Schaltegger, S, Bennett, M, Burritt, RL, Jasch, C (eds.) Environmental Management Accounting for Cleaner Production. In series Tukker A (ed): Eco-Efficiency in Industry and Science vol. 24, pp. 91–102. Springer Netherlands, Dordrecht (2009)
24. Klöpffer, W: Life-cycle based methods for sustainable product development. Int. J. Life Cycle Assess. **8**, 157–159 (2003)
25. Finkbeiner, M, Reimann, K, Ackermann, R: Life cycle sustainability assessment

(LCSA) for products and processes. In: SETAC Europe 18th Annual Meeting; 25–29 May 2008. Warsaw, Poland (2008)

26. Lozano, R: Envisioning sustainability three-dimensionally. J. Clean. Prod. **16**, 1838–1846 (2008)

27. Kloepffer, W: Life cycle sustainability assessment of products (with comments by Helias A. Udo de Haes, p. 95). Int. J. Life Cycle Assess **13**, 89–95 (2008)

28. Hischier, R, Baitz, M, Bretz, R, Frischknecht, R, Jungbluth, N, Marheineke, T, McKeown, P, Oele, M, Osset, P, Renner, I, Skone, T, Wessman, H, de Beaufort, ASH: Guidelines for consistent reporting of exchanges from/to nature within life cycle inventories (LCI). Int. J. Life Cycle Assess. **6**, 192–198 (2001)

29. Postawa, AB: Personal communication about alternator parts and their replacement probability. Department for Machine Tools and Factory Management. Technische Universitaet Berlin, Germany (2010)

30. Lindow, K: Personal communication about materials, weights and replacement probabilities in alternative design of alternators. Chair Industrial Information Technology, Department of Machine Tools and Factory Management. Technische Universität Berlin, Germany (2010)

31. PE, LBP: GaBi 4 - Software-System and Databases for Life Cycle Engineering., Stuttgart, Echterdingen (2009). ver. 4.4.137

32. CML: CML-IA Characterisation Factor; version 3.6. Leiden University, Leiden (2009). http://cml.leiden.edu/software/data-cmlia.html. Accessed 17 Nov 2010

33. Norris, CB, Aulisio, D, Norris, GA: Working with the Social Hotspots Database - methodology and findings from 7 Social Scoping Assessments. In: Dornfeld, D.A., Linke, B.S. (eds.) Leveraging Technology for a Sustainable World, pp. 581–586. Springer, Berlin (2012)

34. Benoît, C, Norris, G: Social Hotspots Database. http://www.socialhotspot.org/. Accessed 10 Jan 2011

35. Traverso, M, Finkbeiner, M, Jørgensen, A, Schneider, L: Life cycle sustainability dashboard. J. Ind. Ecol **16**, 680–688 (2012)

36. Sundin, E, Björkman, M, Jacobsson, N: Analysis of service selling and design for remanufacturing. In: Proceedings of the 2000 IEEE International Symposium on Electronics and the Environment, pp. 272–277. ISEE - 2000, San Francisco, CA (8–10 May 2000)

37. Benoit, C: Development of a screening tool for social LCA: the Social Hot Spots Database part 1. In: Life Cycle Assessment IX 'toward the global life cycle economy'. ACLCA, Boston (29 Sept–2 Oct 2009)

38. Benoît, C, Norris, G, Valdivia, S, Ciroth, A, Moberg, A, Bos, U, Prakash, S, Ugaya, C, Beck, T: The guidelines for social life cycle assessment of products: just in time! Int. J. Life Cycle Assess. **15**, 156–163 (2010)

39. Statistics Sierra Leone: UNICEF-Sierra Leone: Sierra Leone - Multiple Indicator Cluster Survey 2005. Statistics Sierra Leone and UNICEF-Sierra Leone. Freetown, Sierra Leone (2007)

40. UNICEF: Childinfo: statistics by area/child protection - percentage of children aged 5–14 engaged in child labour http://www.childinfo.org/labour_countrydata.php. Accessed 30 March 2011

41. International Labour Organization: Minimum Age Convention, (No. 138). ILO, Geneva (1973). http://www.ilo.org/ilolex/cgi-lex/convde.pl?C138. Accessed 30 March 2011

42. World Bank: World Development Indicators & Global Development Finance., (2011). http://databank.worldbank.org/. Accessed 20 Feb 2012

43. The World Bank: Poverty Headcount Ratio at a Day (PPP) (% of population). http://data.worldbank.org/indicator/SI.POV.2DAY (2011). Accessed 29 Jan 2011

44. ILC: International Comparisons of Hourly Compensation Costs in Manufacturing. (2008). http://www.bls.gov/news.release/pdf/ichcc.pdf. Accessed 17 Dec 2010

45. ILO: Wages in Manufacturing (Table five B). (2010). http://laborsta.ilo.org/. Accessed 17 Dec 2010

46. Sincavage, JR, Haub, C, Sharma, O: Labor costs in India's organized manufacturing sector. Mon. Labor Rev. **133**, 3–22 (2010)

47. Paycheck.in: Minimum Wages India - Current Minimum Wages Rate India. (2011). http://www.paycheck.in/main/officialminimumwages. Accessed 22 March 2011

48. Paycheck.in: Minimum Wages in Maharashtra. (2011). http://www.paycheck.in/main/officialminimumwages. Accessed 22 March 2011

49. 2008 Human Rights Report: Sierra Leone. http://www.state.gov/g/drl/rls/hrrpt/2008/af/119023.htm. Accessed 16 Dec 2010

50. Statistics Sierra Leone: 2008 Annual Economic Survey. Statistics Sierra Leone, Economic Statistics Division, Freetown, Sierra Leone (2009)

51. Traverso, M, Finkbeiner, M: Life Cycle Sustainability Dashboard. In: Proceedings of the 4th International Conference on Life Cycle Management, pp. 6–9. Cape Town, South Africa (September 2009)

52. Bradfield, M: Improving alternator efficiency measurably reduces fuel costs. In: Improving Alternator Efficiency Measurably Reduces Fuel Costs. Remy Inc, Pendleton (2008)

53. Ajayi, OA: Application of automotive alternators in small wind turbines. TU-Delft, Delft (2012)

54. Kuppers, S, Henneberger, G: Numerical procedures for the calculation and design of automotive alternators. IEEE T. Magn. **33**, 2022–2025 (1997)

55. Ijomah, WL, McMahon, CA, Hammond, GP, Newman, ST: Development of design for remanufacturing guidelines to support sustainable manufacturing. Robot. Cim-Int. Manuf. **23**, 712–719 (2007)

56. Östlin, J, Sundin, E, Björkman, M: Product life-cycle implications for remanufacturing strategies. J. Clean. Prod. **17**, 999–1009 (2009)

57. Liu, S-c, Shi, P-j, Xu, B-s, Xing, Z, Xie, J-j: Benefit analysis and contribution prediction of engine remanufacturing to cycle economy. J. Cent. S. Univ. Technol. **12**, 25–29 (2005)

58. Lund, RT: Comment on "Remanufacturing and Energy Savings". Environ. Sci. Technol. **45**, 7603–7603 (2011)

59. Gutowski, TG, Sahni, S, Boustani, A, Graves, SC: Remanufacturing and energy savings. Environ. Sci. Technol. **45**, 4540–4547 (2011)

60. Hunkeler, D: Societal LCA methodology and case study. Int. J. Life Cycle Assess **11**, 371–382 (2006)

3

Strategic decision making method for sharing resources among multiple manufacturing/ remanufacturing systems

Shinsuke Kondoh[1*†] and Timo Salmi[2†]

Abstract

Purpose: To reduce products' environmental impact over their entire life cycle, adequate reuse and recycling of products and their components are indispensable. In this context, it is important to establish efficient closed-loop manufacturing systems (CMS), where products are made from post-use as well as new materials. However, the establishment of economically and environmentally efficient CMS is difficult due to the uncertainty associated with the return flows of post-use products. Since product usage conditions and lifetimes differ from user to user, there are significant fluctuations in product flows' quantity and quality. This results in insufficient utilization of manufacturing/remanufacturing resources (e.g., labor and equipment) and high investment costs for CMSs, which hinder proper reuse and recycling of post-use products.
The objective of this study is to propose a strategic decision-making method for sharing resources among multiple CMSs to reduce the cost of product reuse and recycling.

Methods: We first discuss the benefits and difficulties of sharing production resources among multiple CMSs. Then, a transferability benefit index (TBI) is introduced to help identify the most promising resources to be shared among multiple systems.

Results: A simplified example calculation is provided as an illustration of the method. Two disassembly systems with the similar structure are considered as a case study. As a result, we successfully applied the index to determine the most promising resources in a case study.

Conclusions: We find that TBI is useful because it provides a simple and easily understandable decision criterion for identifying the resources to be transferred and shared among multiple CMSs to reduce the cost for reuse and recycling of used products. TBI also screens outs the promising resources which should be redesigned and modified before sharing among multiple CMSs. Development of practical redesign methods and modification guidelines for these resources will be included in our future work of this study.

Keywords: Closed-loop manufacturing system (CMS), cost for reuse and recycling, sharing resource, transferability benefit index (TBI), design structure matrix (DSM)

1. Introduction

Due to growing concern about environmental problems, it is becoming important for manufacturers to add more value while causing less environmental impact. In order to reduce the environmental impact of products over their entire life cycle, adequate reuse and recycling of products and their components are quite promising [1,2]. In this context, it is quite important for manufacturing firms to establish efficient closed-loop manufacturing systems (CMS) [3] in which products are made from used components and materials as well as new ones. Some firms have successfully established quite efficient CMS from both environmental and economical viewpoints. CMSs for one-time-use cameras [4],

---
* Correspondence: kondou-shinsuke@aist.go.jp
† Contributed equally
[1]National Institute of Advanced Industrial Science and Technology (AIST), 1-2-1 Namiki, Tsukuba, Ibaraki, Japan
Full list of author information is available at the end of the article

photocopying machines [5], and automobile components [6] are typical examples.

However, establishment of an environmentally and economically efficient CMS is not easy, mainly due to high uncertainty associated with the return flow of post-use products. Since product usage conditions and lifetimes differ from user to user and cannot, in general, be controlled by manufacturers, there are significant fluctuations in the quality and quantity of product return flows [7,8]. In addition, the return flow of post-use products may contain different product models in different conditions, each of which requires different remanufacturing operations (e.g., some may need cleaning and inspection while others may need disassembly into their components). Therefore, CMS should have higher flexibility and redundancy than conventional production systems to adapt these significant fluctuations.

Both of these requirements are quite expensive to meet. Flexible machines and labours are generally more expensive (sometimes less effective) than fixed purpose ones. In addition, the differences in necessary operations for each used product need frequent reprogramming and set up for manufacturing equipment. This hinders the automation of CMSs and results in higher operation cost, especially in developed countries where labour cost is expensive. The high redundancy in production resources also leads to their less efficient utilization and causes higher investment cost than conventional ones.

In order to solve these problems, many studies have been conducted in recent years. Examples include, Holonic Manufacturing Systems (HMS) [9], Biological Manufacturing Systems (BMS) [10], cellular manufacturing systems [11], and SOCRADES (Service Oriented Cross-layer infRAstructures for Distributed smart Embedded deviceS) [12] based on Service Oriented Architectures (SOA) [13]. Some of these [9-11] focus on the development of completely new conceptual (sometimes ideal) flexible manufacturing systems, while others [12-14] concentrate on enabling technologies (e.g., XML-based communication protocols for embedded devices and semantic webs for realizing SOA).

However, most of the studies assumed complete replacement of existing systems, which might require prohibitive investment at the beginning. There is a lack of systematic and practical methods for improving the flexibility of existing systems by gradually introducing these concepts. This is a major reason for that many of these concepts have not spread widely into industry.

The objective of this study is to propose a strategic decision making method for designing environmentally and economically efficient CMS while maintaining the flexibility and the redundancy to adapt the significant fluctuations in product return flows. Especially, this paper deals with the investment reduction of a CMS

through effective sharing of its resources across multiple production systems.

To this end, we introduce a transferability benefit index (TBI), the ratio of the benefits to difficulties, to identify the most promising resources for sharing among multiple production systems. We also provide a simplified example calculation to illustrate the method and discuss its result and the future development needs of the methods.

2. Transferability Benefit Index (TBI)

2.1 Benefits of sharing production resources
The wide fluctuations in the return flow of used products cause inefficient utilization of resources in a CMS. Thus, sharing idle resources among multiple CMSs may significantly reduce the initial investment over these systems.

Generally speaking, utilization rate of each resource is given by the ratio of actual working time to the whole working hours (e.g., 8 hours or 24 hours etc.) of the system. The resources with low utilization rate (long idle time) have great possibility for sharing across multiple CMSs to reduce the total number of the same kind of resource over these systems. Theoretically, each resource can be transferred to the other systems and utilized until the summation of its utilization rate over different systems reaches 1. Therefore, the benefit potential for its sharing is evaluated by Equation 1, assuming that the same (or similar) resources in different CMSs have the same initial investment cost.

$$b_i^j = \left(1 - u_i^j \right) \cdot c_i \tag{1}$$

where i, j, b_i^j, u_i^j, and c_i, denote the index for each resource, the index for each CMS, the benefit potential for sharing the resource i in the CMS j with other systems, the utilization rate of resource i in the CMS j, and initial investment cost for the resource i, respectively.

When the resource i is shared across n_i production systems, the actual benefit for the sharing b_i is given as follows;

$$b_i = \left(n_i - 1 \right) \cdot c_i \tag{2}$$

where

$$\sum_{j=1}^{n_i} u_i^j < 1 \tag{3}$$

2.2 Difficulty of sharing resources among multiple CMSs
Even if resources have high benefit potential when shared among multiple systems, it is possible that some of them are very difficult to transfer from one system to the others. Thus, the difficulty of sharing should also be

considered in determining which resources hold the most promise for sharing.

Generally speaking, the difficulty of sharing a certain resource among multiple systems that use similar resources depends on the number of its interactions with other elements in a set of production systems. For example, in order to transfer one piece of equipment to another CMS, adjustment and reprogramming of system segments that are connected to that equipment will likely be needed in addition to adjustment and reprogramming of the equipment itself. These additional necessary operations can be regarded as the main source of difficulty in sharing the equipment.

In order to represent the interdependence among multiple resources in CMSs and formulate the difficulty of resource sharing, we used a design structure matrix (DSM) [15]. The DSM, which is sometimes called an interdependency matrix, is a product or project representation tool that is widely used for representing interdependence among all constituent subsystems or activities to improve the structure of a product or project. Table 1 shows a typical DSM. It lists all constituent activities along rows and columns and shows interdependency with a digit number 1 in each cell where the activity in the corresponding row of the matrix depends on the activity in the cell's column in some ways. For example, the number 1 in the 2nd row and the 1st column of the matrix shows that the activity 'b' depends on the activity 'a'.

Since the necessary time or cost is different for each task, it is necessary to weight the difficulty of each task by introducing weighting factors into the DSM.

The difficulty weight assigned to each task is generally evaluated as its necessary labour time or cost. However, it sometimes happens that some of them need special labour skills or conditions that are difficult to evaluate as a function of labour time and cost. In such cases, difficulty weights are determined on an empirical basis considering these factors other than labour time and cost.

Table 1 Example of Design Structure Matrix

		Element activity							Row No.
		a	b	c	d	e	f	g	
Element activity	a								1
	b	1		1			1		2
	c		1						3
	d		1	1					4
	e								5
	f				1			1	6
	g			1			1		7
Column No.		1	2	3	4	5	6	7	

First, all the necessary tasks for transferring a resource (i. e., removal and reinstallation) from the CMS j are listed across the rows and columns of interdependency matrix M_{kl}^j. Each element of the matrix takes a Boolean value of 0 or 1. If task k should be executed whenever task l takes place, M_{kl}^j is assigned to be 1. Otherwise its value is 0.

Then, by using a weighting factor w_k^j, the total difficulty of task l in the CMS j is calculated as shown in Equation 4.

$$d_l^j = \sum_k w_k^j M_{kl}^j \qquad (4)$$

where d_l^j and w_k^j denote the difficulty of operation l in the CMS j and the weighting factor for operation k in the CMS j, respectively.

The total difficulty of transferring resource i in the CMS j is calculated as the sum of the difficulties of necessary tasks associated with it, as given by Equation 5.

$$d_{S_i^j} = \sum_{l \in S_i^j} d_l^j \qquad (5)$$

where S_i^j denotes the set of tasks necessary to transfer resource i in the CMS j.

The total difficulty of sharing a resource among a set of CMSs is formulated as the sum of the difficulties over these systems as follows:

$$d_{S_i} = \sum_{j \in S_i} d_{S_i^j} \qquad (6)$$

where S_i denotes a set of given CMSs among which the resource i is to be shared.

2.3 Transferability benefit index formulation

All resources can be classified into three categories as shown in regions I, II, and III in Figure 1, considering the benefit of and the difficulty for their sharing, which are represented by horizontal and vertical axes of the figure, respectively. Manufacturers should consider the sharing of resources located in region I because their sharing produces larger benefit with relatively smaller difficulty. In addition, the resources located in region II also hold the promise for the sharing, especially when it is possible to reduce the difficulties for their sharing. They should be redesigned and modified to reduce their interdependency on the other resources in CMSs. In other words, these resources should be replaced with more flexible and reconfigurable resources to ease the sharing across multiple CMSs. For the resources located in region III, there are no immediate needs for the sharing.

In order to identify which resources are located in region I, a Transferability Benefit Index (TBI) is

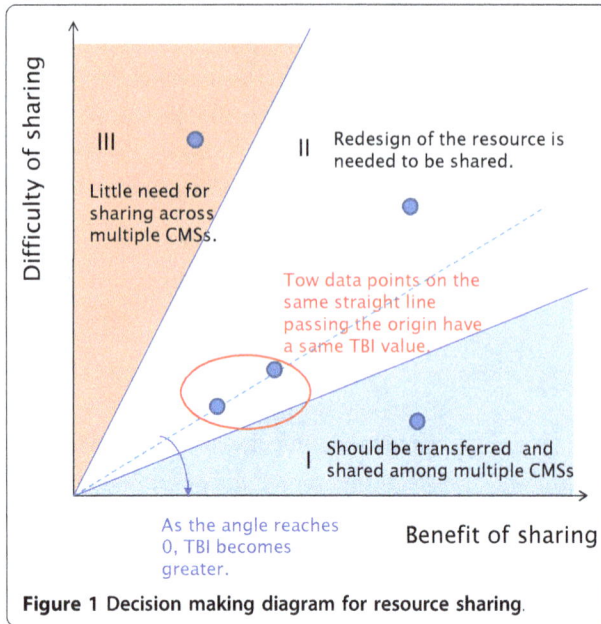

Figure 1 Decision making diagram for resource sharing.

introduced, which can be calculated using Equation 7.

$$TBI = \frac{benefit\ of\ a\ resource\ sharing}{difficulty\ for\ the\ sharing} \tag{7}$$

A high TBI value means that sharing the corresponding resource has a relatively large benefit compared to its difficulty.

Using Equations 1 and 5, the TBI of the resource i in the CMS j is given as follows;

$$TBI_i^j = \frac{(1 - u_i^j) \cdot c_i}{\sum_{l \in S_i^j} d_l^j} \tag{8}$$

As shown in Figure 1, two data points on the same straight line passing by the origin have the same TBI value and the region closer the horizontal axis has higher TBI. Thus, TBI is an adequate index for identifying the most promising resources.

When a set of CMSs S_i among which the resource i to be shared is given, TBI of sharing the resource i across n_i systems from S_i is given by using Equations 2 and 6 as follows:

$$TBI_i^{S_i} = \frac{(n_i - 1) \cdot c_i}{\sum_{j \in S_i} d_{S_i^j}} \tag{9}$$

3. Strategic decision-making procedure for sharing resources among multiple production systems

Step 1: Define a set of CMSs among which the resources are to be transferred and shared

The designer should first define a set of CMSs among which the constituent resources are to be transferred and shared. Then the designer identifies the resources to be considered for sharing taking into account their applicability to their corresponding tasks in each CMS. The cost reduction target by the sharing of the resources is also defined in this step.

Step 2: Estimate sharing benefits

The investment cost and utilization rate are estimated for each resource element identified in the previous step. The designer can then calculate the benefit potential for sharing each element in each CMS using Equation 1. The actual benefit for sharing each element in a set of CMSs given in previous step is also calculated by Equation 2.

Step 3: Estimate sharing difficulties

The tasks necessary to transfer each resource element to each CMS (e.g., mechanical adjustments, reconfiguration of software settings) are identified first. Then, the designer weights each individual task, considering its difficulty in terms of cost, lead-time, necessary tools, and labour skills required. Interdependencies among these tasks in each CMS are also identified and represented by M_{kl}^j. Using interdependency matrix M_{kl}^j and weighting factors w_k^j for each task in each CMS, the difficulty of sharing each element in a given set of CMSs is calculated using Equations 4, 5, and 6.

Step 4: Identify the most promising resources to be shared and transferred

The TBI of each element is calculated using its sharing benefit and difficulty. Then the resources with the highest TBI values are selected one by one until the total benefit of their sharing satisfies the cost reduction target defined in step 1.

Step 5: Evaluate the feasibility of the sharing

Finally, the feasibility of each element sharing is evaluated by considering its summation of utilization rate over a given set of CMSs. Each element sharing is feasible only if it satisfies Equation 3.

Some resources need to be redesigned and modified before sharing across multiple CMSs. For these resources, the feasibility and the possible cost for the redesign and modification should also be evaluated.

If the estimated benefit does not satisfy the cost reduction target defined in step 1, the designer moves to step 4 and selects the resource with the next highest TBI value until the target is satisfied.

4. Case study

In order to illustrate a strategic decision-making method for sharing resources among multiple CMSs, a simplified case study is provided in this section.

4.1 Define a set of CMSs among which the resources are to be transferred and shared

Figure 2 shows a set of two disassembly systems to be considered; disassembly system 1 for used air conditioners and disassembly system 2 for used refrigerators. Each system is assumed to consist of four pieces of equipment and three of them; namely, belt conveyor 'b, ' refrigerant gas collector 'c, ' and crushing machine 'd' can be applicable to both systems. The other equipment, disassembly stations 'a' and 'e' are specialized equipment for each system and cannot be applicable to the different system. The cost reduction target is defined as 5% in this case study.

4.2 Estimate sharing benefit

Initial investment for each piece of equipment and its utilization rate are assumed as shown in Table 2. Since the fluctuation in the volume of returned air conditioners is larger than that of refrigerators, the utilization rate of each piece of equipment in disassembly system 1 is smaller than that of corresponding one in system 2. Substituting these values into Equation 1, the benefit potential for sharing each piece of equipment is calculated as shown in the 4th row in the table. The benefits

for sharing resources 'b', 'c, ' and 'd' among both systems are calculated by using Equation 2 as shown in the 5th row in Table 2. For example, the benefit potential for sharing equipment 'b' is calculated as 1600 [euro] by substituting its initial investment cost (i.e., 2000 [euro]) and utilization rate (i.e., 0.2) into Equation 1. Substituting 2 to n_2 in Equation 2, the benefit for sharing equipment 'b' is calculated as 2000 [euro].

4.3 Estimate sharing difficulty

Since each system contains different equipment from each other, its interdependency pattern also differs from each other. Thus, two interdependency matrix (M_{kl}^1 and M_{kl}^2) are calculated as shown in the tables in Additional Files 1 and 2, respectively.

Necessary tasks for transferring each piece of equipment and their difficulties are first identified as shown in the tables. Interdependence among these operations is assumed as given in the 1st to 10th rows in the tables. For example, three operations (i.e., physical adjustment, software installation, and reprogramming) are required to remove/install a disassembly station 'a' from/to the disassembly system 1. Since the resource is not manually

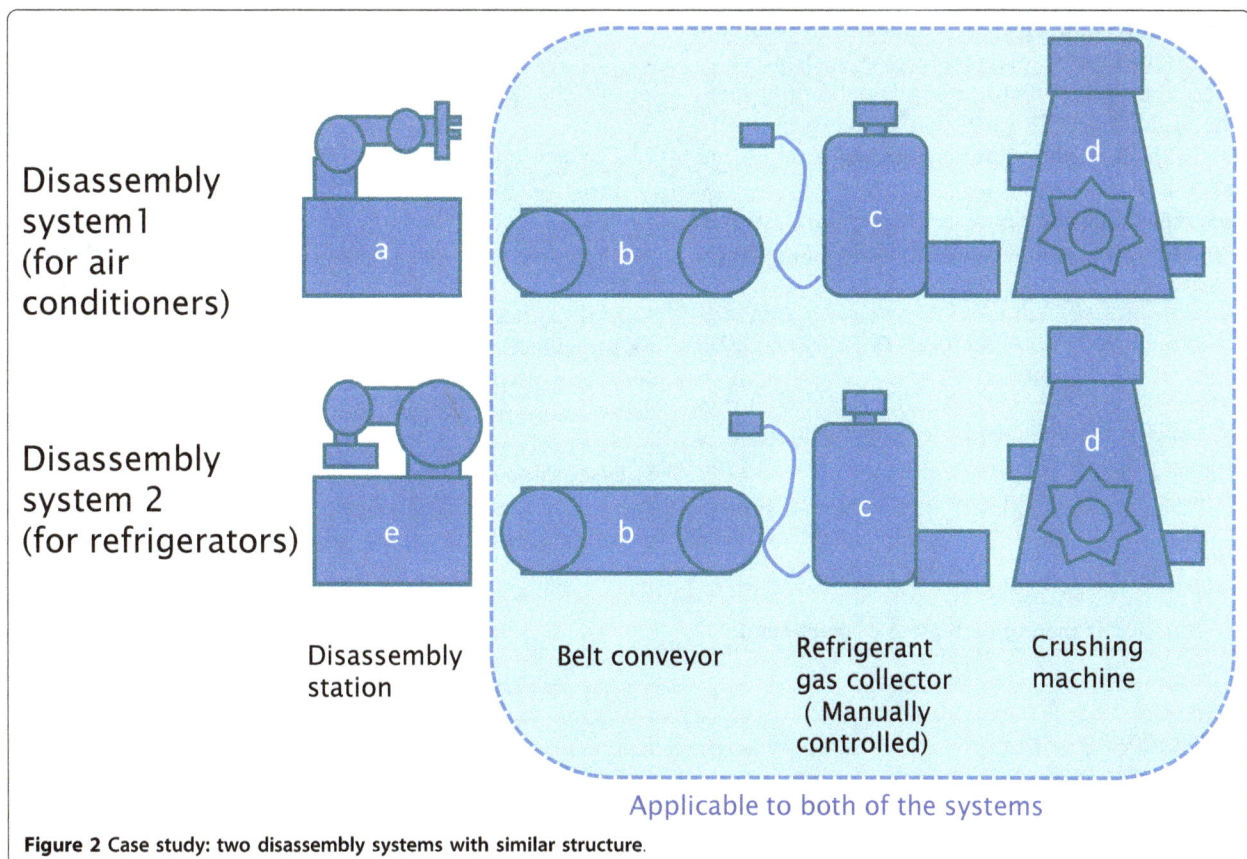

Figure 2 Case study: two disassembly systems with similar structure.

Table 2 Benefit of sharing each piece of equipment

Equipment	Disassembly system 1				Disassembly system 2				Total investment	Row No.
	a	b	c	d	e	b	c	d		1
Initial investment [euro]	8000	2000	5500	40000	7000	2000	5500	40000	110000	2
Utilization rate	0.2	0.2	0.01	0.4	0.5	0.5	0.02	0.9	-	3
Benefit potential	6400	1600	5445	24000	3500	1000	5390	4000	-	4
Actual benefit for the sharing	-	2000	5500	40000	-	2000	5500	40000	-	5
Resource No. i	1	2	3	4	1	2	3	4	-	6
Column No.	1	2	3	4	5	6	7	8	9	-

controlled, its removing and installation also require the physical adjustment and reprogramming of its connected equipment, belt conveyor 'b.'

Then, difficulty weight for each operation (w_k^j) is assigned as shown in the shaded cells in these tables. Among them, physical adjustment of resource 'd' assumed to be the most difficult task since it is too large to transfer (i.e., $w_8^1 = 9$ and $w_8^2 = 9$).

Substituting these values into Equation 4, total difficulty of each task is calculated as shown in the 11th row in the tables. The difficulty of transferring each piece of equipment is calculated by using Equation 5 as shown in the 12th row in the tables. Overall difficulty for sharing each resource among both systems is given by the total of the difficulty for transferring each piece of equipment over two systems. Using Equation 6, overall difficulty of the sharing is calculated as shown in the 3rd row in Table 3, which summarizes the calculation results.

For example, focusing on 'b' in disassembly system 1, the difficulty of operation physical adjustment (d_4^1) is calculated as follows by using Equation 4;

$$d_4^1 = 1 \times 1 + 1 \times 0 + 3 \times 1 + 3 \times 1 + 1 \times 0 + 3 \times 0 + 3 \times 1 + 9 \times 1 + 1 \times 0 + 1 \times 1 \quad (10)$$

Aggregating the difficulties of three operations (i.e., d_5^1, d_5^1, and d_6^1), the difficulty of transferring 'b' from/to disassembly system 1 ($d_{S_1^1}$) is calculated as 35.

Overall difficulty for sharing 'b' across the two systems is the total of transferring difficulty of 'b' for each

system (i.e., $d_{S_2^1}$ and $d_{S_2^2}$), which is given as follows by using Equation 6.

$$d_{S_2} = 35 + 39 \quad (11)$$

4.4 Identify the most promising resources to be shared and transferred

Figure 3 and Table 4 summarizes the result of the TBI calculation. As shown in the figure, crushing machine 'd' has the greatest potential to reduce initial investment cost with relatively small effort compared to other equipment items.

Since the benefit of sharing 'd' is calculated as 40, 000 [euro], which is larger than the target value 5, 500 [euro], 5% of the total investment cost (i.e., 110, 000 [euro]), the designer proceeds to the next step.

4.5 Evaluate the feasibility of the sharing

Although crushing machine 'd' has the highest TBI value, it cannot be shared because the total of its utilization rate over two systems is calculated as 1.3, which is larger than 1. Thus, the equipment with next highest TBI, refrigerant gas collector 'c' is chosen to be shared. As the overall benefit of the sharing is calculated as 5, 500 [Euro], which satisfies the cost reduction target defined in step 1, designer stops the calculation.

5. Discussion

We introduced TBI measure to determine the most promising resources in a case study. Although the

Table 3 Difficulty of sharing each piece of equipment

	Equipment			Row No.
	b	c	d	
Difficulty of transferring resources from/to disassembly system 1	35	3	21	1
Difficulty of transferring resources from/to disassembly system 2	39	3	21	2
Total difficulty for the sharing	74	6	42	3
Resource No. i	2	3	4	

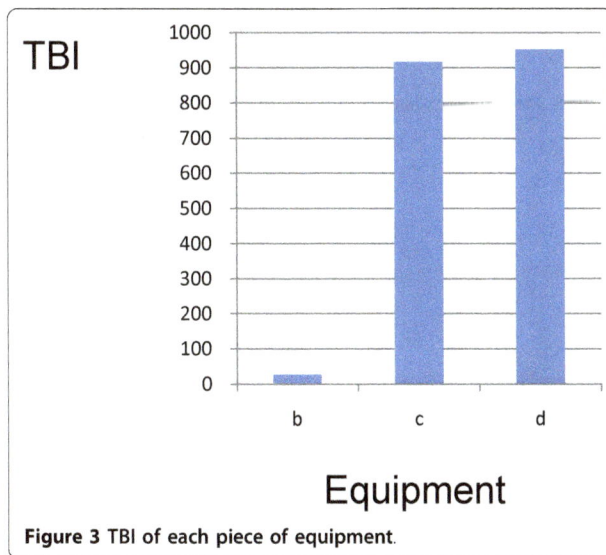
Figure 3 TBI of each piece of equipment.

example consists of a small number of resources, it is also possible to apply this method to the systems consisting of many resources with complicated interdependency patterns. Thus, TBI can be a useful tool for selecting resources to be shared among multiple CMSs.

Although the method proposed here is simple and useful for determining the most promising resources for the sharing, the calculation procedure becomes more complicated (and time consuming) when designers contemplate sharing resources among a large number of CMSs with different interdependency patterns. Simplification of the calculation procedure for production systems with different structures should be undertaken in our future work.

Although in this paper we focus on the sharing of resources with high TBI values, which are located in region I of Figure 1, sharing those located in region II also holds promise if they can be redesigned to reduce the difficulty of necessary tasks to transfer them among systems. Development of a redesign method to improve resources' transferability will also be part of our future work. The DSM, which is used to estimate the difficulty of sharing resources among multiple

CMSs in this paper, can also be used to determine workable structures for such systems and resources.

6. Conclusion

This paper proposes a strategic decision-making method for sharing resources among multiple CMSs, aiming at reducing the cost of reuse and recycling of used products. We introduce a transferability benefit index (TBI) in this paper, and the feasibility and validity of a method for using it is demonstrated by a simplified case study of two disassembly systems with four pieces of equipment. Future work includes the following topics:

Development of a simpler method to calculate resource-sharing difficulties among production systems with different interdependency patterns.

Development of a redesign method to reduce the difficulty of sharing resources among multiple production systems.

More practical case studies to evaluate the effectiveness and feasibility of the methods.

Additional material

Additional file 1: Weighting factors and interdependency among transferring tasks in disassembly system 1.

Additional file 2: Weighting factors and interdependency among transferring tasks in disassembly system 2.

List of abbreviations
CMS: closed-loop manufacturing system; DSM: design structure matrix; TBI: transferability benefit index;

Acknowledgements
This study was supported by Tekes - the Finnish Funding Agency for Technology and Innovation, VTT and Finnish Industry. The study has been part of the project called LIIKU - Transferable and reconfigurable production cells and a part of SISU2010 - Innovative Production research program. The authors thank Dr. Nozomu Misihima, Dr. Mikael Haag, Dr. Markku Hentula, and Dr. Otso Vaatainen for their fruitful suggestions and discussions.

Author details
[1]National Institute of Advanced Industrial Science and Technology (AIST), 1-2-1 Namiki, Tsukuba, Ibaraki, Japan [2]VTT Technical Research Centre of Finland, Metallimiehenkuja 6, Espoo, Finland

Authors' contributions
SK developed a strategic decision making method for sharing resources among multiple manufacturing/remanufacturing system, carried out case study, and drafted the paper. TS also participated in the development of the method and helped to draft the paper. All authors read and approved the final paper.

Authors' information
SK has been a researcher in the National Institute of Advanced Industrial Science and Technology, since 2005. He received his PhD in Precision Machinery Engineering from the Graduate School of the University of Tokyo in 1999. His research interest includes life-cycle engineering (eco-design and life-cycle simulation) and manufacturing system design.
TS has been a researcher in the VTT Technical Research Centre of Finland since 1988 after graduating as a M.Sc. in Tampere University of Technology.

Table 4 TBI calculation results

	Equipment			Row No.
	b	c	d	
Overall benefit for the sharing	2000	5500	40000	1
Total difficulty for the sharing	74	6	42	2
TBI for the sharing	27	917	952	3
Resource No. i	2	3	4	

Now he is working as a senior research scientist and his research interests include production systems design and robotics.

Competing interests
This study was supported by Tekes - the Finnish Funding Agency for Technology and Innovation and VTT. The study has been part of the project called LIIKU - Transferable and reconfigurable production cells and a part of SISU2010 - Innovative Production research program.

References
1. Tomiyama T: **A Manufacturing Paradigm Toward 21st Century.** *Integrated Computer Aided Engineering* 1997, **4**:159-178.
2. Ayres R, Ferrer G, van Leynselee T: **Eco-efficiency, asset recovery and remanufacturing.** *European Management Journal* 1997, **15(5)**:557-574.
3. Steinhilper R: **Recent Trends and Benefits of Remanufacturing: From Closed Loop Businesses to Synegetic Network.** In *Proceedings of EcoDesign 2001: 11-15 December 2001; Tokyo* Edited by: Suga T 2001, 481-488.
4. Kamata M, Uchida S: **Inverse Manufacturing System of One-time-use Camera "QuickSnap".** *FujiFilm Res & Dev* 2000, **45**:28-34, in Japanese.
5. Kerr W, Ryan C: **Eco-efficiency gains from remanufacturing. A case study of photocopier remanufacturing at Fuji Xerox Australia.** *Journal of Cleaner Production* 2001, **9**:75-81.
6. Ferrer G, Whybark DC: **Material planning for a remanufacturing facility.** *Production and Operations Management* 2001, **10(2)**:112-124.
7. Guide VDR Jr, Jayaraman V, Srivastava R: **Production planning and control for remanufacturing: a state-of-the-art survey.** *Robotics and Computer-Integrated Manufacturing* 1999, **15**:221-230.
8. Kondoh S, Soma M, Umeda Y: **Simulation of Closed-loop Manufacturing Systems Focused on Material Balance of forward and inverse flows.** *International Journal of Environmentally Conscious Design & Manufacturing* 2007, **13(2)**:1-16.
9. Babiceanu RF, Chen FF: **Development and applications of holonic manufacturing systems: a survey.** *J of Intelligent Manufacturing* 2006, **17**:111-131.
10. Ueda K: *Biological Manufacturing System* Kogyo Chosakai Publishing; 1994, (in Japanese).
11. Kondoh S, Umeda Y, Tomiyama T, Yoshikawa H: **Self organization of the cellular manufacturing system.** *Annals of the CIRP* 2000, **49(1)**:347-350.
12. Taisch M, Colombo AW, Karnouskos S: **Socreades Roadmap, The Future of SOA-based Factory Automation.**[http://www.socrades.eu/Documents/objects/file1274836528.84], accessed 10.12.2010.
13. McGovem J, Sims O, Jain A, Little M: **Enterprise Service Oriented Architectures.** Springer Netherlands; 2006.
14. Morioka M, Sakakibara S: **A new cell production assembly system with human-robot cooperation.** *Annals of the CIRP* 2010, **59**:9-12.
15. Browning TR, Co LMA, Worth F: **Applying the design structure matrix to system decomposition and integration problems: A review and new directions.** *IEEE Transactions on Engineering Management* 2001, **48(3)**:14.

4

Performance analysis of the closed loop supply chain

Farazee MA Asif[1*], Carmine Bianchi[2], Amir Rashid[1] and Cornel Mihai Nicolescu[1]

Abstract

Purpose: The question of resource scarcity and emerging pressure of environmental legislations has brought a new challenge for the manufacturing industry. On the one hand, there is a huge population that demands a large quantity of commodities; on the other hand, these demands have to be met by minimum resources and pollution. Resource conservative manufacturing (ResCoM) is a proposed holistic concept to manage these challenges. The successful implementation of this concept requires cross functional collaboration among relevant fields, and among them, closed loop supply chain is an essential domain. The paper aims to highlight some misconceptions concerning the closed loop supply chain, to discuss different challenges, and in addition, to show how the proposed concept deals with those challenges through analysis of key performance indicators (KPI).

Methods: The work presented in this paper is mainly based on the literature review. The analysis of performance of the closed loop supply chain is done using system dynamics, and the Stella software has been used to do the simulation.

Findings: The results of the simulation depict that in ResCoM; the performance of the closed loop supply chain is much enhanced in terms of supply, demand, and other uncertainties involved. The results may particularly be interesting for industries involved in remanufacturing, researchers in the field of closed loop supply chain, and other relevant areas.

Originality: The paper presented a novel research concept called ResCoM which is supported by system dynamics models of the closed loop supply chain to demonstrate the behavior of KPI in the closed loop supply chain.

Keywords: Closed loop supply chain, Key performance indicator, Logistics, Operations management, Production management, Performance measurement, Resource conservative manufacturing, Supply chain management, System dynamics, Remanufacturing

Background

Due to worldwide population boost, economic growth, and increase in standards of living, current reserves of natural resources are proven to be insufficient, and the Earth's ecosystems are facing increasing threat. The current growth indicates that the worldwide population will be doubled by 2072 [1]. This double population size will result in a fivefold increase in the gross domestic product (GDP) per capita, with a tenfold increase in resource consumption and waste generation [2]. By contributing 30.7% to the total world GDP and employing a

0.7 billion workforce worldwide (estimated in 2010) [3], the manufacturing industry serves as one of the main driving forces in economic growth. Indeed, the manufacturing industry is consuming resources and generating waste on a large scale at the same time.

It is estimated that if the current consumption rate continues and recycle rate remains the same, then there will be no iron ore left for consumption in the next century [4-6]. Besides, the manufacturing industry is one of the largest contributors to waste generation. In 2008, approximately 363 million tons of solid waste (account for 14% of the total waste) was generated by the manufacturing industry in the EU-27 [7]. In addition to this, through the extended producer responsibility regulation, manufacturers are now fully or partially accountable for

* Correspondence: asif@iip.kth.se
[1]Department of Production Engineering, KTH Royal Institute of Technology, Stockholm, Sweden
Full list of author information is available at the end of the article

End-of-Life (EoL) products that are sold in the market. The problem has become more serious with an increase in tax and restriction on the landfill of solid waste.

Moreover, in the fast-growing and evolving consumer market, products seldom reach EoL when a consumer decides to shift to the next generation of products. In those cases, products end up in scrap yards although they retain some values. Recovering only material from a product when it could be possible to recover other values is not the best practice both from a manufacturing and an environmental point of view.

To summarize, manufacturing industries have to grow in the same proportion as the market demands with limited resources, higher-energy efficiency, and lower emission and wastes. The manufacturing industry needs solutions that can solve entirely, or partially, all the problems. Resource conservative manufacturing (ResCoM) is a novel holistic concept which deals with the conservation of resources through the product's multiple life cycle [8]. ResCoM is defined as follows:

A strategic model which emphasizes conservation of resources through product's multiple life cycle by product design, incorporating supply chain and business model and by integrating OEMs, consumers and other relevant stakeholders. Resources conservative manufacturing system seeks to optimize material and energy usage in manufacturing, use phase and end of use and value recovery from the product at the end of life.

ResCoM proposes to design products in a way that can sustain a number of predefined life cycles. At the end of each predefined life cycle, products are returned to the original equipment manufacturer (OEM) or to the authorized third party; upon return, remanufacturing or other EoL strategies, such as recycling and landfilling, are undertaken. Remanufactured products are then redistributed through the ResCoM closed loop supply chain using the ResCoM business model. As multiple life cycles require the same product to come back and forth in several occasions, a robust closed loop supply chain is vital.

The main objectives of this research are as follows:

- To introduce a novel concept named as ResCoM,
- To demonstrate how key performance indicators (KPI) such as rate (production, assembly, shipment, order), delivery delay, level of inventory, backlog, and capacity (production and assembly) in the closed loop supply chain are affected under variable quantity of product flow at variable times, [9,10], and
- To show how adoption of ResCoM concept improves the robustness of the closed loop supply chain.

Closed loop supply chain: a state-of-the-art review

Designing and managing supply chains to ensure collection of used products (usually addressed as 'core') are two of the essentials for products' multiple life cycles. A supply chain of this kind is usually addressed as a reverse supply chain or closed loop supply chain. A significant difference can be observed when defining these two terms. It is appropriate to address the chain of core collection as the reverse supply chain, if the following conditions are fulfilled:

- The recovered cores do not enter the main stream of the forward supply chain.
- The recovered contents of the original products used by other firms to manufacture products serve a different purpose [11,12].

It should be noted that core collection activities can only be referred to as a closed loop supply chain if the following conditions are fulfilled:

- The core is collected by the OEM or the third party remanufacturer that acts as the supplier to the OEM.
- The core enters (and is used) in the main stream of a manufacturing forward material flow.
- The remanufactured product is sold in the same way as the new one, i.e., the remanufactured product is not considered as a different product variant, and order and supply is not handled separately.

Figure 1a,b,c describes the material flow in different types of supply chains.

The ideal closed loop supply chain, which is essential for the success of the product's multiple life cycle, is shown in Figure 1a. By clarifying the existing misconceptions, the closed loop supply chain management can be defined as follows [13]:

The design, control, and operation of a system to maximize value creation over the entire life cycle of a product with dynamic recovery of value from different types and volumes of returns over time.

In the remanufacturing system, the core acts as raw material, and seamless operation of the system entirely depends on the efficiency of the core collection. It becomes especially challenging as the core is not supplied by one or a few suppliers in a periodic and systematic manner. Instead, the suppliers of the core are the end consumers who own one or a few products and return those products whenever they need or want. In addition to this, the consumers' geographic locations could be anywhere on the globe. The supply chain

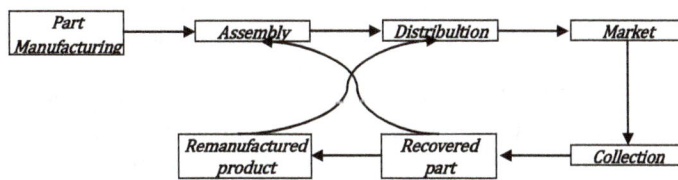

a. Remanufacturing is performed by the OEM or a 3rd party but the product is distributed through the same channel and to the same market (ideal closed loop supply chain)

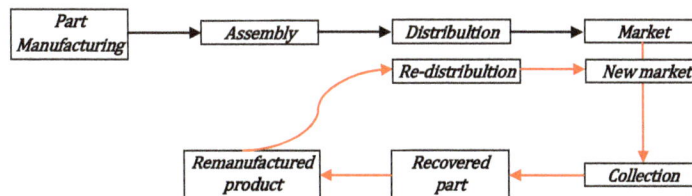

b. Remanufacturing is performed by the OEM or a 3rd party but the product is distributed through a different channel and to a different market (reverse supply chain with an open end, often mistaken as a closed loop supply chain).

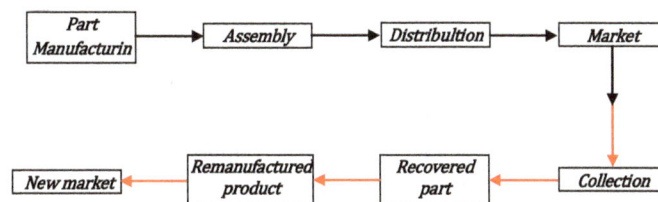

c. Remanufacturing is performed by the 3rd party and the product is distributed to a different market (entirely open system, two forward supply chains, one for the manufacturer and the other for the remanufacturer).

Figure 1 Material flow in different types of supply chain. Different arrows in the figures illustrate direction of product and component flow in the supply chain.

becomes further complicated with product variety, return time, quality of the core, product life cycle, technology life cycle, cost of collection, and so on.

Guide and Van [13] have put together the past 15 years' research development in the closed loop supply chain which provides an overview of relevant research areas, and Lundmark et al. [14] have presented a literature review pointing out industrial challenges within the remanufacturing system. Researchers who have worked with the closed loop supply chain have more or less acknowledged the problem of uncertainty related to timing and quantity of the returned core, quality of the core, and mismatch between the supply and demand of the core and remanufactured product. This problem was mentioned in the early research done by Thierry et al. [15], and the most recent work presented by Guide and Van [13] indicates that the problem still exists. Along the way, these problems have been brought up by several authors; among them, the contributions of Gungor and Gupta [16], Seitz and Peattie [17], and Toffel [18] are worth mentioning. By reviewing several relevant research, the underlying reasons of uncertainty have been identified as follows:

- The return of the core occurs for different reasons in different periods of time [19,20].
- A product's EoL is the result of the complex relationships between age and pattern of use (user conditions, user interactions, levels of service and maintenance, etc.) [21].
- Some products never return as the products move out of the region where legislative or other obligations are not valid and return is not economically feasible.
- The product's information is lost; thus, the core collection from the product is done manually on the basis of trial and error, which often causes destruction of cores. Freiberger et al. [22] have given an example of difficulties in testing and remanufacturing of electronic and mechatronic vehicle components due to lack of information.
- Remanufacturing is treated as a separate business; therefore, demand and supply is tackled independently.
- Products are not designed for efficient recovery [23,24].

However, with the increase in interest to conserve resources, efforts to minimize the uncertainties in core collection are also getting attention. To ensure flawless core returns, some sort of agreement between OEMs, consumers, and remanufacturers is needed. There are several business models that have been adopted by the pioneering OEMs in remanufacturing. Östlin et al. [20] have discussed some of the relationships and core acquisition strategies often used by the OEMs. Kumar and Malegeant [25] pointed out that strategic alliance between the OEMs and eco-non-profit organizations in the collection process not only helps to acquire cores at EoL/end of use (EoU), but also creates value for the firm. Among all these, the most commonly used business models for core collection are ownership-based and buyback. However, from the publication of Lifset and Lindhqvist [26], it is understood that the ownership-based business model is not straightforward, and its success depends upon careful analysis of the profit and loss. In other words, the ownership-based business model is not always feasible. Buyback is not as efficient as it is supposed to be if the consumers are not concerned and motivated.

Moreover, this solves half of the problem. It is true that these kinds of business models bring a certain level of certainty to the timing of the core returns, but uncertainty related to the quality of the core remains unsolved [8-11]. At the same time, the above-mentioned business models aim to bring cores back at the EoL/EoU, at which point, value recovery becomes extremely difficult. It is also important to consider the consumer's perception about newness of the product as it influences return of the core. Most of the business models may fail to fulfill its purpose if the consumers have a negative attitude towards the remanufactured product. The rest of the business models may fail if the customers wish to change brand or manufacturer.

System dynamics and its application in closed loop supply chain

System dynamics is a method to enhance learning in the complex system which is grounded in the theory of non-linear dynamics and feedback control developed in mathematics, physics, and engineering. Basically, the dynamic tendency of any complex system is the result of a system's internal structures, feedback mechanism, and causal relationships among factors that are active in the system [27]. System dynamics was introduced by Jay Wright Forrester and developed at MIT in the mid 1950s. Since then, system dynamics has been applied to a wide range of issues in social, economic, and engineering sciences.

Ilgin and Gupta [28] concluded that the application of simulation, stochastic programming, robust optimization, and sensitivity and scenario analyses has become quite popular in research due to a high degree of uncertainty involved in reverse logistics. They have also mentioned that more studies are needed to better control the effects of uncertainties in the closed loop supply chain. System dynamics is one of such modeling and simulation method which is widely used in the management of production systems, especially in the supply chain (forward) for about five decades. An overview of the frame of the research that applied system dynamics in the supply chain is presented by Angerhofer and Angelides [29], Georgiadis and Vlachos [30], and Vlachos et al. [31]. The trend of using system dynamics in the analysis of the closed loop supply chain is relatively new but growing; at the same time, Kumar and Yamaoka [32] mentioned the lack of system dynamics research in studying the closed loop supply chain. Nevertheless, fair progress has been made in this respect.

Georgiadis and Vlachos [30] studied the long-term behavior of reverse supply chains with product recovery under the influence of various ecological awarenesses. Later, Vlachos et al. [31] examined capacity planning policies of a single product's forward and reverse supply chain transient flows due to market, technological, and regulatory parameters. Georgiadis et al. [33] and Georgiadis and Efstratios [34] developed models of system dynamics to study the closed loop supply chain with remanufacturing both for single-product and two-product types under two alternative scenarios which incorporate a dynamic capacity modeling approach.

In a recent work, Qingli et al. [35] continued the work of Vlachos et al. [31], to some extent, and added the bullwhip effect into their studies. Similar modeling has been done by Schröter and Spengler [36], but their focus was product recovery to obtain spare parts for equipment, when the original equipment is no longer produced. Poles and Cheong [37] modeled the closed loop supply chain to determine factors that influence the return of cores and concluded that customer behavior and the level of service agreement improve control over returns, thus, reducing uncertainty in remanufacturing systems.

The research mentioned above focused mainly on operational issues of the closed loop supply chain. Besiou et al. [38] argued that even though system dynamics has been applied to various environmental problems, business policy, and strategy, few strategic sustainability problems in the closed loop supply chain are reported. Nevertheless, Georgiadis and Besiou [39] combined strategies of environmental sustainability with operational issues of the closed loop supply chain to study their interaction and understand their impact on the environment. In an earlier research, Georgiadis and Besiou [40] examined the impact of innovation and ecological motivation to study the long-term behavior of the closed loop supply chain which can be used as a strategic tool.

These are only few of the publications; besides these, there are a large number of publications available. Apart from applying system dynamics, linear programming models for the closed loop supply chain network design are quite popular among researchers. The review of mathematical models using linear programming has not been included in this paper. The authors recommend the work of Özkir and Basligil [41] for an overview of research done to design the closed loop supply chain network using the linear programming approach.

Critical review of the state-of-the-art

There is a misconception concerning the closed loop supply chain. Supply chains designed to collect cores and developed to sell remanufactured products in a secondary market (bypassing the OEMs) are not necessarily closed. Apart from this fact, it is an established truth that the main problem of the closed loop supply chain is the uncertainty in timing of core return and the quality of the returned cores. A fundamental but rarely discussed truth is that most of the researchers suggest implementing the closed loop supply chain concept where the product, or the business model, is not designed for it.

The scope of applying system dynamics in the closed loop supply chain is large compare to what had been done so far. System dynamics has been applied in both operational and strategic issues of the closed loop supply chain. However, studying and analyzing the performance (or behavior of KPI) of the closed loop supply chain under the influence of uncertain quantity and quality of returned core in unpredictable intervals using system dynamics had not been the main focus of the research up to this point. The influence of the uncertainty in strategic resources such as inventory, capacity, backlog, and demand in the closed loop supply chain had not been extensively covered in most research.

ResCoM: a new paradigm of manufacturing

According to the definition presented in the 'Background' section, ResCoM is a holistic approach that provides a complete solution to maximize resource conservation and minimize waste. ResCoM is built upon the concept of the product's multiple life cycle.

The concept of product life cycle can simply be explained as follows: the life cycle of a product (that contains more than one part) is generally equal to the life cycle of the component that has the shortest life. For example, a product consists of three components X, Y, and Z, and each has the designed life of 1, 2, and 3 years, respectively. Basically, the product will reach its EoL when one of the components fails. Considering that other factors do not affect the component's life, component 'X' will fail at the age of one, and eventually, the

product will be discarded. It is to be noted that components 'Y' and 'Z' have equal and twice as many remaining lives compared with X, respectively. If the entire product is discarded, the potential recoverable values are lost. Instead, if component X can be replaced or upgraded at the age of 1 year and component X is replaced or upgraded again along with Y at the age of 2 years, then the product can sustain three life cycles. Alternatively, the ideal case is to design a product that contains components that have the same duration of life. Of course, in reality, it is not as straightforward as it has been explained. It is important to highlight that the life cycle of a product/component is not time-dependent, which means that the time when a product/component will reach its EoL is not exactly deterministic. The life cycle duration of a product is the result of a complex relationship mainly between age, operating conditions, service, and maintenances during the life cycle and the user locations. It is not possible to determine the exact interval of each life cycle. Therefore, ResCoM proposes to develop a robust design method to reduce the uncertainty in predicting the EoL and to integrate life cycle-monitoring devices to monitor the physical/functional condition of critical components.

In ResCoM, the product is named as resource conservative product (RCP) which is used as a 'brand' name. Each life cycle of RCP is labeled with a resource conservation level (RCL). The concept is illustrated in Figure 2a. In principle, RCL_0 refers to a new RCP that contains only new components, and it is at the start of its first life cycle, having several life cycles ahead. Components at a certain level are called RCL_i (where $i = 0, 1, 2...$) components, such as RCL_0 components, RCL_1 components, and so on. At the end of life cycle 1 (i.e., end-of-resource conservation level 0 ($EoRCL_0$)), when the desired performance reaches the minimum allowable, the product is recalled, upgrading and replacement of complements are done, and remanufacturing is performed. RCL_1, which is the beginning of the second life cycle, contains new components of RCL_0 and upgraded components of RCL_0 and may contain some new components. This approach continues until the product finishes its predetermined number of life cycles. At the end of each life cycle, the product is restored to the desired performance level. This so far explains the life cycle at the product or subassembly level. The life cycle of the component is slightly different and is illustrated in Figure 2b.

Let us assume that a product is assembled with three components represented by X, Y, and Z and that their performance index over time is shown in Figure 2b with red, blue, and green curves, respectively. In this particular case, the life cycle (i.e., three life cycles) of the product is determined based on the component which has

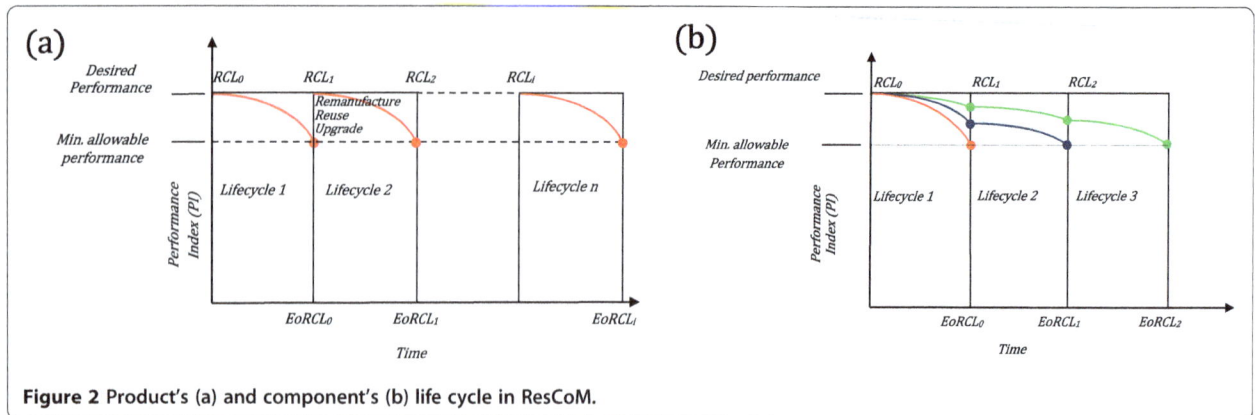

Figure 2 Product's (a) and component's (b) life cycle in ResCoM.

the longest design life, i.e., component Z. At RCL_0, the product contains three new components. At $EoRCL_0$, component X reaches the minimum allowable performance, which is then replaced with a similar component. It means that at RCL_1, the product will have two RCL_0 components and one new component and so on.

So far, the core concept of ResCoM and product's multiple life cycle has been briefly presented. Readers are referred to the work of Asif [8] for further details.

Among the many dimensions of ResCoM research framework, the closed loop supply is an essential element. The innovative approach of managing the closed loop product system in ResCoM is further elaborated in the following sections.

ResCoM closed loop supply chain

As discussed in the preceding section, in the product's multiple life cycle approach, the product will return at several occasions and will go through remanufacturing. To facilitate this, the closed loop supply chain is required. The operational effectiveness of a supply chain mostly depends on the smooth flow of material both in forward and reverse directions without constraining the planned capacity of the manufacturing processes. It means that the manufacturing system for RCL_0 products and the manufacturing systems for RCL_1 to RCL_i should not be over- or under-capacitated. To ensure this, the expected quantity of the product to be manufactured at RCL_0 and RCL_1 to RCL_i needs to be known. As the RCL_0 product refers to a newly manufactured product and follows a standard manufacturing forward supply chain, it is relatively simple to plan. On the other hand, RCL_1 to RCL_i manufacturing significantly depends on the availability of the products (cores) from their previous life cycles. Availability of the returned products and scrap rate of the returned products are the main obstacles to the success of the closed loop supply chain. In ResCoM, the problem of product availability is solved through product design, estimation of life cycle duration, and the number of life

cycle and business model. In ResCoM, the quantity and the timing of the product return are predictable within a certain confidence interval. However, the other problem with the quality of the returned product is not entirely solved but minimized to a large extent.

In the conventional approach, it is estimated that the scrap rate of returned product can be anything from 15% to 85%. The reasons for this large variation are mainly due to age, operating conditions of product, and quality of service that the product receives during the life cycle. In ResCoM, the age of the returned product is known, and the service of the product is managed and controlled by the OEM (or authorized service provider). This means that in normal circumstance, the quality of the returned products is known within a certain confidence interval. The functional condition of critical components in the products will be monitored during operation; if there is any deviation in the desired performance, the product will be recalled earlier. In this way, currently perceived scrap rate can certainly be reduced, which eventually can create a robust closed loop supply chain with minimum uncertainty.

There are other issues, such as designing the network, planning and controlling the logistics, and production at RCL_1 to RCL_i, which are also related to the success of the closed loop supply chain. These issues are greatly influenced by the types of product, size, and periphery of the market. As the concept is presented in a generic context and does not refer to any specific product, discussion around these issues is not within the scope of the research at this point.

ResCoM business model

The ResCoM approach is not well fitted with the ordinary sell-buy-sell business model. It requires a model that goes beyond the conventional business model and establishes a strong relationship among OEMs, consumers, and third parties (if the OEM decides to outsource RCL_1 to RCL_i production). Based on the concept of RCP brand and RCL labeling, the business model of resource conservative

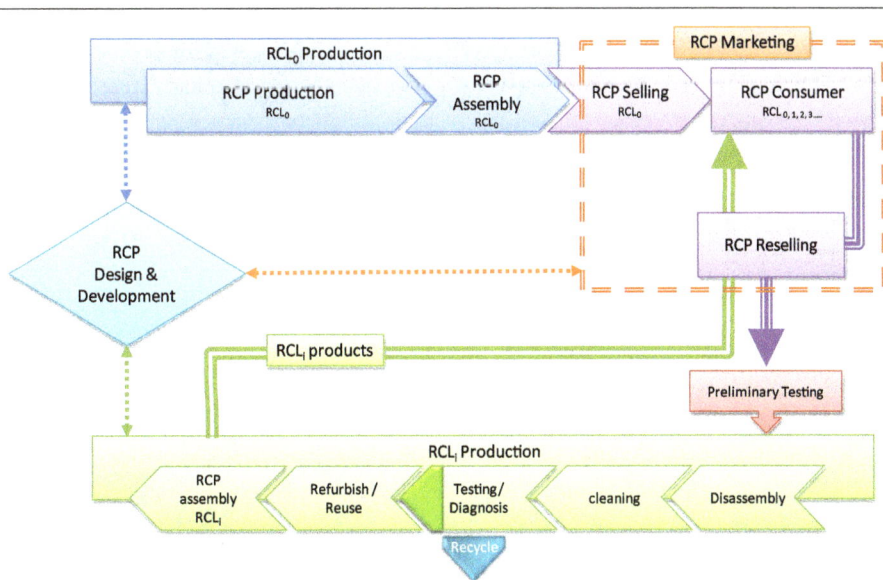

Figure 3 The summarized resource conservative manufacturing business model. RCL_0 is the new RCP product with resource conservation level zero; RCL_i is the RCP with resource conservation level $i = 1, 2, 3$.

manufacturing is illustrated in Figure 3. In this model, the RCP production at RCL_0 and RCL_i are separate functions of the same enterprise. However, RCL_i production can be outsourced by the OEM only if the entire process is controlled by the OEM. In the ResCoM business model, consumers are part of the manufacturing system and mostly responsible for returning the product at the end of each life cycle. As mentioned earlier, consumers are still reluctant towards secondhand products. Therefore, at the beginning, the business model suggests a dedicated RCP reselling unit, which will act as the bridge between the consumer and OEM or third party suppliers. The basis of their relationship and the interest of each stakeholder are determined mainly based on the product type, number of returns, arrangement of returns, and way of reselling. Besides, the RCP reselling unit will also be engaged in promoting RCL_1 to RCL_i product adoption as a social and moral responsibility. Once the business model is established and consumers become comfortable with the product's multiple life cycle and consider product returning as part of their social responsibility, the RCP reselling unit will be abolished. The ordinary product distribution unit will take over both RCL_0 and RCL_i product selling.

The modeling approach and the models of supply chains

The models that are presented in this paper retain different objectives than the publications mentioned in the 'System dynamics and its application in closed loop supply chain' section. The main purpose of the modeling is to study and analyze the performance of the closed loop supply chain. Therefore, the model does not propose

any solution; instead, the model is used to understand the behavior of KPI of the closed loop supply chain in conventional and in the proposed ResCoM context. The models are used to analyze the robustness of the conventional forward supply chain in the settings of the conventional closed loop supply chain and compare it to the proposed one by ResCoM. The aim of the modeling is to see how the KPI in the closed loop supply chain vary with time in different settings. In addition, the aim is to understand the main drivers affecting the KPI as well as the end results, and the behaviors of the strategic resources. Two models have been built, and four different analyses have been made. The behavior of KPI has been analyzed for the following:

1. Conventional forward supply chain.
2. Conventional reverse supply chain.
3. Forward supply chain when reverse supply chain is combined, i.e., the conventional closed loop supply chain.
4. Closed loop supply chain proposed by ResCoM.

The models have been built in three steps. In the first step, forward and reverse supply chains have been modeled without any dependency. In the second step, forward and reverse supply chains have been combined, i.e., the closed loop supply chain where the forward supply chain is influenced by the reverse supply chain. In the third step, the model has been built as how ResCoM proposes. In the following sections, the structure of these models is described. The supply chain is modeled with inventory control

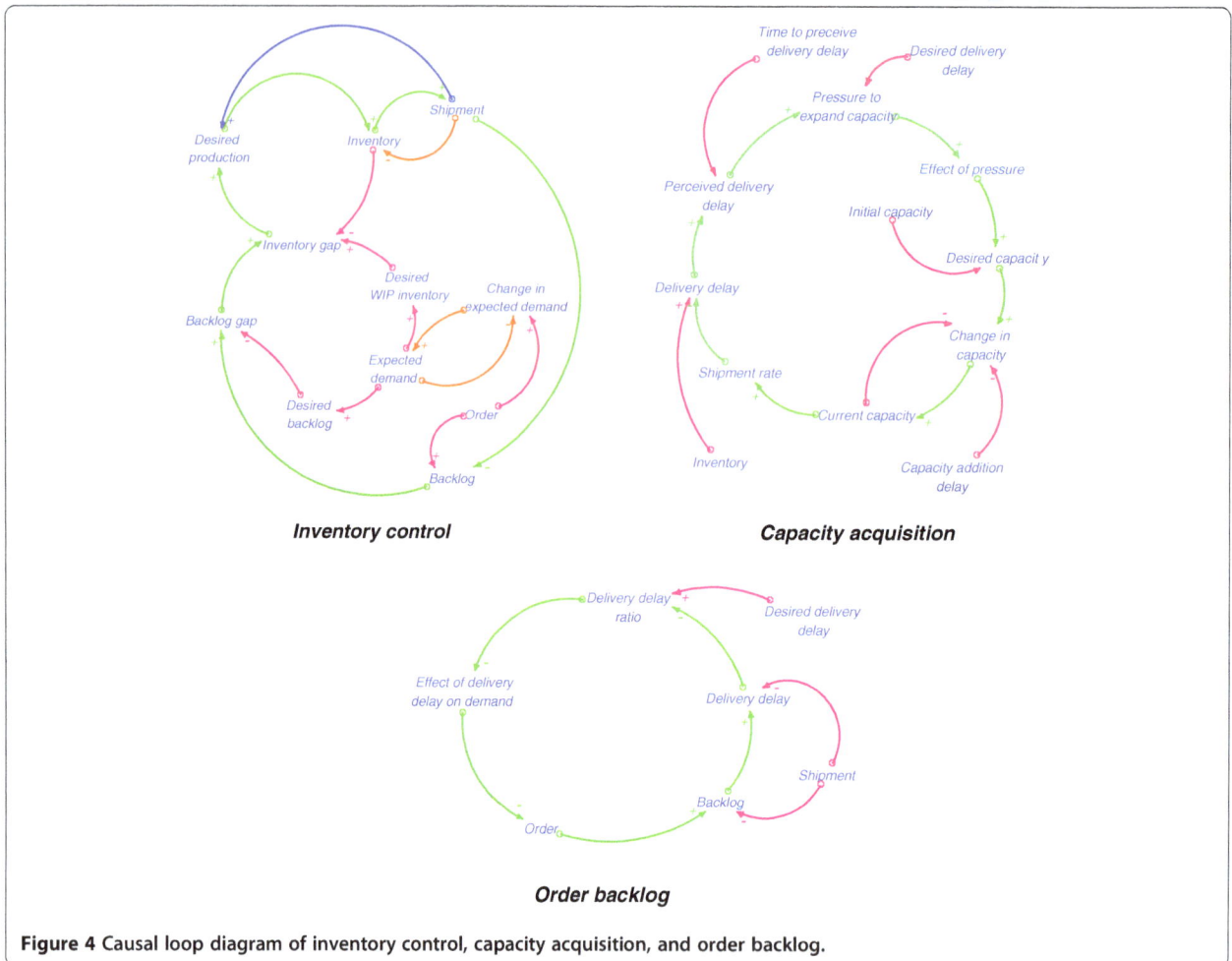

Figure 4 Causal loop diagram of inventory control, capacity acquisition, and order backlog.

mechanism, capacity acquisition, demand backlog, and demand forecasting. The performance of the supply chain is analyzed in respect to the level of inventories, backlogs, rates (production, assembly, shipment, etc.), and delays. The causal structures of the feedback loops used in the models are shown in Figure 4.

Regardless of which model settings are discussed, the performance indicators, drivers, end results, and strategic resources have the same structure and relationships. For example, the end result order (rate) directly influences the strategic resource backlog. The end result is driven by the delivery delay ratio which is influenced by delivery delay. Delivery delay is influenced by the shipment (rate).

Similarly, in case of capacity acquisition loop, the end result is the change in the capacity of the system. This end result is driven by the pressure to expand capacity, which causes the strategic resource capacity to fall or rise. This is directly influenced by the delivery delay.

Finally, in case of inventory, the end result is the desired production rate which is driven by inventory gap and backlog gap. The gaps are influenced by the strategic resource backlog, expected demand, and the inventory itself which are influenced by the delivery delay.

Mathematical formulation

The main mathematical formulations used in the modeling are shown in Additional file 1. However, depending on where in the models these concepts are used, the notation to define the flows, stocks and variables are named accordingly. For detail mathematical formulation of each section in the model the readers are referred to the work of Asif [8].

Forward supply chain

The forward supply chain has been modeled with the sectors named as *production capacity*, *assembly capacity*, *production work in progress (WIP) inventory*, *assembly WIP inventory*, *finished product inventory*, *production backlog*, *assembly backlog*, *sales backlog*, and *demand forecasting*.[a] The stock and flow diagram of the forward supply chain is shown in Figure 5. The following assumptions have been made:

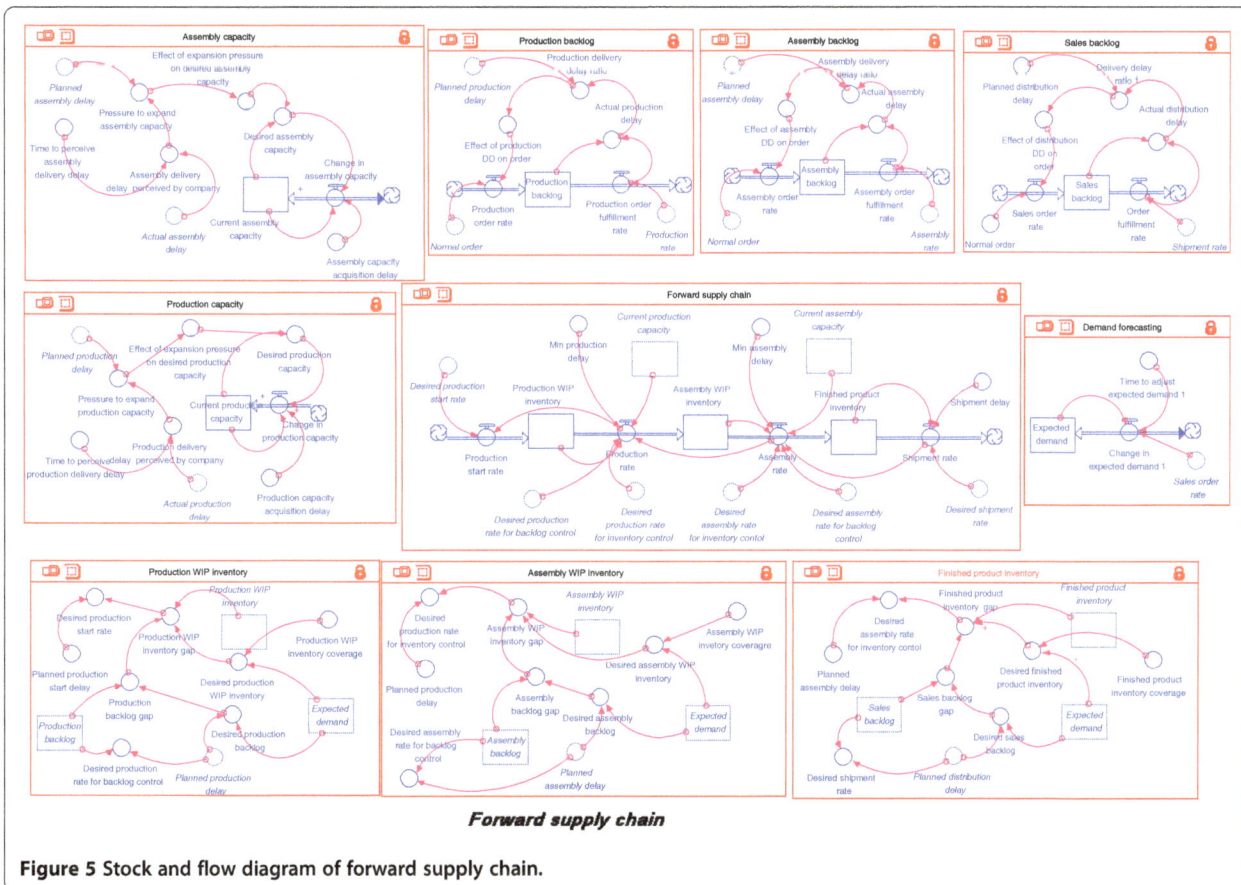

Figure 5 Stock and flow diagram of forward supply chain.

- The models are built for a single product.
- Production starting capacity is infinite.
- Shipment of product is only constrained by availability of product in the *finished product inventory*.
- Order placed by the consumers is constant.

In the *forward supply chain* sector, the stock of *production WIP inventory* is accumulated at the *desired production rate*, and the inventory moved to the next step (*assembly WIP inventory*) at the *production rate*. The *production rate* can be determined in four ways as follows:

- Available *production WIP inventory* starts to move to the next stage after minimum *production delay*.
- Available *production WIP inventory* starts to move to the next stage as much as the *current production capacity* allows.
- Available *production WIP inventory* starts to move to the next stage at a rate that can bring the *production backlog* to the desired level.
- Available *production WIP inventory* starts to move to bring the *assembly WIP inventory* at the desired level.

Current production capacity is an accumulative value of the difference between the *desired and current production*

capacity over time. If the ratio of *actual and planned production delay* becomes larger, then that would create a *pressure to expand capacity*. This pressure causes the *desired capacity* to rise after a predefined delay.

Similarly, in the *sales backlog* sector, the *expected demand* is an accumulative value of the difference between the *expected demand* and *sales order rate* over time. It is to be noted that the *expected demand* represents information, not the physical product. If the ratio of *actual and planned distribution delay* becomes larger, then that would cause a drop in the *order rate*. This causes the *expected demand* to fall. However, the *expected demand* does not rise or fall immediately but after a predefined delay. It is important to note that in the model, *normal order* that is placed by the consumers has been considered as the *order rate* in all steps, i.e., shipment, assembly, and production in the *forward supply chain*.

As mentioned earlier, in the *production backlog* sector, the *production order rate* is considered the same as the *normal order* placed by the consumers. This rate causes the *production backlog* to rise, and backlog decreases with the rate of *production order fulfillment rate*, which is basically the *production rate* (it also reduces *production WIP inventory*). The backlog and the rate at which the order is fulfilled determine the *actual production delay*. The ratio between *planned and actual production delay* would cause the order to fall if the ratio becomes greater than one. Similar to the *expected demand, production backlog* also represents information, not physical product.

In addition, the *production WIP inventory sector* is used to estimate the *desired production WIP inventory* and *desired production backlog*. Based on the *expected demand* and how much inventory to keep, the *desired production WIP inventory* is estimated. Similarly, based on *expected demand* and *planned production delay*, the *desired production backlog* is determined. *Desired production start rate* is estimated based on the gap between the desired and actual inventory and the gap between the desired and actual backlog.

Exactly the same stock and flow structure follows in the *assembly WIP inventory* and *finished product inventory* as of *production WIP inventory*. The *assembly capacity, assembly backlog, assembly WIP inventory, sales backlog*, and *finished product inventory* sectors have exactly the same flow and stock structure as the production part of the model.

Behavior of key performance indicators At the beginning of simulation, the *production WIP inventory* is much less than the desired value, causing a high *production backlog* which results in the *actual production delay* to rise. As soon as the *desired backlog* becomes equal to the actual level, the *actual production delay* becomes equal to the *planned production delay*. For the *desired production backlog* to become equal to the *production backlog*, the *production WIP inventory* level has to rise, and at the same time, the rate at which product is moved to the next stage (*assembly WIP inventory*) also has to rise. The stock of inventory and the backlog are increased with the rate at which products are piling up into the inventory, and the rate of order placed. The inflow and outflow of the inventory and backlog are affected by all other feedback loops that are connected with it. Similarly, with the rise of the *actual production delay*, the capacity side of the model gets alarmed,

causing the *desired production capacity* to rise, which eventually results in the *current production capacity* to adjust. As soon as everything else becomes stabilized, the *current production capacity* also stabilizes. These behaviors are illustrated in the graphs in Figures 6, 7, and 8.

Exactly the same behavior and the same dependency are evident, i.e., after an initial shock, the KPI become balanced, in case of assembly and distribution in the forward supply chain. Therefore, detailed graphical illustration is avoided.

Reverse supply chain

The reverse supply chain consists of sectors namely *reverse supply chain, remanufacturable product inventory, remanufactured product demand forecasting*, and *remanufactured product backlog*. The stock and flow diagram of the reverse supply chain is shown in Figure 9.

The *reverse supply chain* sector consists of the *EoL product inventory* where products accumulate at EoL through three aging chains named as *product in use 1, 2, and 3*. Aging is deterministic; however, the rate at which the product reaches at EoL or to the succeeding stages of *product in use* is probabilistic. It is assumed that the probability of failure increases with age. Products move from *EoL product inventory* to *collected EoL product inventory* after some predefined delay. Products in *collected EoL product inventory* are then inspected, (*inspection rates 1 and 2*) and depending on their physical and functional conditions, products are stored either in *remanufacturable product inventory* or in *non-remanufacturable product inventory*. The physical and functional conditions of returned products are denoted by the *functionality factor*, which is probabilistic and generates any random values between 0.1 and 1. The assumptions made here are as follows:

- There is no capacity constrain in the reverse supply chain.
- The rate, i.e., *shipment rate of manufactured products*, at which product is supplied to the next

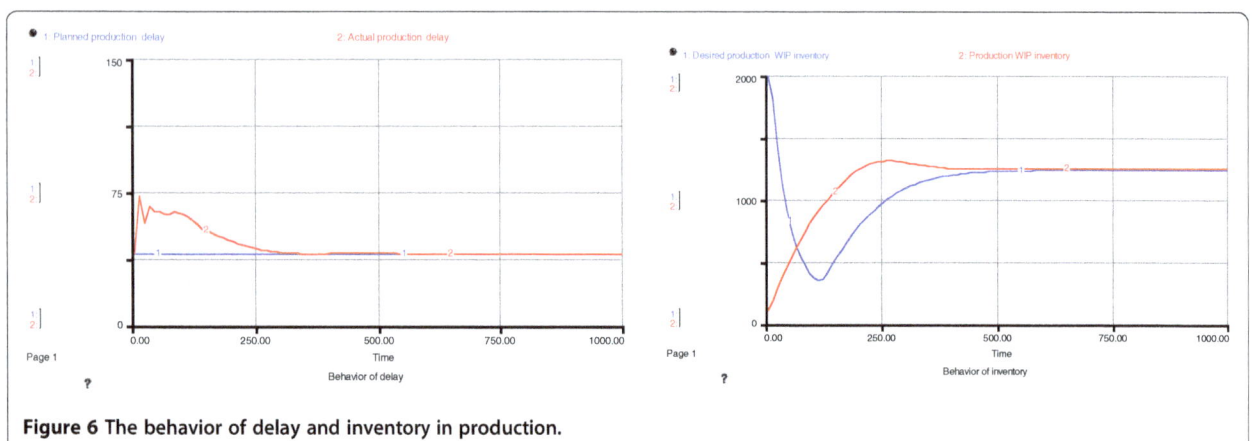

Figure 6 The behavior of delay and inventory in production.

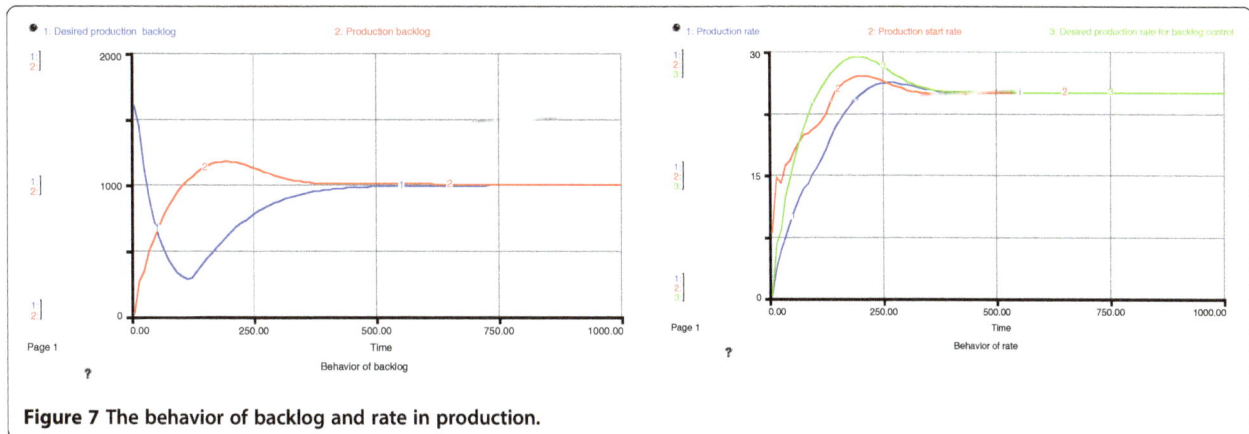

Figure 7 The behavior of backlog and rate in production.

stage is only constrained by the availability of *collected EoL product inventory* and *remanufacturable product inventory* or the *desired shipment rate of remanufactured product*.

- Each product reaching EoL creates a demand, and order is placed immediately.

The stock and flow structure used in the sectors *remanufactured product backlog, remanufactured product demand forecasting*, and *remanufactured product inventory* has the same structure as the backlogs, demand forecasting, and inventory sectors described in the forward supply chain in the previous section.

Behavior of key performance indicators Behavior of KPI in the reverse supply chain is not the same as that in KPI in the forward supply chain. The main reason for inconsistency in the behavior is the random variables that determine different rates in the model. Besides, in

the reverse supply chain model, the demand is considered to be more than the supply. This causes the planned and actual distribution delay, inventory, backlog, and shipment rates never to balance. This hypothesis is a well-known fact in the reverse supply chain. The behavior of KPI is illustrated in the graphs in Figures 10 and 11.

From the above graphs, it can be concluded that the reverse supply chain is unstable in nature. The uncertainty of core arriving time, quantity, and quality causes the feedback loops to suffer. This kind of behavior limits the possibility to create a robust policy. The decision makers usually cannot identify key drivers within the system that can improve the system's performance in such situations.

Conventional closed loop supply chain
In the conventional closed loop supply chain, the above-mentioned two models have been kept the same with two distinct differences. Firstly, *remanufacturable product*

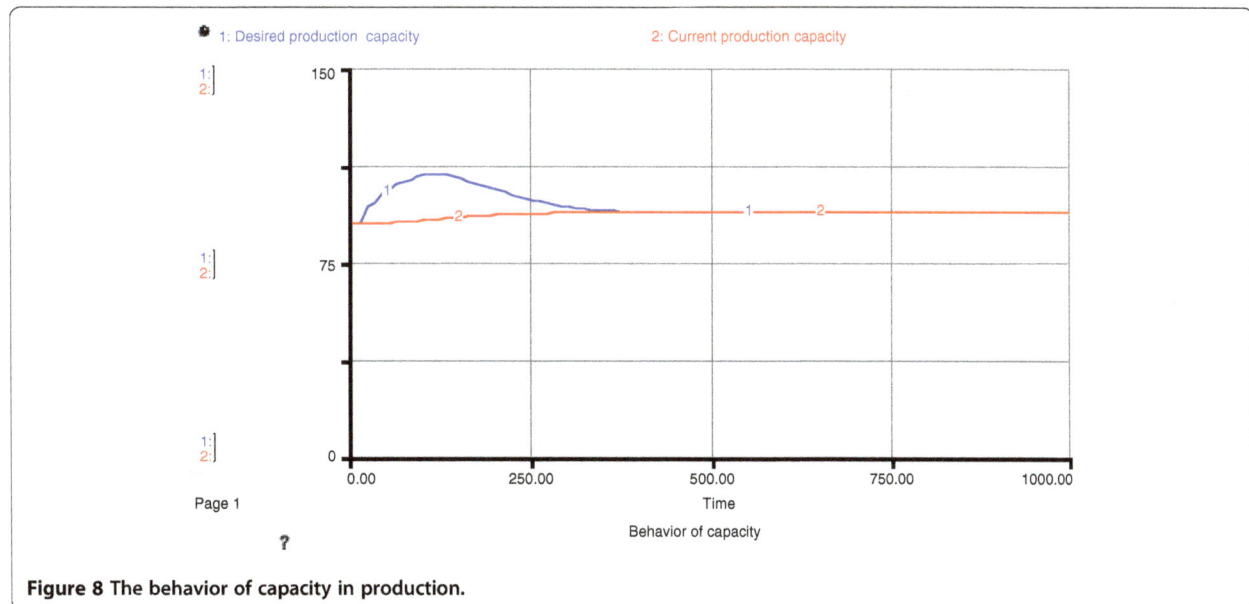

Figure 8 The behavior of capacity in production.

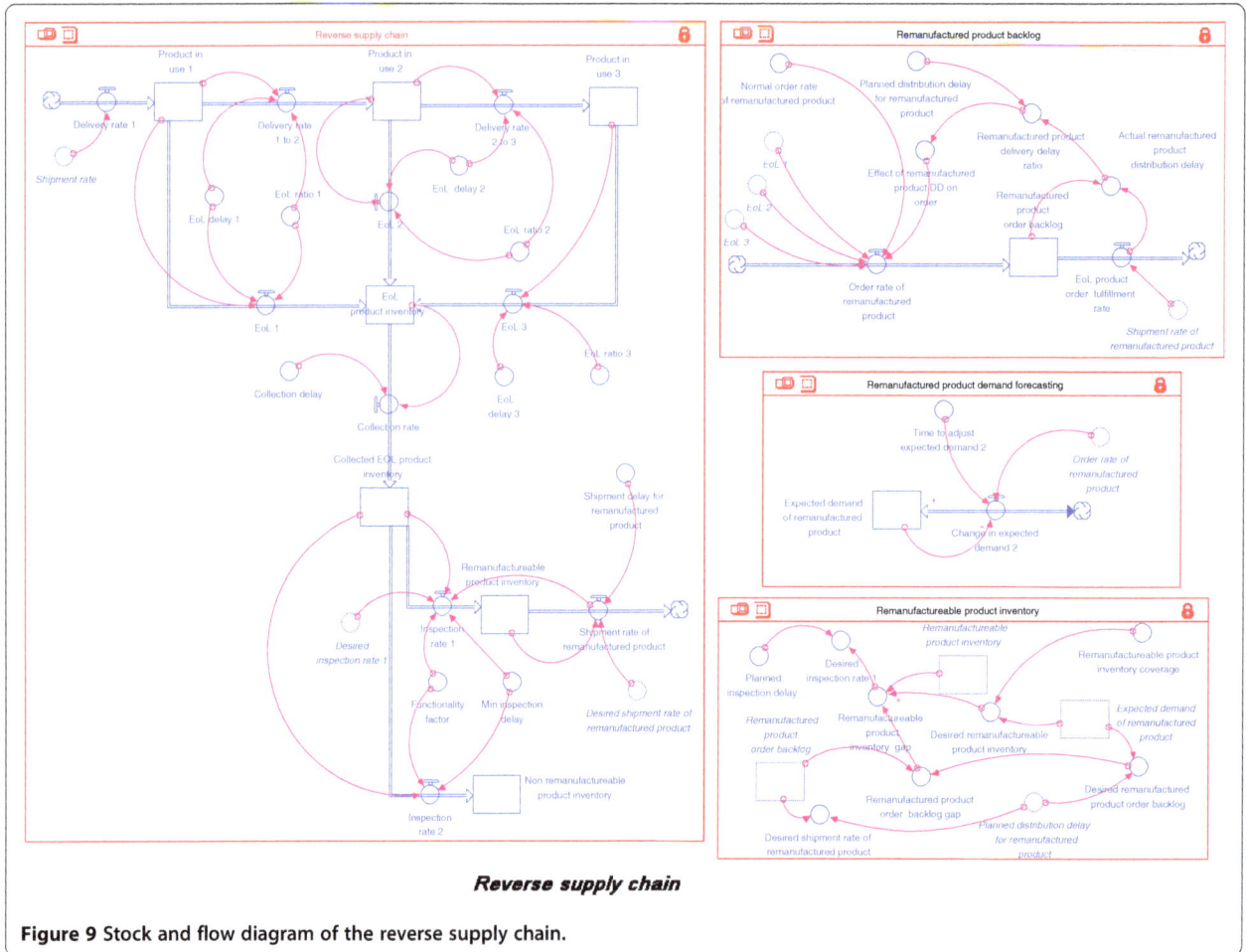

Figure 9 Stock and flow diagram of the reverse supply chain.

inventory has been connected to the *assembly WIP inventory*, i.e., products accumulated in the *remanufacturable product inventory* move to the *assembly WIP inventory* at the *shipment rate of manufactured product*. Secondly, the *order rate of remanufactured product* has been added in the sectors *production backlog, assembly backlog,* and *sales backlog* in the forward supply chain. The changes are shown in the model with 'green'-colored flows and

connections in Figure 12. The main assumptions made here are as follows:

- Both remanufactured and newly manufactured products are sold through the same channel.
- All remanufactured products are as good as the newly manufactured products and can substitute the need for production.

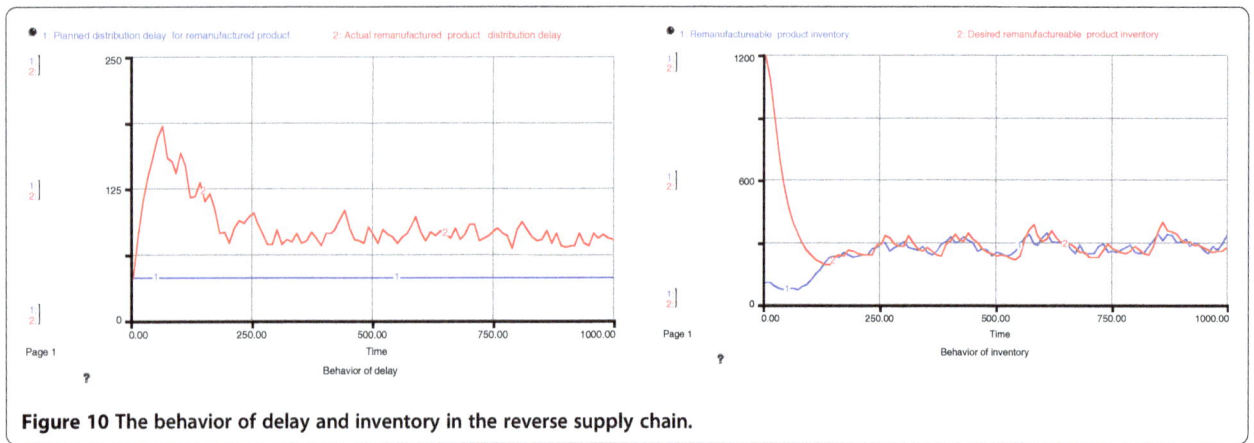

Figure 10 The behavior of delay and inventory in the reverse supply chain.

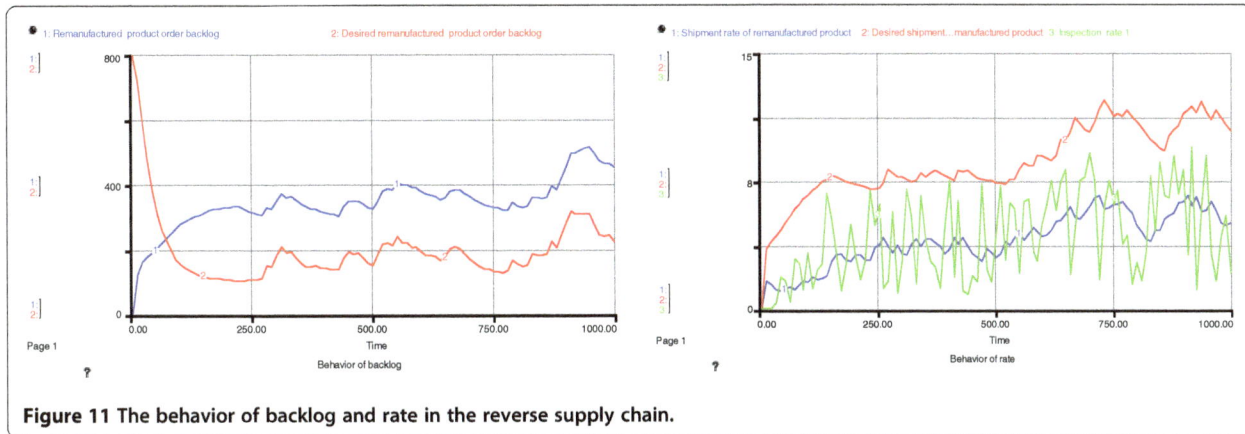

Figure 11 The behavior of backlog and rate in the reverse supply chain.

- The market becomes larger as soon as the firm decides to remanufacture products.
- All remanufacturable products are remanufactured without any delay. It means that the *shipment delay for remanufactured product* is not constrained by other factors such as delay in capacity acquisition and delay in order processing.

Behavior of key performance indicators The behavior of KPI in production (forward supply chain) of the conventional closed loop supply chain is shown in Figures 13, 14, and 15.

Two distinct differences are evident in the behavior of KPI in case of production in the conventional closed loop supply chain compared with the forward

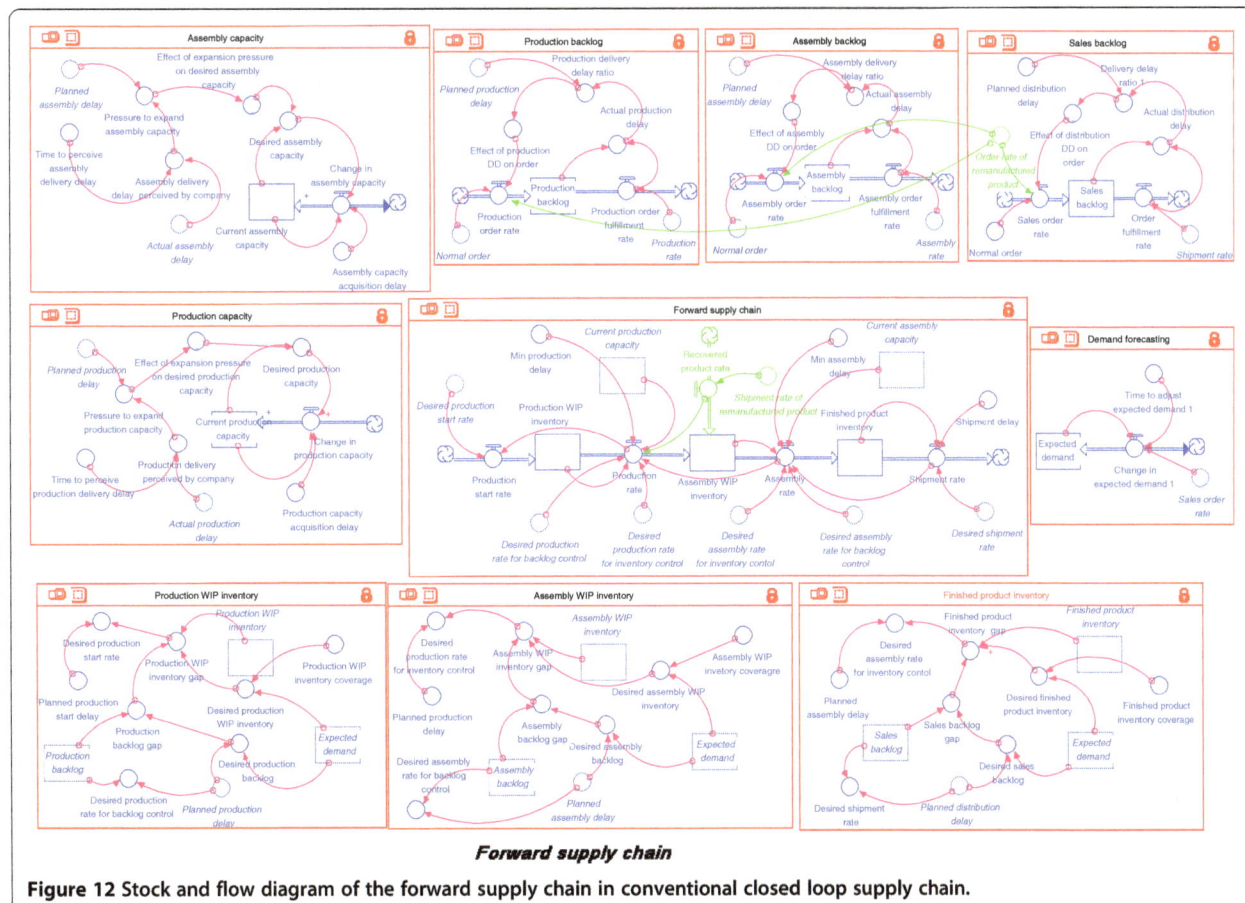

Figure 12 Stock and flow diagram of the forward supply chain in conventional closed loop supply chain.

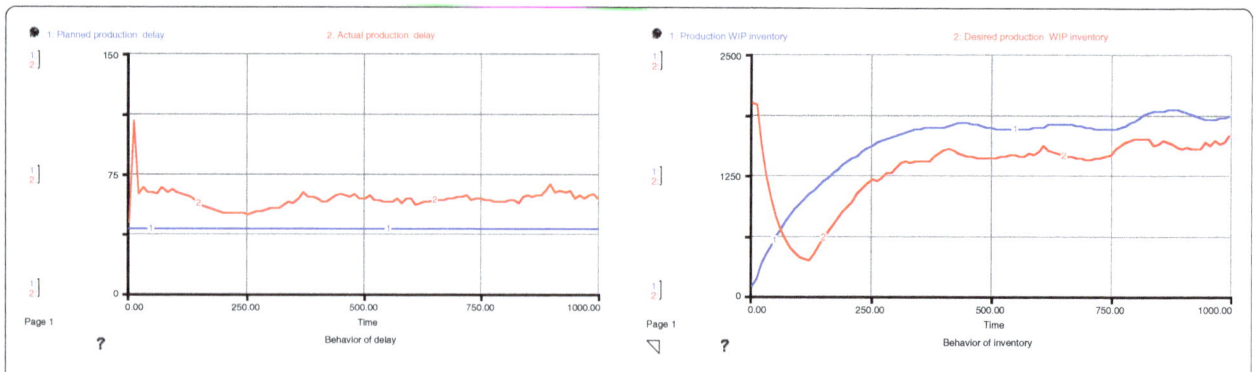

Figure 13 The behavior of delay and inventory in production in the conventional closed loop supply chain.

supply chain discussed in the 'Forward supply chain' section:

- The graphs are not balancing.
- The graphs continuously fluctuate.

In case of other parts of the forward supply chain in the conventional closed loop supply chain scenario, i.e., assembly and distribution exhibit balancing but fluctuating characteristics. The reason of graphs in production not balancing in the closed loop supply chain (both in conventional and ResCoM scenarios) has been mentioned in the 'Discussion' section.

The reverse part of the supply chain in conventional closed loop supply chain shows similar behavior pattern as shown in Figure 10 and Figure 11.

Closed loop supply chain in ResCoM

The closed loop supply chain in ResCoM has a slightly different structure than the conventional closed loop supply chain. As in ResCoM, the time of product return is predetermined; the aging chain does not exist in the model. The only delay to accumulate products from *product in use 1* to *EoL product inventory* is predefined.

In addition to this, all products are assumed to be returned; therefore, there is no random variation in the *EoL ratio*. Moreover, the *functionality factor* that determines *inspection rates 1 and 2* is assumed to be quite high (90% of the products are remanufacturable) and constant. This assumption is in line with the argument made in the 'ResCoM a new paradigm of manufacturing' section, i.e., in the proposed ResCoM approach, the quality of returned products is known (high) to some extent, and almost all of them can be used further (if designed for multiple life cycle). The assumptions made in the models discussed above are valid, and no new assumptions are made. The stock and flow diagram of the reverse part of the ResCoM proposed closed loop supply chain is shown in Figure 16. The stock and flow diagram of the forward part of the closed loop supply chain proposed by ResCoM remains the same as in Figure 12.

Behavior of key performance indicators The behavior of KPI in the forward part in the ResCoM proposed closed loop supply chain is shown in Figures 17, 18, and 19.

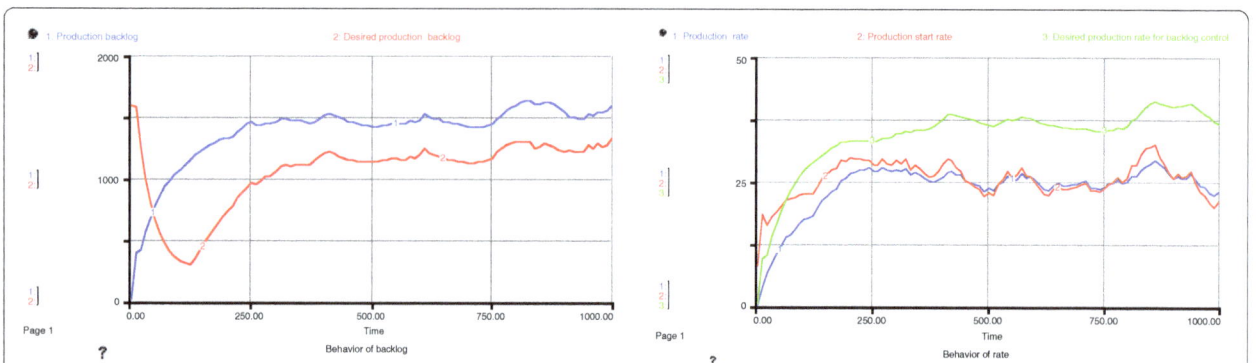

Figure 14 The behavior of backlog and rate in production in the conventional closed loop supply chain.

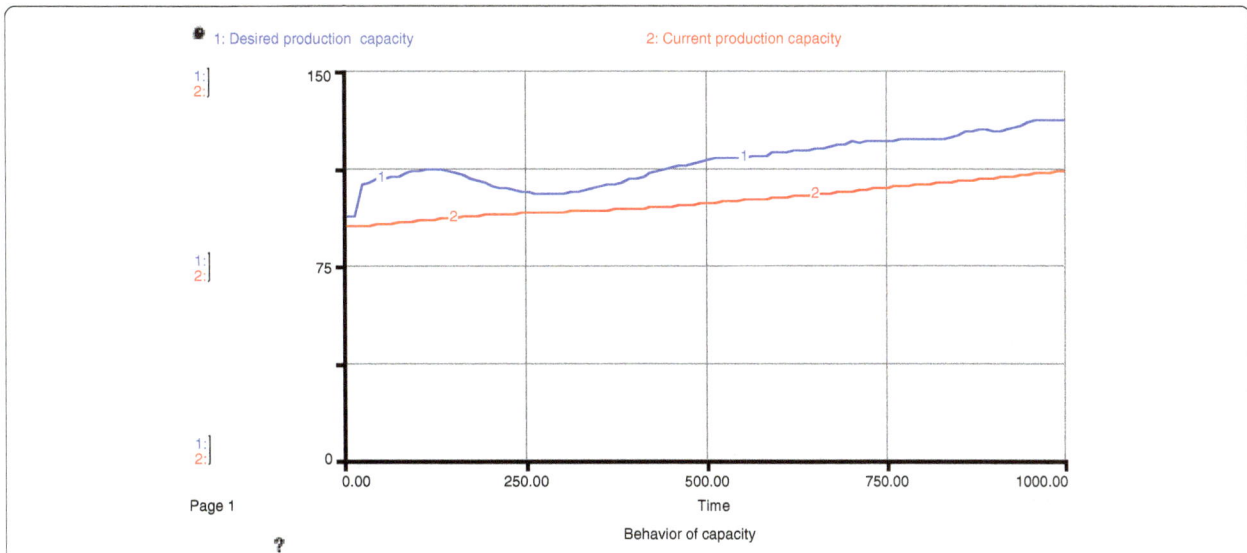

Figure 15 The behavior of capacity in production in the conventional closed loop supply chain.

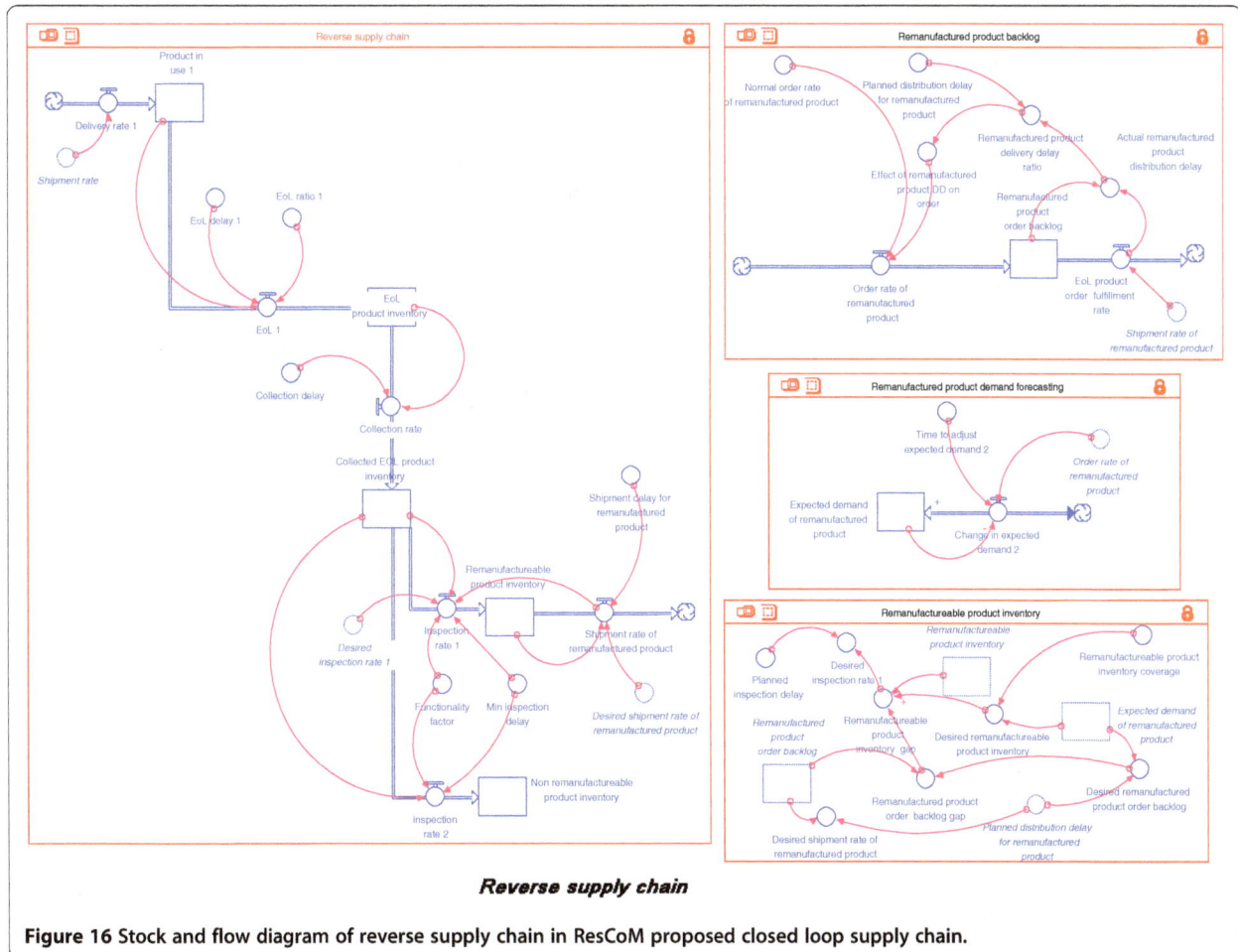

Reverse supply chain

Figure 16 Stock and flow diagram of reverse supply chain in ResCoM proposed closed loop supply chain.

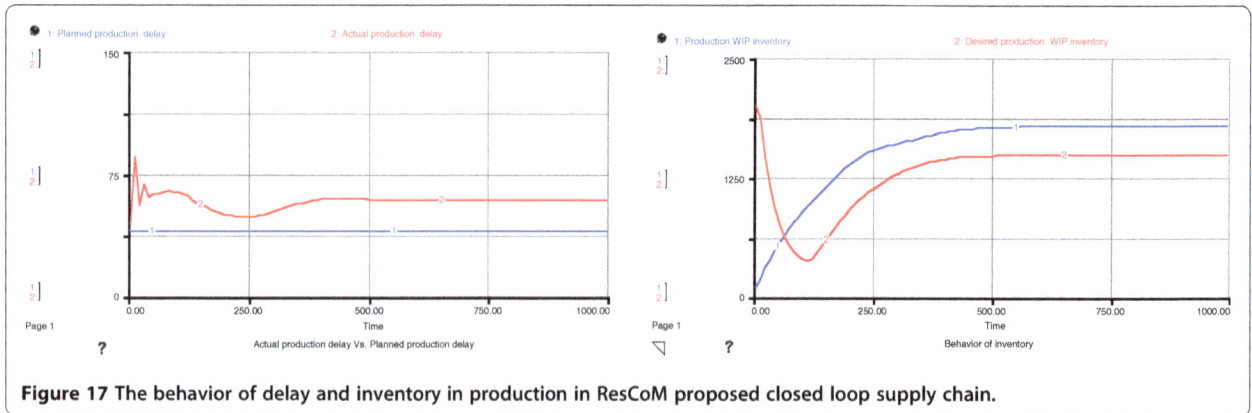

Figure 17 The behavior of delay and inventory in production in ResCoM proposed closed loop supply chain.

The behavior of KPI in the reverse part of the ResCoM proposed closed loop supply chain scenario shows a significant difference compared with that in the conventional closed loop supply chain scenario shown in the 'Reverse supply chain' section. These behaviors are shown in Figures 20 and 21.

Results and discussion
Simulation results
The simulation results have been presented in terms of performance of the supply chain in three different settings. The trend (graphs) of the KPI such as level of inventories, backlogs, rates, and delays are shown in respective sections. The trends clearly depict that the reverse supply chain faces uncertainty due to the availability of cores and the quality of returned cores. The forward supply chain becomes unstable when the reverse supply chain is combined, i.e., the conventional closed loop supply chain. The forward supply chain becomes stable again if the resource ResCoM approach is adopted.

The feedback loop that exists within the dynamics of the supply chain helps decision makers to take actions that are sustainable over time. The simulation helps to understand to what extent the policy is robust and the drivers that affect robustness of the current policy. In

the case of the forward supply chain, this is particularly true and is validated through the model once again. However, in the case of the closed loop supply chain, the conventional supply chain management policies cannot be applied or it is not possible to create a robust policy. Industries that use the reverse supply chain or closed loop supply chain cannot manage their supply chain with traditional thinking and well-established policies. Industries that are planning to incorporate the reverse supply chain with their forward supply chain should, from these models, gain insight that as soon as two supply chains are combined, their policies (that have been in place and working well) will be disturbed, and the robustness will not be within manageable limits. Nevertheless, if the concept of ResCoM is adopted, the closed loop supply chain will behave more or less similarly as how the conventional forward supply chain usually behaves.

Model testing
The models were tested through the initialization of the model in a balanced equilibrium. It means that all stocks in the system remain unchanged despite the variation of time, requiring their inflow and outflow to be equal. The part of the model with random variables could not be initialized as it is; in this case, random variables were

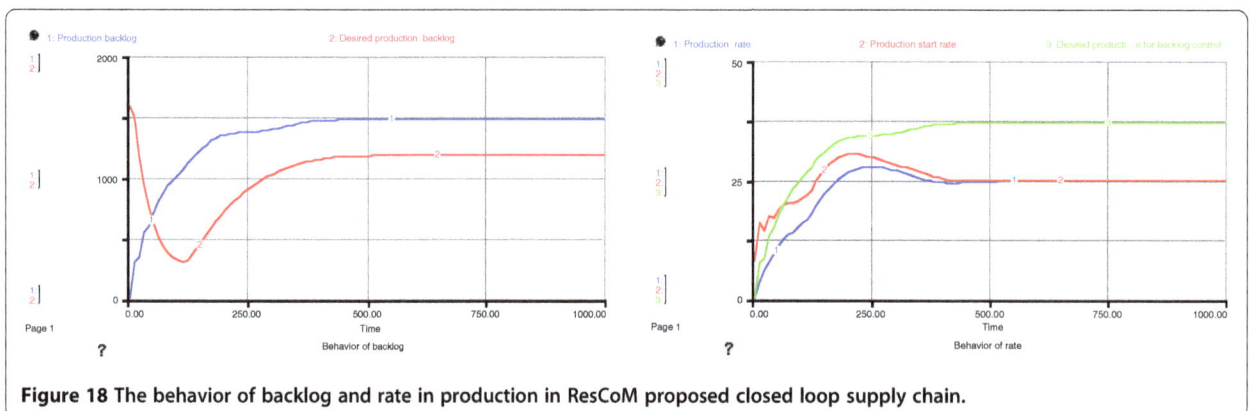

Figure 18 The behavior of backlog and rate in production in ResCoM proposed closed loop supply chain.

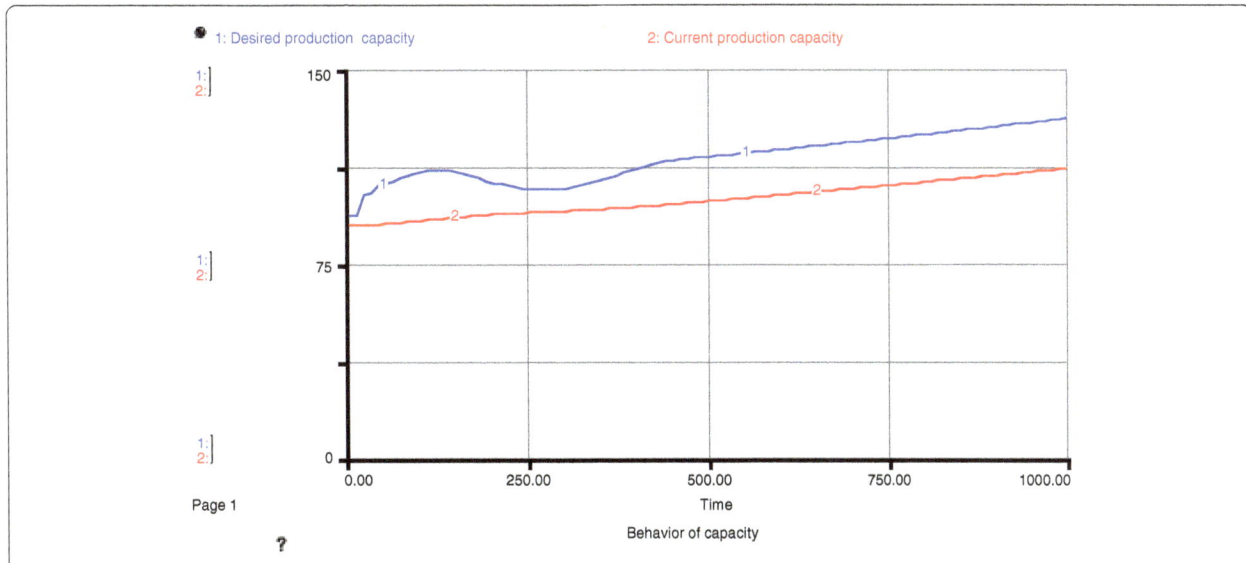

Figure 19 The behavior of capacity in production in ResCoM proposed closed loop supply chain.

replaced by constant values. Initialization confirms that there is no discrepancy in the equations or in the feedback loops.

The models were tested using the extreme condition test [27], where extreme input values were assigned concurrently. The reverse part did not fulfill the condition of the extreme test due to the random variables used in the reverse supply chain.

The simulation time has been extended to test if the model causes any reaction. In this case, the trends (graphs) of KPI remain more or less steady despite the largely varied simulation duration.

Discussion

The models that have been presented are generic models, which do not depict any specific type of product or industry. The boundaries of the models are quite broad; therefore, there is a lack in details in many cases. The

input data of the models are fabricated but correspond to the reality. In the models, some random variables are used, which do not comply with the system dynamics principles as Sterman describes randomness as *a measure of our ignorance, not intrinsic to the system*. In this particular case, randomness could not have been avoided as no research has been found that describes these phenomena otherwise; the span of the analysis is relatively shorter than what system dynamics usually suggests, and finally, there is a lack of empirical data.

The model raised at least two questions related to dynamics of policy and performance of supply chain. This is the first question: when remanufactured products enter (in rate of nondeterministic number) the forward supply chain and the production rate adjusts itself, what are the dynamics and feedback loops acting on it? This explains the behavior (non-balancing trends) of KPI in the production part in the forward supply chain after

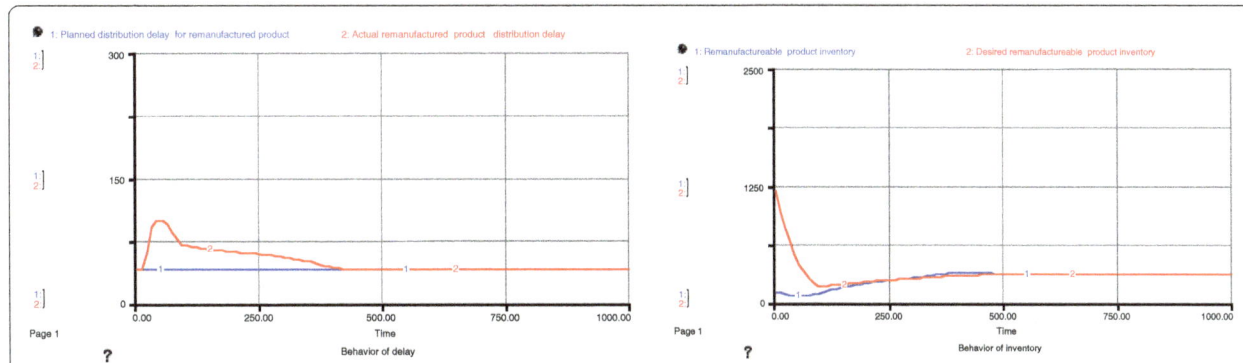

Figure 20 Delay and inventory behavior in reverse supply chain in ResCoM proposed closed loop supply chain.

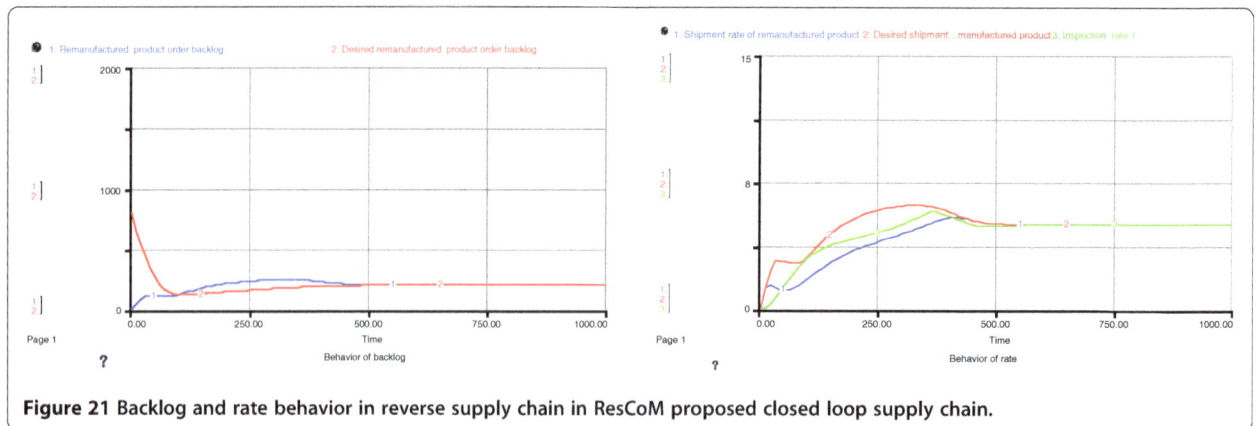

Figure 21 Backlog and rate behavior in reverse supply chain in ResCoM proposed closed loop supply chain.

combining the reverse supply chain with the forward supply chain. The other question is as follows: when a firm decides to enter the remanufacturing (new) market, how do the dynamics of the supply and demand and market share become balanced and what are the feedback loops that cause it to balance? At the same time, it has been realized that environmental benefits, change in societal perception, and level of natural resource conservation are needed to be incorporated in the model to make it complete.

The purpose of the modeling has been different from what is usually expected from system dynamics modeling. Through modeling, it has been shown how the policy and its leverage get affected when there is large uncertainty in any part of the supply chain. Therefore, the descriptions and arguments that are built around the models may not be as they would have been in the case of a conventional system dynamics model.

Referring to the question that usually emerges while choosing between continuous and discrete event simulations, the main factor in deciding which modeling tool to use is the level of aggregation sufficient for a particular object at hand [42]. Morecroft [43] has proven that similar results can be obtained using both system dynamics and discrete event simulation. However, system dynamics is particularly useful in demonstrating the complex dynamic relations of factors that are essential to manage a supply chain. It also helps to visualize the feedback loops and how they influence each other in a supply chain. Moreover, it gives management a base for decision making i.e., in a supply chain, in what degree of freedom one has to change different variables. As the objective of modeling has been to demonstrate performance of the supply chain in different settings and how they influence each other in terms of behavior, no other tool can fulfill the purpose as explicitly as the system dynamics did.

Apart from the modeling, the research presented in this paper tried to collect and summarize the research done in the closed loop supply chain. Moreover, this

work attempted to clarify the misconceptions and problems related to the closed loop supply chain. A novel concept, ResCoM, is presented to show the relevance of the research work with the state-of-the-art research. Finally, through KPI analysis of the closed loop supply chain, it is proven that the closed loop supply chain faces less uncertainty in terms of the supply and demand of products in ResCoM. As a by-product of this research, knowledge base has been created in the field of system dynamics applied in supply chain management.

Conclusions

Based on the review and analysis of the research in the area of closed loop supply chains, it is evident that the prevailing approach to close the loop for product multiple life cycle or remanufacturing is inherent to business thinking and models used for open loop manufacturing. The classic challenges of the closed loop supply chain, i.e., uncertain product returns, create serious problems for the multiple life cycle approach. Only the business thinking unique to closing the loop can solve this problem.

Moreover, it has been observed that isolated research in the areas of product design, closed loop supply chain, and business model has progressed, but the fundamental problems are still unique in the conventional approach. We proposed an alternative approach, which is partially described in this work, called ResCoM. The essential features of the proposed ResCoM model are as follows:

- Products designed for multiple life cycles with predefined life,
- Integration of forward (RCL_0 production) and reverse (RCL_i production) manufacturing functions into a single enterprise, and
- Customer integration as a business function of the enterprise

will ensure enhanced visibility of the products during their entire life cycle as regards to the quality, quantity, and timing of their return to the remanufacturing function; this visibility will minimize the uncertainties in product returns. This work also concludes that for advancement in developing successful product multiple life cycle, the current approach of research on isolated problems and implementation of its results in the industry is inefficient. The ResCoM concept requires a framework for a system level approach integrating four major functions of the manufacturing enterprise: product design and development, supply chain design and management, marketing and consumer behavior, and manufacturing and remanufacturing technologies should be integrated to form a unified research platform.

By reviewing and analyzing the research in the area of closed loop supply chain in stochastic environment, this work concludes that system dynamics has been applied in both operational and strategic issues of the closed loop supply chain. However, there is a need for further research as closed loop supply chain deals with complex issues. Using system dynamics, different researchers have described different phenomena of the closed loop supply chain which are important in creating the knowledge base. Models presented in this paper used system dynamics to demonstrate the robustness of the closed loop supply chain by analyzing the performance in conventional and in the ResCoM proposed approach. Through analysis of the behavior of KPI, it can be concluded that the ResCoM proposed closed loop is much more robust in terms of operations and faces less uncertainty. It is important for the policymakers to understand the behavior of KPI in order to set a robust policy. The behavior of KPI in ResCoM also shows that robust policies can be adopted in this approach as the uncertainty is minimized.

Methods

The methodological approach taken for this research can be best described as the cyclic process explained by Leedy and Ormrod [44] which includes the following:

- Problem identification and setting the research goal,
- Subdividing the problem to smaller elements,
- Introducing hypotheses that might lead to the solution,
- Gathering data and information that the hypotheses and problem lead to,
- Presenting the data in the form of a result to show that the problem has been solved, the question has been answer, or the result support or do not support the hypotheses, and
- Finally, validation and verification of the results.

While research methodology is a systematic way to do research, methods of research is just the means for

conduction of research [45]. The research methodology remains the same throughout the research, while methods can be different at different stages of research. As the research presented in this paper is in conceptual stage, and it is a small part of the ResCoM research paradigm, therefore, all the steps of the cyclic process described above may not be obvious at first glance.

The foundation of the research presented in this paper is mainly based on literature review. Some knowledge and experiences gathered by the authors by attending international conferences have also been reflected in this work. This is to say that the original problem formulation was measured and analyzed against the literature in the topic, and this led to the final problem form. These foundations have motivated the authors to describe by simulation the widely spoken problem of the closed loop supply chain, i.e., uncertainty in quantity and quality and arrival time of core. System dynamics principle has been used to model the closed loop supply chains, and the Stella software has been used to visually demonstrate the behavior of KPI in different scenarios. Finally, the results of simulation have been presented in the form of behavioral comparison of KPI in conventional and ResCoM proposed closed loop supply chain settings. However, no real data has been used to run the simulation as the objective of the modeling was to highlight the particular behavior of the KPI, not to simply quantify them.

Endnotes

[a]Words written in italics from this point forward are the terms used in the simulation models.

Additional file

Additional file 1: Mathematical formulations. Main mathematical formulations used in the models.

Competing interests
The authors declare that they have no competing interests.

Authors' contributions
FMAA contributed in developing the concept of resource conservative manufacturing, carried out the research presented in the paper, did the modeling, and made the draft of the paper. CB contributed in modeling, provided ideas for the modeling approach, and reviewed the modeling part of the research. AR contributed in developing the concept of resource conservative manufacturing, supervised the research, provided ideas for research, and also revised the paper critically for important intellectual content. CMN contributed in developing the concept of resource conservative, supervised the research, revised the paper critically for important intellectual content, and gave the final approval of the version to be published. All authors read and approved the final manuscript.

Authors' information
FMAA is a PhD student at the Department of Production Engineering, KTH Royal Institute of Technology, Sweden. He has been awarded the Technology Licentiate degree in September 2011. Apart from his Licentiate thesis, he has published articles for the Proceeding of Swedish production Symposium, DAAAM Baltic, and DAAAM international conferences. CB is a full professor of

Business and Public Management at the Department of International Studies, University of Palermo (Italy) where he is also the scientific coordinator of the CED4 System Dynamics Group. He is the director of the masters degree course on "Managing business growth through system dynamics and accounting models: a strategic control perspective" and of the international PhD program on "Model based public planning, policy design, and management". He is also the associate editor of the *System Dynamics Review*. His main research and consulting areas are related to small business growth management, entrepreneurial learning, startup, matching system dynamics with accounting models, dynamic scenario planning, dynamic balanced scorecards, business process analysis, and performance management. AR is a researcher and assistant professor at the Department of Production Engineering, KTH the Royal Institute of Technology, Sweden. He has been working in different manufacturing industries until he joined KTH in 2010. His research emphasis has been the analysis and control of machining system dynamics and extending his expertise towards sustainable manufacturing. He is the author of many scholarly articles published in many international journal and highly reputed conference proceedings. He has significant experience in the management of collaborative R&D projects through locally and EC funded projects. CMN is a full professor at the Department of Production Engineering, KTH the Royal Institute of Technology, Sweden. He is the chair of the research division called Machine and Process Technology. Aside from the many publications in different international journal and highly reputed conference proceedings, he has published some books. He has been actively involved in research and teaching since the beginning of his career and had supervised many PhD students.

Acknowledgments
The authors acknowledge the financial support received from the Swedish Institute (www.si.se) through the project Lifecycle Management and Sustainability in the Baltic Region.

Author details
[1]Department of Production Engineering, KTH Royal Institute of Technology, Stockholm, Sweden. [2]Department of Political Sciences, University of Palermo, Palermo, Italy.

References
1. The World Bank: *World development indicators*. 2011. http://data.worldbank.org/data-catalog/world-development-indicators/wdi-2011 (2011). Accessed 18 June 2011.
2. Kumar V, Bee D, Tumkor S, Shirodkar P, Bettig B, Sutherland J: **Towards sustainable "product and material flow" cycles: identifying barriers to achieving product multi-use and zero waste**. In *ASME 2005 International Mechanical Engineering Congress and Exposition*. Orlando, Florida:; 2005.
3. CIA: *The World Factbook*. http://www.cia.gov/library/publications/the-world-factbook/geos/xx.html (2011). Accessed 18 June 2011.
4. Jorgenson JD: *Mineral commodity summaries*. 2011. http://minerals.usgs.gov/minerals/pubs/mcs/2011/mcs2011.pdf (2011). Accessed 20 June 2011.
5. World Steel Association: *World steel in figures*. 2011. http://www.worldsteel.org/media-centre/press-releases/2011/wsif.html (2011). Accessed 20 June 2011.
6. Bureau of Internal Recycling: *World steel recycling in figures 2006–2010: steel scrap - a raw material for steelmaking*. http://www.bir.org/assets/Documents/publications/brochures/aFerrousReportFinal2006-2010.pdf (2010). Accessed 20 June 2011.
7. Eurostat European Commission: *Energy, transport and environment indicators*. http://epp.eurostat.ec.europa.eu/cache/ITY_OFFPUB/KS-DK-10-001/EN/KS-DK-10-001-EN.PDF. Accessed 20 June 2011.
8. Asif FMA: *Resource conservative manufacturing: a new generation of manufacturing*. Licentiate thesis, KTH Royal Institute of Technology; 2011.
9. Akkermans HA, Oorschot KEV: **Relevance assumed: a case study of balanced scorecard development using system dynamics**. *Journal of the Operational Research Society* 2005, **56**:931–94.
10. Bianchi C: **Enhancing Performance Management and Sustainable Organizational Growth through System Dynamics Modeling**. In *Systemic Management for Intelligent Organizations: Concepts, Model-Based Approaches, and Applications*. Edited by Groesser SN & Zeier R. Heidelberg: Springer-Publishing; 2012:(under process)
11. Asif FMA, Nicolescu CM: **Minimizing uncertainty involved in designing the closed-loop supply network for multiple-lifecycle of products**. In *Annals of DAAAM for 2010 and Proceeding of the 21st International DAAAM Symposium: Intelligent Manufacturing and Automation: Focus on Interdisciplinary Solutions, Zadar*. Edited by Katalinic B.; 2010.
12. Hammond D, Beullens P: **Closed-loop supply chain network equilibrium under legislation**. *Eur J Oper Res* 2007, **183**(2):895–908.
13. Guide VDR, Van Wassenhove LN: **The evolution of closed-loop supply chain research**. *Oper Res* 2009, **57**(1):10–18.
14. Lundmark P, Sundin E, Björkman M: **Industrial challenges within the remanufacturing system**. In *Proceedings of Swedish Production Symposium*. Gothenberg:; 2009.
15. Thierry MC, Salomon M, Van Wassenhove L: **Strategic issues in product recovery management**. *Calif Manag Rev* 1995, **37**(2):114–145.
16. Gungor A, Gupta SM: **Issue in environmentally conscious manufacturing and product recovery: a survey**. *Comput Ind Eng* 1999, **36**(4):811–853.
17. Seitz M, Peattie KJ: **Meeting the closed-loop challenge: the case of remanufacturing**. *Calif Manag Rev* 2004, **46**(2):74–79.
18. Toffel MW: **Strategic management of product recovery**. *Calif Manag Rev* 2004, **46**(2):120–141.
19. Parlikad AK, McFarlane DD: *Recovering value from "End-of-Life" equipment: a case study on the role of product information.*: Technical report: Centre for Distributed Automation and Control, University of Cambridge; 2004.
20. Sundin E, Björkman M, Ostlin J: **Importance of closed-loop supply chain relationship for product remanufacturing**. *Int J Prod Econ* 2008, **115**(2):336–348.
21. de Brito MP, Dekker R: *A framework for reverse logistics*. Rotterdam: Erasmus Research Institute of Management; 2003.
22. Freiberger S, Albrecht M, Käufl J: **Reverse engineering technologies for remanufacturing of automotive systems communicating via CAN bus**. *Journal of Remanufacturing* 2011, 1:6.
23. Kerr W, Ryan C: **Eco-efficiency gains from remanufacturing a case study of photocopier remanufacturing at Fuji Xerox Australia**. *J Clean Prod* 2001, **9**(1):75–81.
24. Sundin E, Bras B: **Making functional sales environmentally and economically beneficial through product remanufacturing**. *J Clean Prod* 2005, **13**(9):913–925.
25. Kumar S, Malegeant P: **Strategic alliance in a closed-loop supply chain: a case of manufacturer and eco-non-profit organization**. *Technovation* 2006, **26**(10):1127–1135.
26. Lifset R, Lindhqvist T: **Does leasing improve end of product life management?** *J Ind Ecol* 1999, **3**(4):10–13.
27. Sterman JD: *Business Dynamics: Systems Thinking and Modeling for a Complex World*. Boston: McGraw-Hill/Irwin; 2000.
28. Ilgin MA, Gupta SM: **Environmentally conscious manufacturing and product recovery (ECMPRO): a review of the state of the art**. *J Environ Manage* 2010, **91**(3):563–591.
29. Angerhofer BJ, Angelides MC: **System dynamics modelling in supply chain management: research review**. In *Proceedings of the 32nd Conference on Winter Simulation*. Orlando:; 2000.
30. Georgiadis P, Vlachos D: **The effect of environmental parameters on product recovery**. *Eur J Oper Res* 2003, **157**(2):449–464.
31. Vlachos D, Georgiadis P, Iakovou E: **A system dynamics model for dynamic capacity planning of remanufacturing in closed-loop supply chains**. *Comput Oper Res* 2007, **34**(2):367–394.
32. Kumar S, Yamaoka T: **System dynamics study of the Japanese automotive industry closed loop supply chain**. *J Manuf Technol Manag* 2007, **18**(2):115–138.
33. Georgiadis P, Vlachos D, Tagaras G: **The impact of product lifecycle on capacity planning of closed-loop supply chains with remanufacturing**. *Prod Oper Manag* 2006, **15**:514–527.
34. Georgiadis P, Athanasiou E: **The impact of two-product joint lifecycles on capacity planning of remanufacturing networks**. *European Journal of Operations Research* 2010, **202**(2):420–433.
35. Qingli D, Hao S, Hui Z: **Simulation of remanufacturing in reverse supply chain based on system dynamics**. In *IEEE, Service Systems and Service Management, 2008 International Conference*. Melbourne:; 2008.
36. Schröter M, Spengler T: **A system dynamics model for strategic management of spare parts in closed-loop supply chains**. In *The 23rd International Conference of the System Dynamics Society*. Boston:; 2005.

37. Poles R, Cheong F: **A system dynamics model for reducing uncertainty in remanufacturing systems**. In *PACIS 2009 Proceedings*. Hyderabad:; 2009.

38. Besiou M, Georgiadis P, Van Wassenhove LN: **Official recycling and scavengers: symbiotic or conflicting**. *European Journal of Operations Research* 2012, **218**(2):563–576.

39. Georgiadis P, Besiou M: **Environmental and economical sustainability of WEEE closed-loop supply chains with recycling: a system dynamics analysis**. *Int J Adv Manuf Technol* 2010, **47**:475–493.

40. Georgiadis P, Besiou M: **Sustainability in electrical and electronic equipment closed-loop supply chains: a system dynamics approach**. *J Clean Prod* 2008, **16**(15):1665–1678.

41. Özkir V, Basligil H: **Modelling product recovery processes in closed loop supply chain network design**. *Int J Prod Res* 2012, **50**(8):2218–2233.

42. Semere DT: *Configuration design of a high performance and responsive manufacturing system*.: Doctoral thesis, KTH Royal Institute of Technology; 2005.

43. Morecroft J: *Strategic modelling and business dynamics: a feedback system approach*. West Sussex: John Wiley & Sons Ltd.; 2007.

44. Leedy PD, Ormrod JE: *Practical Research Planning and Design*. New Jersey: Merrill/Pearson Education, Inc.; 2010.

45. Kothari CR: *Research Methodology: Methods and Techniques*. Delhi: New Age International Limited; 2004.

5

Design for remanufacturing in China: a case study of electrical and electronic equipment

Gillian D Hatcher[1*], Winifred L Ijomah[1] and James F C Windmill[2]

Abstract

As global demand for consumer goods continues to rise, the problem of waste electrical and electronic equipment (or e-waste) increases. E-waste is of particular concern to the world's governments and environmentalists alike, not just because of the sheer quantity that is being produced annually, but also because e-waste often contains both hazardous materials and scarce or valuable materials. Much research is now focused upon how this waste can be treated safely, economically, and in an environmentally sound manner. This paper presents the findings from a literature review and case study research conducted as a small part of the Globally Recoverable and Eco-friendly E-equipment Network with Distributed Information Service Management (GREENet) project. The GREENet project aims to share knowledge and expertise in e-waste treatment across Europe (in this case, the UK) and China. The focus of this particular study was upon 'design for remanufacture' and e-waste in China: as a remanufacturing industry begins to emerge, are Chinese original equipment manufacturers (OEMs) prepared to design more remanufacturable products and could electrical and electronic products become a part of this industry? Findings presented in this paper suggest that design for remanufacture could become more relevant to Chinese OEMs in the near future, as environmental legislation becomes increasingly stringent and a government remanufacturing pilot scheme expands. However, findings from case studies of Chinese e-waste recyclers would suggest that electrical and electronic products are not presently highly suited to the remanufacturing process.

Keywords: E-waste, Design for remanufacture, Recycling, China

Background

Design for remanufacture

Remanufacturing is the process of returning a used product to a like-new condition through inspection, disassembling, cleaning, reprocessing, reassembling, and testing. Components which cannot be reused in this way are replaced with new components, and the final remanufactured product may be sold at a lower price than a newly manufactured equivalent, but with an equal warranty [1]. Remanufacturing differs from traditional recycling in that the used products are 'recycled' at a component level, as opposed to a raw material level (see Figure 1). Remanufacturing is often confused with reconditioning, when the used product is returned to a working condition but will not have an equal warranty to a newly manufactured equivalent. 'Repair' typically involves simply the correction of specific faults in the product. While remanufacturing requires more work (including energy and expense) than reconditioning or repairing, the resultant product will be of a higher quality with a further extended life in use. Therefore, remanufacturing can often be considered more energy-saving and cost-effective when compared to other end-of-life processes [2].

However, not all products are suitable for remanufacture. As a general rule, the product must be durable (able to withstand multiple lifecycles) and contain high-value parts (worth investing in). Also, there must be market demand for the remanufactured products. Products typically remanufactured in the UK include automotive products, pumps and compressors, and off-road equipment [4]. As well as product characteristics such as high-value parts or a return flow of used products (i.e., factors beyond the control of the designer), the efficiency and effectiveness of the remanufacturing process can also greatly depend upon how the product has been

* Correspondence: gillian.hatcher@strath.ac.uk
[1]Design Manufacture and Engineering Management, 4th Floor Architecture Building, University of Strathclyde, Glasgow G4 0NG, UK
Full list of author information is available at the end of the article

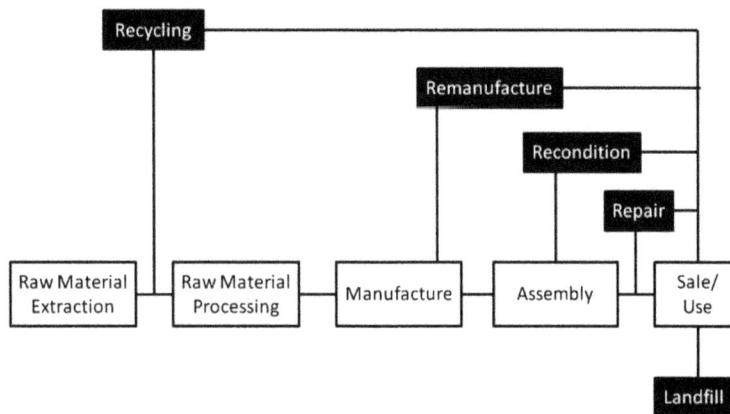

Figure 1 End-of-life treatment 'loops'. Unlike recycling, remanufacturing does not require raw material processing or component manufacture [3].

designed (factors within the designers' control): features such as fastening and joining methods, product architecture, and material choice can have an effect upon ease of disassembly, ease of reprocessing, and so on [5,6]. This understanding has led to the concept of 'design for remanufacture' or 'DfRem', and the development of DfRem guidelines [5]. According to the literature, very few companies currently design for remanufacture [7]. However, in the USA and Europe, examples can be found of companies with successful remanufacturing operations, maximising the potential of their products through DfRem [8,9].

E-waste in the UK and China

The term 'e-waste' refers to waste electrical and electronic equipment, such as computers, wireless devices, and white goods. It is most commonly used within the context of discarded consumer products. E-waste is a growing global problem, with million tonnes annually discarded by UK households alone. Some of this waste may be reused or recycled, but the significant proportion of this mass that ends up in landfill or incineration can have a highly detrimental environmental impact [10]. Furthermore, as our consumer culture shows little sign of abatement, including populous, emerging economies such as China [11], the situation is likely to get worse before it improves. Therefore, many countries, including the UK (European Union (EU)) and China have made attempts to implement legislation with the intention of reducing the volume of e-waste generated and ensuring that producers of such products take responsibility for the environmentally sound treatment and disposal of their waste.

Waste electrical and electronic equipment legislation in the UK and China

The European Union was among the first to attempt to legislate e-waste, with the introduction of the waste electrical and electronic equipment (*WEEE*) *Directive* [12]. Although all member states were expected to comply by 2004, it was not until 2007 that the UK finally implemented the WEEE Directive [13]. The EU WEEE Directive promotes 'extended producer responsibility' (EPR) with which original equipment manufacturers (OEMs) are obliged to contribute financially to the treatment of all WEEE in their given country. In the UK, OEMs of electrical and electronic equipment (which fall within the ten categories outlined in the Directive) must pay into a 'producer compliance scheme' which will finance the take-back and recycling of all electronic products which qualify, regardless of brand. Producer compliance schemes must be approved by the Environment Agency in the UK, and the funds they raise will be used by approved companies for e-waste recycling. Furthermore, OEMs are obliged to provide free take-back of e-waste, either through a take-back scheme or through retailers. Collection targets, however, are placed on the member states, as opposed to the OEMs: at least 4 kg of household waste per capita per year [14]. This target is surprisingly low, considering electronic waste accounts for an estimated 4% of European household waste, which would mean that around 18 kg of e-waste is discarded per capita in the UK [15]. However, a revision of this target has been proposed which would mean that by 2016, 65% of the average weight of e-waste must be collected [14].

In addition to the WEEE Directive, the EU's 'Restriction of Hazardous Substances' (RoHS) legislation is also applicable to waste electrical and electronic products in the UK. RoHS, implemented in 2006, restricts the use of six materials which are considered hazardous to the environment: lead, mercury, cadmium, hexavalent chromium, polybrominated biphenyls, and polybrominated diphenyl ether. These restrictions only apply to products sold within the EU; however, OEMs are expected to pass these requirements to their supply chain [16]. This means that OEMs operating in other countries must also comply with both the WEEE Directive and RoHS if the products are destined to be sold in the EU.

While the UK struggles to control its own e-waste, the problem in China also continues to grow at a rapid pace: a 13% to 15% increase each year [17]. Therefore, in 2007 and 2011, respectively, the China RoHS (*Management Measure for the Prevention of Pollution from Electronic Information Products*) and China WEEE Directive (*Management Regulations on the Recycling and Disposal of Waste Electronic and Electrical Products*) were introduced (however, WEEE legislation is still to be fully implemented). The China RoHS is very similar to its European equivalent, with the same six substances being banned from use in electrical and electronic products intended for Chinese consumer and commercial and industrial markets [18]. It is possible that this list will be extended to include more hazardous substances in the future [16]. An addition to the China RoHS is that the regulation requires producers to include an 'environmental expiry date' with their products (e.g., in the user manual) to inform customers of the safe usage period of their products, before hazardous substances pose a risk to health and the environment.

In the past, and ongoing today, recycling in China was an informal industry sector, unregulated and often damaging to the local environment [11]. The China WEEE Directive, like its EU equivalent will insist upon 'extended producer responsibility'. Formal recycling companies, approved by the government, will receive subsidies to operate safe, environmentally sound recycling for consumer and industrial e-waste, the funds for which will have been provided by the OEM producers [18]. Slightly different from the EU WEEE Directive is the definition of which products apply to the legislation. While in Europe there are ten categories of e-waste which are subject to the WEEE Directive, in China there is a catalogue outlining in detail all products which apply [18]. Furthermore, due to particular problems in this area, the China WEEE Directive also states that informal recyclers, who are not government-approved and are not operating in an environmentally sound manner, can be punished. However, how this will be enforced and the nature of punishment remain unclear [18]. The Chinese government has an overall recycling target of 70% by 2015, considerably more ambitious than the European WEEE Directive's '4 kg' target, but comparable to the proposed new '65%' target for 2016 [19].

E-waste challenges in China

As mentioned in the previous section, informal (and now illegal) recycling is an ongoing problem in China due to its detrimental effect on local environments, as well as employee health and safety. This long-established informal recycling industry also poses several other challenges upon an EPR system in China. First, of all consumers are accustomed to being paid good prices for their e-waste from informal collectors [20]. This means that in order to remain competitive, formal recyclers must also pay the same price for their e-waste, ruling out the recycling fee strategy currently in operation in Japan, for example [14]. Formal recyclers have greater costs than informal, because they must ensure employee health and safety plus environmental protection. Hence, government subsidies are essential for competitiveness. Some pilot studies in China had previously failed because the formal recyclers could not collect enough e-waste for these reasons [21].

Another challenge that the China WEEE Directive must address is 'orphan products'. Orphan products are those which cannot be paid for by the original producer for a variety of reasons. This is a problem in any country, but will be a particularly problematic issue in China because firstly, the sheer number of manufacturing companies in the country combined with a fast-moving economy means that producers are regularly going out of business, often without trace [21]. Also, China has a particular problem with imitation or counterfeit products, an underground operation which is again difficult to hold to account for recycling [21]. While the export of e-waste to China is now illegal, it is still an ongoing problem for the country, which creates further volumes of 'orphaned' e-waste [20].

Another, social problem regarding e-waste in China is collection. According to Li et al. [11], it is common for Chinese consumers to hold on to their e-waste for much longer than is the norm in other countries, due to a belief that the products could somehow become useful again in the future, or be sold to the second-hand market.

For these reasons, Li et al. [20] argue that simply copying the EU's extended producer responsibility system will not provide a solution to the problem of China's e-waste; China needs a system that is specifically designed to accommodate these country-specific issues. Yu et al. [22] suggest that integrating the informal recyclers with the formal may be an option for the future, for example paying informal collectors to supply formal recyclers.

Methods

Research aim

This research was conducted as part of the 'Globally Recoverable and Eco-friendly E-equipment Network with Distributed Information Service Management' project, or 'GREENet'. The project is a 3-year collaboration between various institutes in the EU and China. Part of the programme involves the study of the status of remanufacturing and e-waste across the two global locations. This particular study was conducted during a short visit by UK researchers to Tsinghua University and the Research Centre for Eco-Environmental Sciences in Beijing.

Specifically, the research discussed in this paper is focused upon the status of *design for remanufacture*. Therefore, the research questions addressed in this paper are:

a) How does the 'design for remanufacture' situation in China compare to the UK?
b) How suited are electrical and electronic products to the remanufacturing process with regards to product design?

The first research question was explored primarily through the literature, searching for information on the prevalence of both remanufacture and DfRem in China for comparison with the authors' experiences from the UK. The second question was explored through case study research, as outlined in the next section.

Case studies

Information was gathered during visits to three case study companies in the Beijing, Shanghai and Qingdao regions of China:

Company 1: One of the largest e-waste treatment facilities in Shanghai, it is involved in the collection, sorting, treatment, recycling, and safe disposal of a variety of electrical and electronic products including used televisions, ink cartridges, and white goods.

Company 2: This is a global e-waste recycling company with several facilities in China, Beijing, and Shanghai, where a recycling centre is located. The company treats all manner of e-waste, yet at the visited facility, the focus was upon the recycling of PCB boards to extract precious metals (copper and gold).

Company 3: This company operates a large 'eco-industrial park' which has become an all-inclusive provider of waste solutions, ranging from hazardous and medical waste treatment to soil remediation and recycling. The company's e-waste recycling operations include products such as refrigerators, televisions, air conditioners, washing machines, and computer equipment.

The companies were all part of the 'formal' Chinese recycling industry, and therefore received government subsidies. They have a waste management licence which enables them to collect waste products from a variety of client sources, for example, factory rejects and community collection boxes. They are then responsible for the reverse logistics, treatment, and safe disposal or resale of raw materials (see Figure 2). The companies' waste recycling activities produce recycled ferrous materials (iron steel, manganese, chromium), non-ferrous materials

(gold, aluminium, copper, zinc), and non-metal materials (plastic, cardboard, glass, foam). An example of a case study recycling process is illustrated in Figure 3. The companies will then sell on their reclaimed materials to various manufacturing plants, depending on where the best price could be negotiated. The case study companies each have their own research and development centres and hold patents for state-of-the-art automated recycling equipment.

The three case study companies always recycle: they are not involved in any remanufacturing, refurbishment, or repair activities. At least one of the case study companies collected used products which contain remanufacturable components (for example motors in white goods). However, the lack of a remanufacturing infrastructure in China means that even these parts are currently recycled.

Case study questions and protocol

The following questions were asked to the case study companies in order to gain information on the suitability of e-waste for both remanufacture and design for remanufacture in China:

a) What is the recycler's connection/relationship to the OEMs of their e-waste?
b) What product design-related problems and barriers do they currently face?
c) What other problems and challenges are they currently facing?

Understanding the recycler's relationship with OEMs not only provides context to the case studies but may also reveal any inter-organisational barriers to design for remanufacture, for example, access to intellectual property or design collaboration. This will compliment any findings regarding the technical problems and barriers a recycler may face.

Case studies involved a visit to the company's recycling facilities, where e-waste products could be observed being disassembled and where materials are segregated and re-processed. This tour was followed by an interview with one or more members of site management. The findings reported in this paper represent only a small part of the overall GREENet project, and the authors have identified scope for future contributions from project researchers under this particular theme.

Literature survey
Remanufacturing in China

Unlike in Europe where remanufacturing has been a common industry practice for many decades, remanufacturing is a fairly new concept in China, with recent interest due to the dramatically rising number of vehicles on Chinese roads in recent years [23]. In 2008, the National

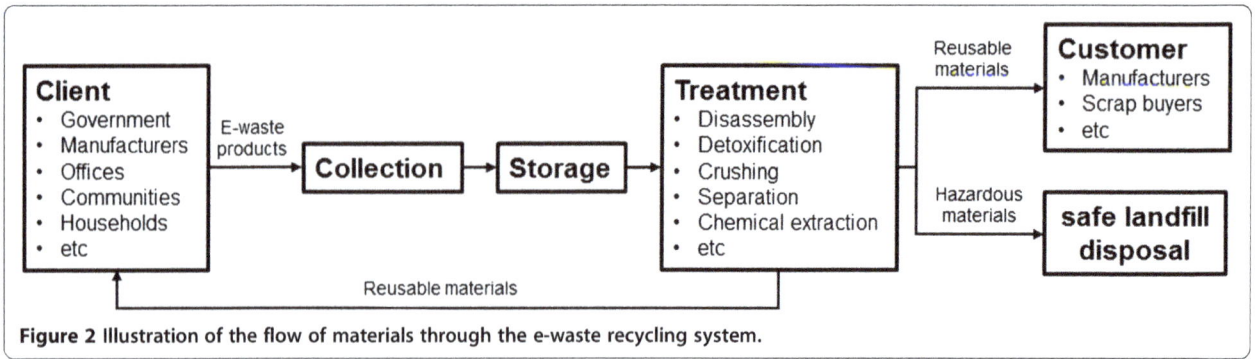

Figure 2 Illustration of the flow of materials through the e-waste recycling system.

Development and Reform Committee launched *Regulations of Remanufacturing Pilot of Automotive Parts*, a pilot study of 14 vehicle remanufacturers, with a view to encourage more remanufacturing activities in the future. This pilot study, which is ongoing at the time of writing, has been the first time vehicle engines, transmissions, starting gears, starters, and generators can be legally remanufactured in China [24]. Remanufacture has also been listed in the *National Long and Medium Term Program of Science and Technology Development Planning* as a key future manufacturing field in China [24]. Both Caterpillar (Illinois, USA) and IBM (NY, USA) have also recently launched remanufacturing ventures in China, another reason to suggest that remanufacturing is becoming a lucrative business venture [25,26]. However, for the time being, there is very little information in the literature regarding remanufacture in China.

Some potential barriers to future remanufacturing expansion are similar to those found in other parts of the world, namely public perceptions that remanufactured products are 'second hand' and, therefore, inferior in quality. Others are more specific to China. For example, Chinese manufacturing often prefers to rely upon cheap labour rather than investing in the latest technologies, something which could slow the development of a strong remanufacturing industry in the country [23]. Local governments can also be wary of registering vehicles with remanufactured parts [24]. Zhang et al. [24] also noted that intellectual property laws in China could prevent third party remanufacturers from establishing themselves (third party remanufacturers play a major role in the European remanufacturing industry). As the kind of products typically remanufactured do not fall under the category of e-waste, there is currently no specific standards or regulations regarding remanufacture in China [24]. As there is very little remanufacturing currently happening in China, it can also be concluded that there is very little, if any, DfRem happening within Chinese OEMs today.

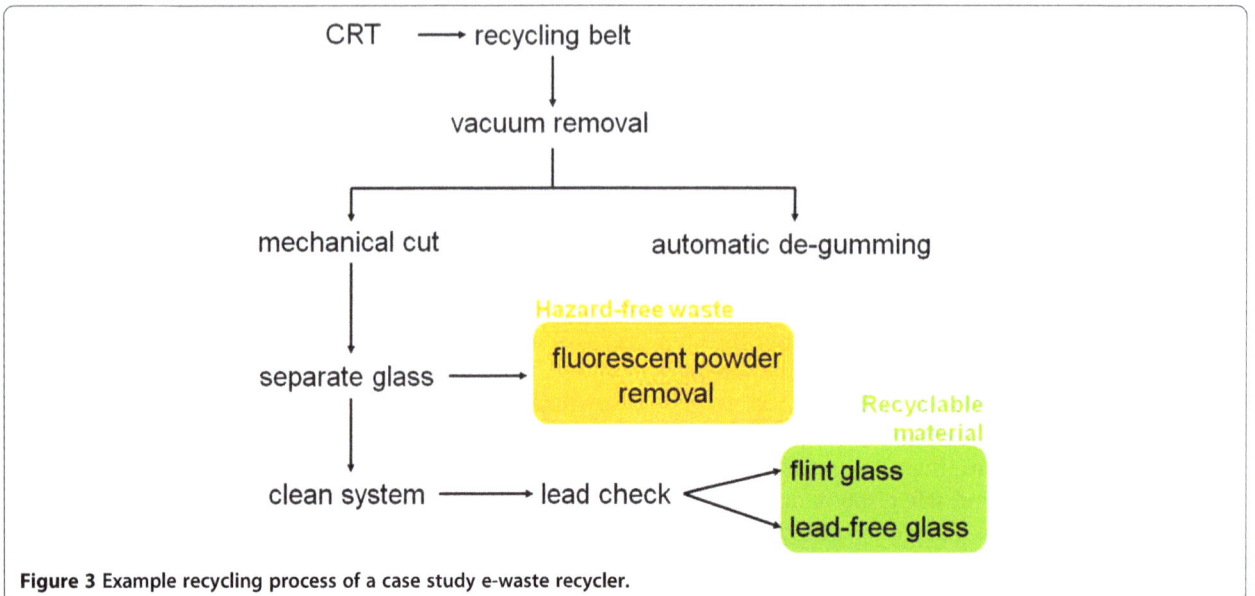

Figure 3 Example recycling process of a case study e-waste recycler.

Design for remanufacture in China

While there may be little or no DfRem happening in China, it is possible to look to ecodesign for some insight. Although ecodesign research and DfRem research are not entirely inter-changeable, there are some similarities between the two, which mean it is worth exploring ecodesign for a comparison [27]. Something that the EU and China have in common is that their environmental legislation (Article 4 of the EU WEEE Directive and Articles 9 and 10 of China RoHS) both call for ecodesign from producers of electronic products; however, neither directive offers measurement for compliance [16].

While ecodesign is also a relatively new concept in China, it is likely to be of rising interest to Chinese designers as the country moves from simply manufacturing products to taking part in the innovation as well [28,29]. If the remanufacturing industry were to expand in China over the next few years, the same could be said of DfRem.

Lindahl [30] conducted a survey of Asian electrical and electronic companies, including those operating in China, to study designers' and managers' awareness and abilities in ecodesign. At the time of the survey, most people interviewed were aware of the need to comply with European environmental legislation when selling to those countries, but not China (the directives were very much in their infancy at the time). Larger Chinese companies surveyed said they had implemented ecodesign measures to reduce energy consumption in their products and had eliminated hazardous substances. Some companies had made attempts to reduce materials usage, but in general, end-of-life considerations were not a major concern in these firms. The survey found that most designers and managers had some basic knowledge of ecodesign; however, the researchers admitted that the research could be biased. It is likely that the companies with no knowledge of ecodesign would not choose to take part in the study [30]. Similar to research conducted in the EU, for example [31], designers interviewed in China did not use ecodesign tools during their design work and felt that any tools implemented in the future would have to be quick to learn and low-cost to implement [30].

Yu et al. [16] conducted a similar survey with Chinese electrical and electronics companies. While they suspected their respondents were over-positive, the results suggested that 58% of companies have responded to the China RoHS legislation with ecodesign, although most of this has simply involved the removal of toxic substances. The survey found that, in general, Chinese companies have limited knowledge of ecodesign at present. However, Chung and Zhang [18] point out that China, in general, has better e-waste and ecodesign awareness than most developing countries.

These surveys highlight a point that is also made by Behrisch et al. [32] following an internet survey of Chinese design consultancy websites. Because the Chinese environmental legislation is so new (and in the case of WEEE Directive still to be fully put into practice), any ecodesign occurring in Chinese companies today is likely to be motivated instead by European legislation, for products destined for those countries. This is likely to change in years to come as Chinese environmental legislation is implemented. However, as with the EU, it is important that future amendments to environmental legislation include not just a mention of the importance of ecodesign/DfRem, but practical guidance and measurements that design engineers can follow, e.g., joining and fastening methods, disassembly targets, and material choices.

Lindahl [30] points out that for a developing country such as China, it will be important to the government that any move towards environmental improvement and increased ecodesign does not hinder economic development. An advantage DfRem has over ecodesign in this respect is that remanufacturing decisions are typically motivated by profitability, not environmental impact [5].

Results

This section outlines the preliminary findings and observations made from data collected during the three case study visits.

Company relationship with OEMs

The case study companies (from now on referred to as 'the recyclers') had many partner OEMs in China, including Panasonic (Osaka, Japan), Sony (Tokyo, Japan), Kodak (New York, USA), and Lenovo (North Carolina, USA). Most of the recycler's e-waste is collected directly from these manufacturers. Some of this e-waste will be used or faulty products returned by the end customer or the plant's own used office goods, but most of the products are factory rejects. OEMs in China are increasingly obligated to take responsibility for the recycling of this waste, due to EPR policies being introduced through the China WEEE and RoHS Directives.

If the collected e-waste contains hazardous waste, which the recycler must treat and safely dispose, WEEE regulations insist that the OEM must pay the recycler for this service. This service enables OEMs to comply with China RoHS legislation and, therefore, cannot legally be avoided. However, most often the precious metals and other materials of value contained within e-waste mean that the waste is considered to be of high market value. In these circumstances the recycler must purchase the e-waste from the OEM. This relationship is in some ways similar to an independent third-party remanufacturing scenario and also to contract remanufacturing. The relationship between the recycler and the OEMs of their e-waste is likely to be complex as, in many ways, the OEM is the supplier of e-waste, but is also the recycler's customer as the

OEM relies upon the recycler to ensure environmental legislation is complied with.

Communication between the recycler and the OEMs of e-waste comes firstly at point of collection and also during auditing processes as specified by the OEM. Because the OEM is legally obliged to be environmentally responsible for their e-waste, it is important that the company can be sure that their products are being treated in the correct manner. If the recycler has a contractual agreement with an OEM, that OEM will provide strict recycling instructions. The recycler is then required to provide the OEM with regular reports, providing evidence that these instructions are being followed. The recycling managers interviewed for this study stated that they would not provide information in these reports regarding common faults, or proposed design changes, mostly because product design is not of concern to them, and because they are not in a position of power to request design changes from an OEM. However, if the recycler is facing difficulties in recycling a particular product, they may be able to consult the OEM for advice, and interviewees in one of the companies interviewed stated that their OEMs will sometimes provide details of the hazardous waste contained within their products (i.e., a materials list) and occasionally will request reusable parts be sent back to their manufacturers.

Even if there is some design-related communication with the OEMs, the recyclers are never provided with any design information for their collected e-waste due to intellectual property concerns. This is also the case for contract remanufacturers. Therefore, it is necessary for the recyclers to 'reverse engineer' any new product that arrives at the plant to determine the proportions of precious metals present and the best course of action for retrieving them. Unlike reverse engineering for remanufacture, it is not necessary for recyclers to consider the disassembly of the new products, as discussed in the next section.

Product design barriers and challenges

Unlike remanufacturing, where the used product must be disassembled without damage to valuable parts, e-waste arriving at the recycling plants can be disassembled destructively. For example, a plastic casing for an air conditioner may be cracked open, because the materials will be shredded for recycling anyway. This advantage eliminates many of the design-related problems found in remanufacturing processes; therefore, the recyclers involved in this study did not recognize product design as a barrier to their recycling process.

Another design issue that creates problems in some remanufacturing processes is the disassembly of components containing hazardous wastes. Almost all disassembly for remanufacture is manual; therefore, the safe removal of these components can at times provide a

challenge to remanufacturing staff [7]. However, at the Chinese recycling companies, the process is almost entirely automated, often using patented equipment. Ink cartridges, for example, can be fed into machinery that will crush and clean the products, removing hazardous waste automatically at no risk to recycling staff. Figure 3 illustrates a similarly automated process, where cathode ray tubes are fed onto a moving belt; several automated steps later, non-hazardous, hazardous and recyclable materials are produced.

The recyclers interviewed stated that there was little difference between new and old models of products; as product design has changed over the years, their recyclability has changed very little. This means that the technological stability of a product will have little effect on recycling effectiveness. Remanufacturing, on the other hand, is very much affected by technological stability, because the products are being re-sold as opposed to being broken down to their base materials. For example, as mobile phone technology is changing rapidly, by the time a used mobile phone is returned and remanufactured, there may be very little market demand for such an out-of-date model. This is simply not an issue for companies breaking down the mobile phones to their raw materials for recycling.

Other challenges

A great challenge for the recyclers is being able to predict profitability. In remanufacturing, this is also the case because it is very difficult for remanufacturers to predict the rate of used product (called 'cores') collection and the condition of the cores once collected (cores in poorer condition will be more expensive to return to an as-new condition). The predictability issues at recycling companies are slightly different; it is a result of varying market prices for the various materials they deal with. The recyclers must be very selective over the e-waste they choose to deal with and the companies they choose to work with to ensure the company maintains profitability. The interviewed recyclers stated that sometimes they would take on a contract that they know will be a loss-maker in order to build long-term relationships with customers and suppliers that will prove profitable in the future.

Some of this uncertainty is alleviated by government subsidies. As 'formal' recycling companies (as opposed to the now illegal informal recycling plants common across China), the companies involved in this study qualified for government financial assistance. The funds for these subsidies are provided by the OEMs under new WEEE legislation and are essential to ensure that formal recycling companies can remain competitive. A problem with informal recycling in China is its lack of worker safety and detrimental effect on the local environment.

In order to obtain the required certification (e.g., ISO 9001 and ISO 14001), recyclers must invest in technologies and practices that would ensure that their recycling operations are both safe for employees and to the environment. These additional costs would reduce the competitiveness of formal recyclers without government subsidy [21].

Discussion

Remanufacture of E-waste in China

With the onset of WEEE regulations in China, which insist that OEMs of electrical and electronic products ensure the responsible end-of-life treatment of their e-waste (and take financial responsibility too), the role of formal recycling companies will become increasingly important in the industry. Whether China's up-and-coming remanufacturing industry will also play a role in 'extended producer responsibility' in the future remains unclear, at least until the outcomes of the Chinese government's current pilot studies are known in more detail.

While the recycling industry in China is primarily driven by environmental legislation, the remanufacturing industry in Europe is driven by profit. Many of the companies involved in remanufacture are not currently subject to any end-of-life legislation that would promote this choice; instead, they have realized that there is customer demand for affordable remanufactured products and spare parts [33]. Despite this contrast, both the recycling and remanufacturing industries will be just as interested in improving the efficiency and effectiveness of their processes, and often, this can be assisted through product design. The recognition that product design can affect the efficiency, effectiveness, and therefore profitability of product end-of-life processing is one reason why 'design for recycling', 'design for remanufacture', 'design for environment', and 'ecodesign' have become familiar terms in the engineering literature.

However, this study has found that many of the product design concerns commonly associated with remanufacturing are simply not important when considering recycling. This becomes clear when considering some points of the 'RemPro' matrix [6], a model which highlights different product design considerations affecting remanufacture. Ease of access, handling, disassembly, and reassembly are irrelevant in the modern recycling technologies used by the Chinese formal recyclers. These recyclers are only interested in the extraction of hazardous substances and the separation of recyclable materials. The most effective way to achieve this is to quickly disassemble the products by the easiest means possible, which may involve breaking casings and snapping joining components. Products that could only be disassembled through breakage would be a major problem for a remanufacturer, but makes no difference to the recycler. After initial separation, the products are essentially crushed and separated by entirely automated means.

This does not confirm that design for recycling is entirely unnecessary. Firstly, as contract recyclers, the case study companies were unable to discuss the product design of their e-waste in any great detail. A 1995 paper on design for recycling guidelines [34] contains many guidelines familiar to remanufacturing research: modularization, minimal joints, easily accessible hazardous parts, etc. However, as recycling technology has progressed to become increasingly sophisticated and increasingly automated, many of these guidelines do not apply to a product destined for a modern recycling facility. At the same time, many of the OEMs that work with the recyclers involved in this study do claim to optimise their products for recycling. For example, on their website, Panasonic claims to be involved in 'green design'. The company claims to be committed to reducing materials such as PVC, which is difficult to recycle; the company also states it is committed to designing more 'recycling-oriented products' [35]. Chinese OEM Lenovo claims to aim to comply with the European WEEE Directive by 'designing equipment with consideration to future dismantling, recovery and recycling requirements' [36].

Considering the suitability of e-waste for remanufacture, the information gathered to date would suggest that, in general, electrical and electronic products are not currently suitable candidates for remanufacture; the process is not a feasible alternative end-of-life solution to current recycling practices. The key reasons for this conclusion, based on the findings from this study, are as follows:

- At present, electrical and electronic products are not typically designed for non-destructive disassembly because modern recycling technologies have deemed this previous requirement unnecessary.
- Many electrical and electronic products contain high volumes of precious metals, which, once recovered through recycling, have a higher market value than a remanufactured product most likely would.
- Most of the used products that arrived at the case study recycling facilities would be considered technologically 'out-of-date', even if they are only a few years old. There would be little market demand for a remanufactured laptop or mobile phone for this reason (hence, low market value).
- A very important part of the e-waste recycling process is data wiping. While this can also be carried out for electronic remanufacture or

refurbishment, unlike with recycling, the sensitive components would be reused, not crushed and recycled. It is possible that many customers would be untrusting, potentially reducing the number of used products that can be collected and then resold.

Design for remanufacture in China

Because environmental legislation has so recently been introduced, ecodesign (and related concepts such as DfRem) is still a very new concept to Chinese producers. Findings from the three case study recyclers would suggest that these new ideas have not yet trickled down to those working in the product end-of-life phase. Design for remanufacture is a relatively uncommon concept in UK companies, and it would appear that China is even further behind. However, companies in China are beginning to make changes to the way they design products (for example the reduction of hazardous materials), and as legislation matures and China expands its research and development sectors, ecodesign can be expected to become increasingly relevant to Chinese design engineers. As discussed in this paper, the highly automated technology used in China's state-of-the-art recycling facilities may render many traditional ecodesign guidelines unnecessary; yet, design considerations that facilitate effective collection and transportation (for example design for stacking or product-service design) remain important. The collection of e-waste was a significant part of the three case study recyclers' businesses and is likewise a significant factor in the remanufacturing industry.

At present, remanufacturing in China is more or less limited to 14 pilot studies recently launched by the government. These studies, which are focused upon the automotive industry, are intended to gather knowledge and experience of the remanufacturing process to enable the development of guidance, standards, and strategies that will expand China's remanufacturing industry. It will be interesting to learn the findings from these pilot studies, which include product design guidelines or requirements. However, these pilot studies are concerned with automotive (not electronic) products. It is most likely that a move towards DfRem in China will begin in the mechanical/electromechanical industries, which has been the case in the UK.

To date, there is no specific environmental legislation encouraging companies to remanufacture in China. As such, DfRem is virtually unknown, and very little remanufacturing research can be found in the literature. However, if the pilot studies prove successful and remanufacturing industry expands in years to come, it is possible that DfRem will follow a similar path to ecodesign in the future. Research conducted in the UK would suggest that there are a number of steps OEMs can take to help ensure that DfRem considerations are included in the design process. Designers must be motivated to design for remanufacture and have knowledge and understanding of their company's remanufacturing processes and capabilities. Management must also be committed to optimising remanufacturability. Key factors which may influence DfRem integration include the presence of remanufacturing requirements in design specifications and regular and quality communication between design engineering teams and the remanufacturers of their products [9].

Conclusion
Summary

This paper has presented the findings from a literature survey and case study research conducted in three Chinese recyclers of waste electrical and electronic products. Recent progress in environmental legislation in China has meant that manufacturers are under increasing pressure to ensure the environmentally sound treatment of their e-waste, and the changes have also led to an increase in formal recycling industrial activity in China. Remanufacturing, on the other hand, remains relatively unknown and un-tried as an end-of-life solution in China; as a result, the concept of 'design for remanufacture' is also virtually unknown. However, it is possible that DfRem could become more relevant to Chinese OEMs in the future, if a government pilot scheme for automotive remanufacturing proves a success, and ecodesign concerns in general become a more common knowledge among Chinese product designers.

Considering e-waste specifically, however, evidence gathered from the three case study companies would suggest that electrical and electronic products are not highly suited to the remanufacturing process due to a combination of technical design and market factors. Also, the remanufacturing process is not as suited to e-waste, in comparison to the hi-tech recycling processes now utilized by China's formal recyclers. Considering that much of the DfRem literature published to date has focused upon electrical and electronic products [27], the issues raised in this paper should be taken into consideration when discussing the future directions of both the remanufacturing industry and future e-waste solutions.

Future work

The findings presented in this paper represent an outlook on design for remanufacture and e-waste in China, from a European (UK) perspective. However, to gain a full understanding of this subject, more in-depth research is required which covers not only the recyclers' perspective but also the perspective of the OEM (specifically design engineering). More evidence of DfRem practice in China is required to further address the research aims stated in this paper. Not only will this enhance our understanding of the relationships between OEMs and their waste treatment facilities but studying those who are responsible for the design of electrical and electronic products may also reveal a wider set of barriers, challenges, and opportunities regarding DfRem and e-waste which must be taken into account during any discussion of the subject. Therefore, future case study work in this area should include relevant design engineers, design managers, and aftermarket management working within electrical and electronics OEMs. An understanding of the existing design process and organisational issues in Chinese OEMs will enable the development of DfRem guidelines that are appropriate for China's organisational structures and cultures, as well as the country's existing product end-of-life technologies and infrastructure.

Due to the nature of the GREENet project, this research was conducted with a specific focus upon electrical and electronic products. However, studies from the UK would suggest that Chinese remanufacturing and DfRem practice is most likely to develop in the mechanical/electromechanical industries first [4]. Therefore, to gain an improved early understanding of emerging DfRem challenges and requirements, more insight into these industries is required.

If the Chinese government's remanufacturing pilot scheme is a success and the remanufacturing industry of China is set to expand, the need for DfRem within Chinese OEMs will increase, and research in this subject will become more relevant. Aside from the development of DfRem guidelines, another important part of this research will be determining how these guidelines may best be integrated into the design processes of Chinese OEMs: a country- and culture-specific study of the operational factors that would enable successful DfRem integration.

Finally, this research has revealed that the design of electrical and electronic products in China is not highly suited to the remanufacturing process, partly because the existing state-of-the-art recycling processes do not require disassembly. However, many of the other challenges to e-waste remanufacture identified through this research were market-driven: in the opinion of waste treatment facilities, customers simply do not want remanufactured electronic goods; therefore, recycling is a more economically lucrative strategy. Further research is required to understand these issues from both sides: What can designers do to increase the emotional longevity of electronic products in an era of rapid technological advancement, and how can society challenge the view that 'new' is always desirable in a world of limited resources?

Abbreviations
DfRem: design for remanufacture; EPR: extended producer responsibility; EU: European union; GREENet: Globally Recoverable and Eco-Friendly E-Equipment Network with Distributed Information Service Management; OEM: original equipment manufacturer; RoHS: reduction of hazardous substances; WEEE: waste electrical and electronic equipment.

Competing interests
The authors declare that they have no competing interests.

Authors' contributions
GDH drafted the manuscript and acquired, analysed, and interpreted the data. WLI made critical revisions for the intellectual content of the manuscript. JFCW also contributed to the critical revision for the intellectual content of the manuscript. All authors read and approved the final manuscript.

Authors' information
GDH graduated from the University of Strathclyde, Glasgow (UK) with a BSc (Hons) degree in Product Design and Innovation from the Department of Design, Manufacture and Engineering Management (DMEM). She is a member of the University of Strathclyde's remanufacturing research group, currently completing a Ph.D. in product design for remanufacture, and is also currently a Teaching Associate in the DMEM Department. WLI is one of the UK's leading Remanufacturing researchers with elements of her work incorporated in British Standards Institute (e.g. BS 8887–2:2009- Terms and definitions). She is a member of national and international committees established to help industry meet international environmental legislation (e.g. TDW/004/0-/05 Design for MADE BSI, The UK Energy Minister's WEEE Advisory Group). She is the initiator and the Editor-in-Chief of Springer's International Journal of Remanufacturing. She is a founding member of the IEEE Robotics and Automation Society technical committee on "sustainable production and service". She created and heads the University of Strathclyde remanufacturing research group, which specialises in interdisciplinary and practitioner-based research, and consists of 14 active researchers with 21 industrial partners. Including current research, she has been actively involved in over 17 industry-focused remanufacturing projects at Ph.D. or post doctorate level (6 as researcher and 10 as supervisor). With key EU and Chinese research institutions, she is developing Information and Communication Technologies-based strategies to enhance global product end-of-life management. She has developed and teaches MSc classes on Sustainable Design and Manufacture. She undertakes knowledge transfer to industry, academia and the public. JFCW received the degrees BEng (Electronic Engineering) in 1998 and Ph.D. (Electromagnetic Force Microscopy) in 2002, both from the University of Plymouth, Plymouth (UK). After his Ph.D. research, he became a post-doctoral Research Associate at the School of Biological Sciences, University of Bristol, Bristol, UK, in 2003, working on insect auditory systems. In 2008, he moved to the University of Strathclyde, Glasgow, UK, to take up the position of Lecturer in the Centre for Ultrasonic Engineering, in the Department of Electronic and Electrical Engineering, where he undertakes research into biologically inspired acoustic systems, with a focus on non-destructive testing and sustainable engineering. In 2011, he was appointed to the position of Senior Lecturer.

Acknowledgements
The authors would like to thank the School of Environment, Tsinghua University and the RCEES in Beijing for their accommodation and support of this research. We also thank the three case study companies involved in this research. This work is carried out as a part of GREENet which was supported by a Marie Curie International Research Staff Exchange Scheme Fellowship within the 7th European Community Framework Programme under grant agreement No. 269122. The paper only reflects the author's views and the Union is not liable for any use that may be made of the information contained therein.

Author details
[1]Design Manufacture and Engineering Management, 4th Floor Architecture Building, University of Strathclyde, Glasgow G4 0NG, UK. [2]Electronic and Electrical Engineering, R3.40a, Royal College Building, University of Strathclyde, Glasgow G1 1XQ, UK.

References

1. Ijomah, W: A model-based definition of the generic remanufacturing business process. The University of Plymouth, Plymouth (2002)

2. Lindahl, M, Sundin, E, Ostlin, J: Environmental issues within the remanufacturing industry. In: 13th CIRP International Conference on Life Cycle Engineering., Leuven (31 May–2 June 2006)

3. King, A, Burgess, S, Ijomah, W, McMahon, C: Reducing waste: repair, recondition, remanufacture or recycle? Sustainable Dev 14, 257–267 (2006)

4. Chapman, A, Bartlett, C, McGill, I, Parker, D, Walsh, B: Remanufacturing in the UK. Centre for Remanufacturing and Reuse, UK (2009)

5. Ijomah, W, McMahon, C, Hammond, G, Newman, S: Development of robust design-for-remanufacturing guidelines to further the aims of sustainable development. Int J Prod Res 45, 4513–4536 (2007)

6. Sundin, E, Bras, B: Making functional sales environmentally and economically beneficial through product remanufacturing. J Cleaner Prod 13, 913–925 (2005)

7. Charter, M, Gray, C: Remanufacturing and Product Design: Designing for the 7th Generation. The Centre for Sustainable Design, Surrey (2008)

8. Kerr, W, Ryan, C: Eco-efficiency gains from remanufacturing: a case study of photocopier remanufacturing at Fuji Xerox Australia. J Cleaner Prod 9, 75–81 (2001)

9. Hatcher, G, Ijomah, W, Windmill, J: Integrating design for remanufacture into the design process: the operational factors. J Cleaner Prod 39, 200–208 (2012)

10. Scottish Environment Protection Agency: Waste electrical and electronic equipment. (WEEE). http://www.sepa.org.uk/waste/waste_regulation/producer_responsibility/weee.aspx. Accessed 12 November 2012

11. Li, J, Tian, B, Liu, T, Liu, H, Wen, X, Honda, S: Status quo of e-waste management in Mainland China. J Mater Cycles Waste Manage 8, 13–20 (2006)

12. Europa: Waste Electrical and Electronic Equipment (WEEE) Directive. http://eur-lex.europa.eu/LexUriServ/LexUriServ.do?uri=CELEX:32002L0096:EN:NOT. Accessed 20 May 2012

13. Turner, M, Callaghan, D: UK to finally implement the WEEE directive. Comput Law Secur Rep 23, 73–76 (2007)

14. Ongondo, F, Williams, I, Cherrett, T: How are WEEE doing? A global review of the management of electrical and electronic wastes. Waste Manag 31, 714–730 (2011)

15. Recycling. http://www.environment-agency.gov.uk/cy/ymchwil/llyfrgell/data/34425.aspx. Accessed 12 November 2012

16. Yu, J, Hills, P, Welford, R: Extended producer responsibility and eco-design changes: perspectives from China. Corp Soc Responsibility Environ Manage 15, 111–124 (2008)

17. He, K, Li, L, Ding, W: Research on Recovery Logistics Network of Waste Electronic and Electrical Equipment in China. In: 3rd IEEE Conference on Industrial Electronics and Applications., Singapore (3–5 June 2008)

18. Chung, S, Zhang, C: An evaluation of legislative measures on electrical and electronic waste in the People's Republic of China. Waste Manag 31, 2638–2646 (2011)

19. Freyberg, T: China Takes Charge. http://www.waste-management-world.com/articles/print/volume-12/issue-6/regulars/from-the-editor/china-takes-charge.html. Accessed 18 May 2012

20. Li, B, Du, H, Ding, H, Shi, M: E-waste recycling and related social issues in China. Energy Procedia 5, 2527–2531 (2011)

21. Kojima, M, Yoshida, A, Sasaki, S: Difficulties in applying extended producer responsibility policies in developing countries: case studies in E-waste recycling in China and Thailand. J Mater Cycles Waste Manage 11, 263–269 (2009)

22. Yu, J, Williams, E, Ju, M, Shao, C: Managing E-waste in China: policies, pilot projects and alternative approaches. Resour Conserv Recycling 54, 991–999 (2010)

23. Xiang, W, Ming, C: Implementing extended producer responsibility: vehicle remanufacturing in China. J Cleaner Prod 19, 680–686 (2011)

24. Zhang, T, Wang, X, Chu, J, Cui, P: Remanufacturing mode and it's reliability for the design of automotive products. In: 5th International Conference on Responsive Manufacturing - Green Manufacturing (ICRM 2010)., Ningbo (11–13 January 2010)

25. Caterpillar: Caterpillar announces remanufacturing joint venture with China Yuchai to promote China's sustainability and environmental preservation initiatives. http://www.caterpillar.com/cda/files/2501386/7/121409%20Caterpillar%20Announces%20Remanufacturing%20Joint%20Venture%20with%20China%20Yuchai.pdf. Accessed 18 May 2012

26. IBM: IBM Opens the First Server Remanufacturing Center in China. http://www-03.ibm.com/press/us/en/pressrelease/36976.wss. Accessed 18 May 2012

27. Hatcher, G, Ijomah, W, Windmill, J: Design for remanufacture: a literature review and future research needs. J Cleaner Prod 19, 2004–2214 (2011)

28. Wong, S, El-Abd, H: Why electronics in China - a business perspective. IEEE Trans Components Packaging Technol 26, 276–280 (2003)

29. Altenburg, T, Schmitz, H, Stamm, A: Breakthrough? China's and India's transition from production to innovation. World Dev 36, 325–344 (2007)

30. Lindahl, M: The State of Eco-Design in Asian Electrical and Electronic Companies: A Study in China, India. Thailand and Vietnam. The Centre for Sustainable Design, Surrey (2007)

31. Lofthouse, V: EcoDesign tools for designers: defining the requirements. J Cleaner Prod 14, 1386–1395 (2006)

32. Behrisch, J, Ramirez, M, Giurco, D: Representation of ecodesign practice: international comparison of industrial design consultancies. Sustainability 3, 1778–1791 (2011)

33. Seitz, M: A critical assessment of motives for product recovery: the case of engine remanufacturing. J Cleaner Prod 15, 1147–1157 (2007)

34. Kriwet, A, Zussman, E, Seliger, G: Systematic integration of design-for-recycling into product design. J Cleaner Prod 38, 15–22 (1995)

35. Panasonic: Green design. http://www.panasonic.co.uk/html/en_GB/About+Panasonic/Green+Design/1555006/index.html#anker_1556569. Accessed 20 May 2012

36. Lenovo: Lenovo Statement Concerning WEEE. http://www.lenovo.com/social_responsibility/us/en/sustainability/Lenovo_WEEE_statement.pdf. Accessed 20 May 2012

6

An entropy-based metric for product remanufacturability

Monsuru O Ramoni[*] and Hong-Chao Zhang

Abstract

Manufacturing contributes heavily to environmental life cycle measures such as energy, material use, and water consumption through depletion and pollution. To lessen the environmental impacts, a number of initiatives have been developed. One of such initiatives is the used product take back, a process through which manufacturers collect used products and remanufacture them to like-new condition. However, remanufacturing of the used products at a modest cost is becoming a daunting task for many manufacturers due to the increasing complexity in many products. To mitigate this remanufacturing challenge, this paper develops a metric to quantify the remanufacturability incorporated into the new product at the design stage. The metric is based on entropy, a phenomenon well known in engineering.

Keywords: Remanufacturing, Entropy, Metric

Background

Remanufacturing is the process of restoring durable used product to like-new condition (in terms of product functions) with only a modest investment. The process involves the complete disassembly of a product, during which each component is thoroughly cleaned, examined for damage, and reprocessed to its original equipment manufacturer specifications [1-3]. A remanufactured product often comes with a warranty, another major criterion that differentiates remanufacturing from other end-of-life strategies [4] (Figure 1). It is worthy to always emphasize the differences between remanufacturing and other product recovery processes due to the lingering confusion about the characteristics of different product recovery processes.

Remanufacturing differs from recycling in that value added for original manufacturing including labor, energy, and equipment expenditure is conserved. The added value is lost during recycling, which reduces the product to its material constituents and requires additional labor, energy, and machinery [3,5,6]. On the other hand, remanufacturing preserves the product's (or the part's) identity and performs the required operations in order to

bring the product to a desired level of quality like that of a new product.

Remanufacturing also differs from repairing, a process limited to making a product operational as opposed to thoroughly restoring it to like-new condition [4,5]. If the remanufacture of the product is not extensive, i.e., few parts are replaced, either of the terms reconditioning or refurbishing are more suitable. Reconditioning typically refers to the restoration of parts to a functional and/or satisfactory condition by surfacing and painting. Remanufacturing typically involves greater work content than other product recovery processes, and as a result, its products tend to have superior quality and performance [2,3,6,7].

Prominent among remanufacturing problems is the poor remanufacturing potential of many products as designs have typically focused on functionality and cost at the expense of environmental issues. Moreover, designers may lack remanufacturing knowledge because there is a paucity of remanufacturing knowledge, design, and research. Design-for-recycling has received more attention among design and manufacturing engineers than design-for-remanufacture (DfRem) [8,9], even though remanufacturing may provide greater environmental and financial benefits than recycling.

As environmental awareness is gaining more ground, manufacturers have begun to emphasize remanufacturing

* Correspondence: olalekan.ramoni@ttu.edu
Department of Industrial Engineering, Texas Tech University, Lubbock, TX 79409, USA

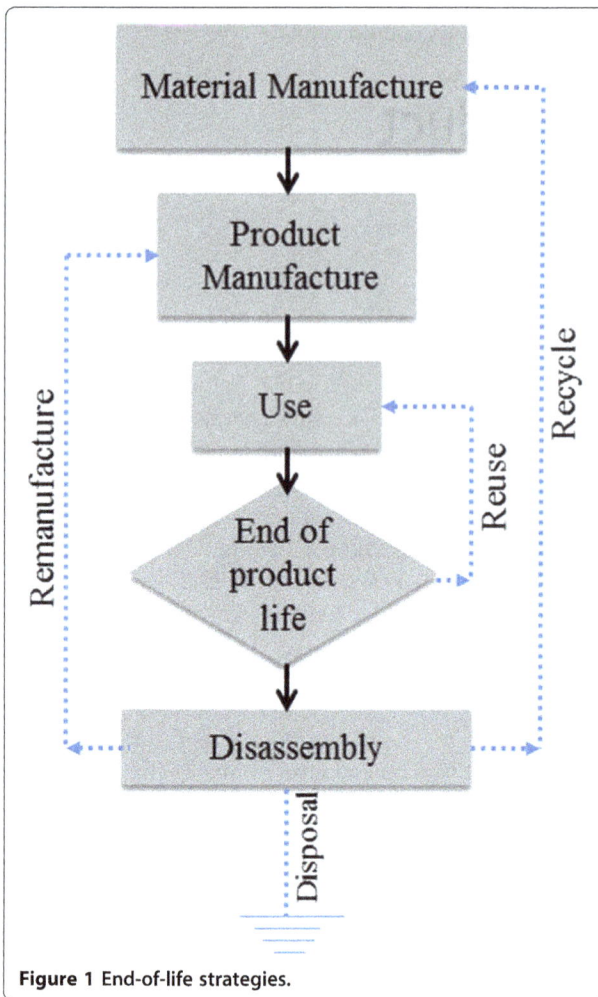

Figure 1 End-of-life strategies.

of post-consumed products. Many manufacturers, such as Xerox, HP, and Caterpillar, now have programs to collect their products after first life cycle and remanufacture them for sale. Moreover, environmental legislations are becoming strict in many countries, as reflected in legislations such as Waste Electronics and Electrical Equipment and End of Life Vehicle directives [10,11]. Remanufacturing can help original equipment manufacturers meet their commitment to environmental issues as well as legislations in a profitable manner [12]. Also, Lund indicated in his studies the total energy required for remanufacturing compared to energy used in initially producing a product, the ratios of which are in the order of 1:4 and 1:5, though these ratios are being disputed by Gutowski et al. [1,3,4]. Still, these gains, coupled with evidence from the auto industry [4,13,14], indicate remanufacturing to be a great idea for both economical and sustainable advantages. However, the increasing complexity of many products nowadays [8,15], as customers are demanding multifunctional and complex products, is hindering the venture of

making remanufacturing a relatively easy task with modest investment (Figure 2). This new trend of increasing complexity in products necessitates the need for more research in remanufacturing.

The opportunity to implement remanufacturing at the onset of product design culminates into one particular research area in the field of remanufacturing, which is called 'design-for-remanufacture' [3,16]. This is the area that requires a lot of work and research to achieve efficiency in remanufacturing without diminishing the functional deliveries of a product. Research has indicated that whether a product is suitable for remanufacture or not greatly depends upon decisions made during the design stage. The importance of considering remanufacturing issues in product designs has frequently been stated in many literatures [17,18], and it is well concluded that the largest gain in enhancing remanufacturability could be made in the product design [4,19]. Zwolinski et al. [4,20] argued that remanufacturing must be considered as early as possible in the design process, ideally as the 'concept generation phases'. However, many of the tools offered so far are too complex and technical to be used at a very early stage of the design process. Given the issues and in order to assist the product designers in developing products which can be remanufactured efficiently at the end of its life cycle, there is a need to have a less fuzzy quantitative methodology through which the designers can systematically assess the degree of remanufacturability incorporated into the new product at the design stage.

Therefore, the aim of this paper is to develop a metric through which the product development team at the onset of design can quantify the degree of remanufacturability incorporated into a new product before actual manufacturing of the product takes place. The metric, however, must rely on uncomplicated methodology to assess the design. The methodology in turn must rely on theoretical basis that could be easily understood by

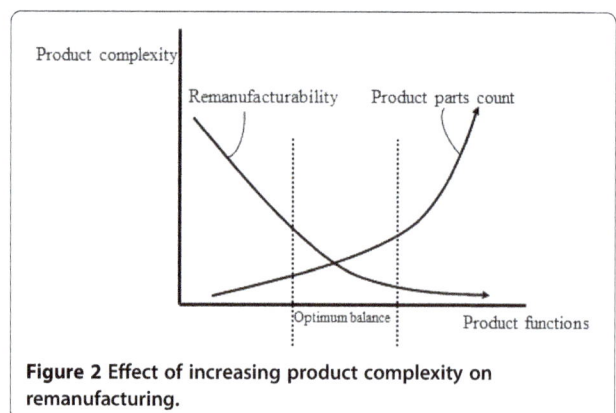

Figure 2 Effect of increasing product complexity on remanufacturing.

designers and engineers who are involved in the product development. The metric will allow the designers to translate properties of a design into quantitative scores in terms of product remanufacturability and thus provide a means of identifying weaknesses in the design and comparing alternatives.

The metric proposed in this paper is based on the property of thermodynamics. Thermodynamics has been used for many years as a foundation of methodologies for assessments of systems or resources. The underlying hypothesis in this study is that thermodynamics offers a comprehensive basis for the development of such metric as entropy for assessment.

Literature review

A number of studies has been done to measure product remanufacturability or to determine end-of-life strategy for a product. Prominent among such studies are the works of Rose and Ishi [7] Lee [21], Sundin and Bras [5], Sundin et al. [6], Brasand Hammond [22], Ijomah et al. [3], and Hatcher et al. [4].

Sundin and Bras [5] studied which product properties are important to facilitate remanufacturing by looking at properties that are suitable for the different remanufacturing steps (inspection, cleaning, disassembly, storage, reprocess, reassembly, and testing); a matrix called RemPro was created to rank and correlate the properties to appropriate strategies. Rose and Ishi [7] developed an end-of-life design advisor (ELDA) based on product characteristics to determine the end-of-life strategies; reuse, service, remanufacture, or recycle. The ELDA makes the designers aware of the impacts of their decision on the end-of-strategy and provides designers with a guide to appropriate end-of-life strategy. The ELDA works at the early stage of product design, which might require substantial knowledge from the designer in determining the appropriate end-of-life strategy [1]. Therefore, finding the balance between designer knowledge and effective end-of-life strategy might be subjective and even problematic.

Lee [21], in her paper, provided an ontology-based design for processing product information to provide decision supports in determining product end-of-life performance. Boothroyd and Dewhurst developed a metric for product assembly; their method involved simple part counts involved in the assembly of the product, which provide close measure for the assemblability of a product [17,18,22].

Bras and Hammond [22] developed a metric for assessing product remanufacturability based on Boothroyd and Dewhurst design-for-assembly as a foundation for remanufacturability assessment metrics based on product design features. Their method also involved aggregation of defined remanufacturing criteria such as disassembly, reassembly, cleaning, testing, and inspection. These criteria were assigned with different weight to measure product remanufacturability.

Review of most of these studies indicates that the tools and metrics provided are only suitable for use later in the product development when most of the major decisions have already been made. A detailed review provided by Hatcher et al. [4] indicates that many of these studies remain largely within the academic realm due to issues such as the complexity of the design aids and metrics and the lack of life cycle thinking in their designs. These issues prompt the need to have a metric that will be less cumbersome and whose methodology will be more familiar with the designers.

Some studies have been done on the use of entropy as a measure to assess systems of either product or service and provide feedback on their complexity for the purpose of making improvements. Frizelle and Woodcook [23] defined an entropy measure to quantify complexity in the supply chain based on Shannon's information entropy. Their entropy measure was qualitatively used to classify complexity as structural and operational (dynamic). Structural complexity deals with variety (schedule), and operational complexity deals with uncertainty (deviation from the schedule).

A static measure of complexity based on entropy measure was used for part mix in job shop scheduling [23,24]. Karp and Ronen [25] developed an approach which includes a formula for the determination of a lot location based on entropy measurement [23]. Fujimoto et al. [8] introduce a complexity measure based on product structure using different stages of process planning. ElMaraghya et al. [26] applied entropy function to quantify the complexity of manufacturing systems and their configurations with examples in machining processes. Guenov [12] uses the fundamentals of architectural design and entropy to introduce a system design metric for a comparison of alternative base cost, value, and performance and technical risk or complexity.

Entropy as a statistical mechanics of thermodynamics

From the perspective of statistical mechanics, entropy is viewed as the probability that certain events may occur within the framework of all possible events. By observing the behavior of large numbers of particles, statistical mechanics has succeeded in providing equations for the calculation of entropy as well as justification for equating entropy with a degree of disorder.

Shannon [27-29] looked at information as a function of a priori probability of a given state or outcome among the universe of physically possible states. He considered entropy as equivalent to uncertainty.

Thus, information theory parallels the second law of thermodynamics in expressing that the uncertainty in the world always tends to increase [27,28]. In Boltzmann's definition, entropy is a measure of the number of possible microscopic states (or microstates) of a system in thermodynamic equilibrium consistent with its macroscopic thermodynamic properties (or macrostate).

Shannon's measure of entropy

Consider a situation where there are n possible outcomes (the functional requirements), each with a probability of occurrence of p_i (design parameters). Let $p = (p_1, p_2 \ldots, p_n)$ be the probability distribution, such that $p_i \geq 0$ for all i and $\Sigma_i pi = 1$ for all i and for i between 1 and n.

The measure of entropy for this distribution is given by

$$S = \sum_i p_i \ln p_i$$

for i between 1 and n and whose function decreases from infinity to 0 for p_i ranging from 0 to 1. This measure is derived by using the axiomatic method of Euclid to quantify the concept of the uncertainty of a probability distribution.

Shannon used the following properties:

- S depends on all probabilities $(p_1, p_2 \ldots, p_n)$.
- $S (p_1, p_2 \ldots, p_n)$ is a continuous function of $p_1, p_2 \ldots, p_n$.
- $S (p_1, p_2 \ldots, p_n)$ is permutationally symmetric. It does not change if $S (p_1, p_2 \ldots, p_n)$ are reordered among themselves. This property is desirable since the labeling of outcomes should not affect the value of entropy.
- $S (1/n, 1/n \ldots, 1/n)$ should be a monotonic increasing function of n. As the number of outcomes increases, then, entropy increases.
- The maximum value of $S = (p_1, p_2 \ldots, p_n)$ occurs when all the outcomes have an equal probability of occurring. This maximum value is indicated as equal to $\ln n$ from the statistical description [28-31].

Statistical description of entropy

Examining the behaviors of the statistical definition of entropy as regards randomness, a uniform probability distribution reflects the largest randomness; a system with n allowed states will have the greatest entropy when each state is equally likely. In these probabilities where $n =$ total number of microstates, the entropy is thus $S = n \ln n$.

Therefore, the summation can be summarized as S is maximum when $\ln n$ is maximum, which permits many states, hence much randomness; equally, S is minimum when $\ln n$ is minimum. For instance, $n = 1$, the randomness will be zero, and S will be zero. For the additive property of entropy with respect to probabilities, if there is a quantum state when A is in its state x and B in its state y, it would be $p_{Ax} * p_{By}$ since the two probabilities are independent. The number of probabilities for the combined system is thus defined as $n = n_A * n_B$. The entropy of the combined $S = \ln (n_A * n_B)$ is $S = S_A + S_B$.

Entropy and remanufacture structure matrix

Entropy is defined as the measurement of system disorder. In this study, the function called entropy is used to evaluate the remanufacturing sequences, which are systemically generated by design alternatives of a product, and to select the design with the lowest entropy value. A high degree of entropy indicates a significant disorder and a high degree of symmetrical effect, while low levels of entropy represent an orderly state and a high degree of asymmetrical effect [16,30]. Therefore, in designing for remanufacturing, it is important to understand the entropy level and to control or adjust the design when the product design shows result towards a disadvantaged state of remanufacturability.

Axiomatic design is used to create a design structure matrix that assesses how the product, after its life cycle, passes all the different remanufacturing operation [16,32]. The relationship between the design structures can be determined and used to assess the extent of ease the entire remanufacturing operations would take place. The idea here is to study each stage required in remanufacturing, determine the functional requirement (FRs) of the product at a particular stage of remanufacturing, and transform the requirements into design parameters (DPs). The design parameters are affected by the remanufacturing operation variables (RVs) which might cause the transformation process to incur some degree of uncertainty. This uncertainty in this paper is interpreted as the unlikelihood that functional requirement will be achieved after the remanufacturing. Meanwhile, the goal of DfRem is to achieve FRs, and thus, any uncertainty in accomplishing the goal is considered to incur complexity. To solve or reduce the complexity, what is required is information. Therefore, information is an effective measure of uncertainty since it is what is required to resolve any uncertainty. In that sense, complexity should be proportional to the information. Axiomatic design theory has the quantity called information content, which is quite similar to that of Shannon's. Since axiomatic design complexity is explicitly defined in terms of uncertainty, it is natural to relate complexity to information indicated in the design structure [29,30,32].

To evaluate the adequacy of the product design to meet remanufacturing goal, the information content, via FRs and DPs, of a product can be calculated by Shannon's entropy. The high entropy value indicates a high degree of uncertainty, which provides deviation from the expected state in the process of remanufacturing the product.

The advantages of the metric developed in this paper lies in the familiarity of the product development team with the axiomatic designs whose approach focused on the low-level structure of the product to be remanufactured [32]. Axiomatic design (AD) would help in providing information on technical solution and expected performances about the product after its remanufacturing.

AD approach is akin to designers finding information about what the customer would require from the product after its first life cycle and transform the knowledge into design parameters that would be embedded into the product. Design parameters are collections of physical/non-physical entities that co-operatively deliver overall functional requirements of a product. Both functional knowledge and design parameters could be quantified to assess how easy the remanufacture of the product could be carried out. Hence, the metric on this approach provides a systematic assessment for deriving and optimizing designs for remanufacture and helps avoid traditional design-build-use-remanufacture cycles for remanufacture solution search.

For instance, the cleaning stage of remanufacturing is the process of removing anything that is not intended to be present in the part; it involves removing any substance like oil, sand, and other foreign materials. Let us assume that a customer wants a pipe (used pipe) attached to a pump machine (Figure 3) to serve as conduit for drawing water. The used pipe needs to be cleaned as part of remanufacturing the pump machine, and both soil and grease debris could be cleaned by a solution of cleaning agents and/or mechanical brushing.

Figure 3 A pump machine.

$$C = conduit$$
$$\begin{Bmatrix} remove\ soil \\ remove\ grease \end{Bmatrix} = \begin{pmatrix} a11 & a12 \\ 0, & a22 \end{pmatrix} \begin{Bmatrix} mechanical\ brush \\ chemical\ solution \end{Bmatrix}$$

The mechanical brush will only remove the soil, and the solvent will both dissolve the grease and rinse away the soil. Thus, it is more appropriate to use the chemical solution for the cleaning operation of the pipe remanufacture [16].

Hence, in the design for such pipe or conduit, the materials should not be made of anything susceptible to chemical solution. New design will enhance the efficiency of remanufacturing operations.

At each stage of remanufacturing operation, a design hierarchy is created. The relations (the dependencies) between the FRs and the DPs can be represented in an equation of the form:

FR = [A] DP
Also, the design parameters are affected by the RVs.
DP = [B] RV

By substituting, the two matrix equations can be combined into a single relation, linking requirement with remanufacturing operation.

FR = [A] [B] R

FR = [C] RV

where [C] = [A] [B]. The multiplication orders reflect the chronological order of the design and all remanufacturing operations. In theory, if the resulting matrix [C] is diagonal, then the design is uncoupled, and all design parameters and remanufacturing operation variables satisfy the functional requirements, as well as Axiom 1. The same iteration process would be done for all other process [7,12,16,30].

However, in reality, obtaining this formation might not be easy; either [A] or [B] has to be adjusted to meet the customer's needs. Minimizing the information content of design parameters would increase the chance of satisfying a function and meeting the customer's needs in the used product.

Illustrations

A new product is being developed; assuming that after the first life cycle, the manufacturer would want to remanufacture the product for reuse. The product part (pipe) will have to go through a series of processes: cleaning, reprocessing, and reassembly.

Let us assume that the functional requirements of the product part (pipe) after the cleaning process are: (1) strength, (2) aesthetic appearance, (3) attachment to the machine, (4) ability to draw liquid, and (5) ability to provide direction [30]. The functional requirements are mapped into design parameters as follows:

Table 2 The design matrices and coupling for reprocessing process

	DP1	DP2	DP3	DP4	DP5
FR1	×				×
FR2		×			
FR3			×		
FR4		×		×	
FR5				×	×

From the structure above, Table 1 is the derived design matrices and coupling for the cleaning process matrices, {FR} = [A]{DP}. Table 2 represents the design matrices and coupling for the reprocessing process. Table 3 represents the design matrices and coupling for reassembly process.

Remanufacture and entropy-based analysis

Definition 1: (Function entropy): For a function requirement 1(FR$_1$), its entropy is defined as

$$S (FR) = - \Sigma_i^n FRi \ln FRi$$

where FRi is the number of all functional requirements i assigned to the design parameters, and n is the number of design parameters corresponding to functional requirement FRi. S (FR) is meant for the determination of all customer requirements in the product after the remanufacturing of the product has taken place.

To normalize the value of functional requirement, the logarithm of the number of design parameters is taken.

S (cleaning/FR) = 1ln1 + 1ln1 + 1ln1 + 1ln1 + 1ln1 = 0

S (cleaning process/FR) = Log$_5$ (0) = 0

$$\begin{Bmatrix} FR \\ strength \\ aesthetic\ appearance \\ attachment\ to\ the\ machine \\ ability\ to\ draw\ liquid \\ ability\ to\ provide\ directions \end{Bmatrix} = \begin{pmatrix} \times & & & & \\ & \times & & & \\ & & \times & & \\ & & & \times & \\ & & & & \times \end{pmatrix} \begin{Bmatrix} DP \\ structure \\ color \\ joining\ process \\ hole \\ label \end{Bmatrix}$$

Table 1 The derived design matrices and coupling for cleaning process matrices

	DP1	DP2	DP3	DP4	DP5
FR1	×				
FR2		×			
FR3			×		
FR4				×	
FR5					×

Table 3 The design matrices and coupling for reassembly process

	DP1	DP2	DP3	DP4	DP5
FR1	×				
FR2		×			
FR3		×	×		
FR4		×		×	
FR5		×			×

S (reprocessing/FR) = 1ln1 + 2ln2 + 1ln1 + 2ln2 + 2ln2 = 4.16
S (reprocessing/FR) = Log_5 (4.16) = 0.89
S (reassembly/FR) = 1ln1 + 4ln4 + 1ln1 + 1ln1 + 1ln1 = 5.5
S (reprocessing process/FR) = Log_5 (5.5) = 1.06

Definition 2: (Design parameter entropy): For a design parameter (D), its entropy is defined as the average entropy of all design parameters for functional requirement. Since not all the customer requirements might be met in the remanufactured products, it is necessary to determine the entropy of design parameters that would fulfill customer requirements.

S (DP) = $(\Sigma_i^k DPi)/k$
S (cleaning/DP) = (1 ln1)/1 = 0
Log_5 (0) = 0
S (reprocessing/DP) = (2ln2 + 2ln2 + 2ln2) = (4.16) / 3 = 1.38
Log_5 (1.3) = 0.16
Similarly, S (reassembly/DP) = 2.77
Log_5 (2.77) = 0.62

Definition 3: The literature on remanufacturing indicates that the processes are sequential for almost all remanufacturing operation [30]. For a given product, from its design matrices of remanufacture, we can define the entropy metric as the sum of the values of its entropy value of functional requirements and design parameters from all the processes proposed for the product's remanufacturing.

S (remanufacturing) = S (cleaning, reprocessing, reassembly)
S (remanufacturing/FR) = 0 + 0.89 + 1.06 = 1.95
S (remanufacturing/DP) = 0 + 0.16 + 0.62 = .78

Σ S (processes in remanufacturing) $= 1.95 + 0.78$

Entropy metric of remanufacturability = Log $\sum$ (entropy of all required stages, via FRs and DPs, in the remanufacturing) = 1.95 + 0.78 = 2.73.
To normalize between (0, 1), take log_{10} of the value.
Entropy metric value for the illustrated product's remanufacturability = log_{10}(2.73) = 0.44.

Discussion and conclusion

The design matrix has been identified as tools to express the dependence and information flow between various stages in designing a product for remanufacture. Moreover, the main point offered in this paper is that the existing tool of thermodynamics such as entropy could

be engaged to develop a metric for the analysis of a product's remanufacturability via its design matrices. Considering the illustrations provided in the paper, it is shown that by finding functional requirements from the product after its life cycle, its corresponding design parameters, and establishing design matrices for potential processes for the product remanufacturing, one can determine the degree of remanufacturability existing in the product at the product development stages.

The main advantage of using this metric is that its theoretical basis can easily be understood by the product designers and engineers. Subsequently, this metric may be associated with monetary value owing to the fact that remanufacturing must be done with minimum investment, especially with contentious arguments of some authors that economic consideration must be at the forefront of DfRem. Quantifying remanufacturability on the product labeling may serve as an incentive for the customers to buy the product, just like energy consumption is nowadays labeled on many products.

The functionality of the metric is contingent on developing information about the status of product end-of-life and functional deliverables after product remanufacturing. Therefore, it might not be sufficient to conclude that the metric developed is far better than other metrics, but the metric reinforces life cycle thinking and early use in the design stage of the product development, both critical components missing in many other metrics.

However, a design having a very high entropic value for its product remanufacturability might not necessarily mean it would not be remanufacturable but require more information to understand what is happening in the design. Therefore, developing methods for acquiring information needed to conclude designs will require more research.

Competing interests
The authors declare that they have no competing interest.

Authors' contributions
MOR and HCZ made contributions to all parts of the manuscript. Both authors read and approved the final manuscript.

References
1. Gutowski, T, Sahni, S, Boustani, A: Remanufacturing and energy saving. Environ. Sci. Technol. **45**, 4540–4547 (2011)
2. Gungor, A, Gupta, SM: Issues in environmentally conscious manufacturing and product recovery: a survey. Comput. Ind. Eng. **36**, 811–853 (1999)
3. Ijomah, W, Childe, S, McMahon, C: Remanufacturing: a key strategy for sustainable development. In: Proceedings of the 3rd International Conference on Design and Manufacture for Sustainable Development (2001)
4. Hatcher, G, Ijomah, W, Windmill, J: Design for remanufacturing: a literature review and future research needs. J. Cleaner. Prod. **19**, 2004–2014 (2011)
5. Sundin, E, Bras, B: Making functional sales environmentally and economically beneficial through product remanufacturing. J. Cleaner. Prod. **13**(9), 913–925 (2005)
6. Sundin, E, Lindahl, M, Ijomah, W: Product design for product/service systems-design experiences from Swedish industry. J. Manuf. Technol. Manage. **20**(5), 723–753 (2009)

7.   Rose, C, Ishii, K: Product end-of-life strategy categorization design tool. J. Electron. Manuf. **9**, 41–51 (1999)
8.   Fujimoto, H, Ahmed, A, Iida, Y, Hanai, M: Assembly process design for managing manufacturing complexities because of product varieties. Int. J. Flexible. Manuf. Syst. **15**(4), 283–307 (2003)
9.   Fukushige, S, Yamamoto, K, Umeda, Y: Lifecycle scenario design for product end-of-life strategy. J. Remanuf. **2**, 1 (2012)
10. European Parliament and Council: Directive 2002/96/EC of the European Parliament and of the Council of 27 January 2003 on waste electrical and electronic equipment (WEEE). Official. J. Eur. Union. **L37**, 24–38 (2003)
11. European Parliament and Council: Directive 2000/53/EC of the European Parliament and of the Council of 18 September 2000 on end-of-vehicle (ELV). Official J. Eur. Union. **L269**, 34–42 (2000)
12. Guenov, MD: Complexity and Cost Effective Measures for Systems Design. Cranfield University, Cambridge (2002)
13. Seitz, M: A critical assessment of motives for product recovery: the case of engine remanufacturing. J. Cleaner. Prod. **15**, 1147–1157 (2007)
14. Smith, V, Keoleian, G: The value of remanufactured engines: lifecycle environmental and economic perspectives. J. Ind. Ecol. **8**, 193–221 (2004)
15. Freiberger, S, Alberct, M, Kaufl, J: Reverse engineering technologies for remanufacturing of automotive systems communicating via CAN bus. J. Remanuf. **1**, 6 (2011)
16. Shu, L, Flower, W: A structured approach to design for remanufacture. In: (ed.) DE-Vol 66, Intelligent Concurrent Design: Fundamental, Methodology, Modeling and Practice. ASME, Boston (1993)
17. Amezquita, T, Hammond, R, Bras, B: Characterizing the remanufacturability of engineering systems. In: ASME Advances in Design Automation Conference, Ed – Vol. 82, pp. 271–278. ASME, Boston (1995)
18. Amezquita, T, Hammond, R, Bras: Design for remanufacturing. In: (ed.) 10th International Conference on Engineering Design (ICED 95), pp. 1060–1065. Praha, Czech Republic (1995). Heurista, Zurich, Switzerland
19. US Congress Office of Technology Assessment: Green Products by Design: Choices for a Cleaner Environment. United States Government, Washington (1992)
20. Zwolinski, P, Lopez-Ontiverous, M, Brisssaud, D: Integrated design of remanufacturable products based on product profiles. J. Cleaner. Prod. **14**, 1333–1345 (2006)
21. Lee, HM: An ontology-based product design adviser for assessing end-of-life (EOL) performance. Nanyang Technological University, Singapore (2008)
22. Bras, B, Hammond, R: Towards remanufacturing—metrics for assessing remanufacturability. In: 1st International Workshop on Reuse, Eindhoven, pp. 35–52., Eindhoven (1996)
23. Frizelle, G, Woodcock, E: Measuring complexity as an aid to developing operational strategy. Int. J. Oper. Prod. Manag. **15**, 26–39 (1994)
24. Calinescu, A, Efstathiou, E, Bermejo, J, Schirn, J: Assessing decision-making and process complexity in manufacturing through simulation. In: Proceedings of 6th IFAC Symposium on Automated Systems Based on Human Skill, pp. 159–162, Kranjska Gora (1997)
25. Karp, A, Ronen, B: Improving shop floor control: an entropy model approach. Int. J. Prod. Res. **30**(940), 923–38 (1992)
26. ElMaraghya, HA, Kuzgunkayaa, O, Urbanic, RJ: Manufacturing systems configuration complexity. Annals CIRP **54**, 445–448 (2005)
27. Bakshi, B, Gutowski, T, Sekulic, D: Thermodynamics and the Destruction of Resources. Cambridge University Press, New York (2009)
28. Dusan, P: An entropy generation metric for non-energy systems assessments. Energy **34**, 587–592 (2009)
29. Zhang, Z., Xiao, R: Empirical study on entropy models of cellular manufacturing systems. Progr. Nat. Sci. **19**, 389–395 (2009)
30. Lee, T: Complexity theory in axiomatic design. PhD dissertation, Massachusetts Institute of Technology (2003)
31. Wang, S, Capretz, A: Dependency and entropy based impact analysis for service-oriented system evolution. In: International Conferences on Web Intelligence and Intelligent Agent Technology, Lyon (2011). 22–27 August
32. Taglia, A, Campatelli, G: Axiomatic design and QFD: a case study of a reverse engineering system for cutting tool. In: Proceedings of ICAD 2006, 4th International Conference on Axiomatic Designs, Florence, (2006). 13–16 June

7

Coping with disassembly yield uncertainty in remanufacturing using sensor embedded products

Mehmet Ali Ilgin[1*], Surendra M Gupta[2] and Kenichi Nakashima[3]

Abstract

This paper proposes and investigates the use of embedding sensors in products when designing and manufacturing them to improve the efficiency during their end-of-life (EOL) processing. First, separate design of experiments studies based on orthogonal arrays are carried out for conventional products (CPs) and sensor embedded products (SEPs). In order to calculate the response values for each experiment, detailed discrete event simulation models of both cases are developed considering the precedence relationships among the components together with the routing of different appliance types through the disassembly line. Then, pair-wise t-tests are conducted to compare the two cases based on different performance measures. The results showed that sensor embedded products improve revenue and profit while achieving significant reductions in backorder, disassembly, disposal, holding, testing and transportation costs. While the paper addresses the EOL processing of dish washers and dryers, the approach provided could be extended to any other industrial product.

Keywords: disassembly line, experimental design, sensor embedded products, cost-benefit analysis, discrete event simulation

1. Background

Remanufacturing is an industrial process involving the conversion of used products into like-new condition. This process starts with the collection and transportation of EOL products to a remanufacturing plant where they are disassembled into parts. Following the cleaning and inspection of disassembled parts, repair and replacement operations are performed to deal with defective and worn-out parts. Finally, all parts are re-assembled into a remanufactured product which is expected to function like a new product. In addition to repair and replacement, some parts or modules may also be upgraded while remanufacturing a product.

New and stricter government regulations on EOL product treatment and increasing public awareness towards environmental issues have forced many manufacturers to establish specific facilities for remanufacturing operations. Being the most environment-friendly and profitable product recovery option, remanufacturing has many advantages over other recovery options such as recycling,

repairing or refurbishing. In remanufacturing, majority of labor, energy and material values embedded in an EOL product are recovered because the disassembled parts are used as is in the remanufacturing process. On the other hand, in recycling, only the material is recovered because the EOL products are simply shredded in a recycling facility. Remanufactured products provide superior performance due to replacement of worn-out parts and upgrading of some key parts. That is why many manufacturers are willing to give consumers the same warranty provisions as with the new products. Although replacement of some parts may occur during the repair or refurbishment option, there is no upgrading. Therefore repaired or refurbished products may not provide a superior performance and their warranty provisions are inferior to those of the remanufactured or new products.

Although remanufacturing is more sustainable than the traditional way of manufacturing where we only use virgin materials to produce new products, it involves more uncertainty. In a traditional manufacturing system, there are strict requirements to be obeyed by suppliers regarding the quality, quantity and arrival time of components. On the other hand, in remanufacturing, such strict requirements can not be imposed on the quality, quantity

* Correspondence: mehmetali.ilgin@deu.edu.tr
[1]Department of Industrial Engineering, Dokuz Eylul University, Buca 35160, Izmir, Turkey
Full list of author information is available at the end of the article

and arrival time of EOL products. That is why, determination of the condition, type and quantity of a component before actually disassembling it is not possible. This increases the uncertainty associated with the used component yield.

Sensor embedded products which involve sensors embedded into their critical components during the production process can solve this problem by providing information on the condition, type and number of components before actually disassembling them. In this study, we consider the application of SEPs in disassembly of components from EOL appliances for remanufacturing. The impact of SEPs on system performance is analyzed by performing separate experimental design studies based on orthogonal arrays for conventional products (CPs) and SEPs. Detailed discrete event simulation (DES) models of both cases are used to calculate various performance measures under different experimental conditions. Then, the results of pair-wise t-tests comparing the two cases based on different performance measures are presented.

The paper is organized as follows. In Section 2, a review of the issues considered in this study is presented. In Section 3, characteristics of the appliance disassembly line are explained. Section 4 and Section 5 explain the details and results of the design of experiments study, respectively. Finally, some conclusions are presented in Section 6.

2. Literature Review

Heuristics, tools or methodologies developed for manufacturing systems can not directly be applied to remanufacturing systems in most cases due to unique characteristics of remanufacturing process. Hence, researchers developed novel techniques considering different issues in remanufacturing including logistics [1,2], operations and production management [3,4], design for remanufacturing [5-7] and disassembly [8]. A complete and up-to-date overview of these studies can be found in the reviews by [9] and [10]. Being a crucial step in remanufacturing, disassembly has received increasing attention of researchers. Many studies have been presented on different domains of disassembly including sequencing [11,12], scheduling [13], disassembly line [14,15], disassembly line balancing [16,17], disassembly-to-order systems [18] and design for disassembly [19]. Researchers have also addressed the issues related to the disassembly of different type of products e.g., vehicles [20], electronics [21] and consumer appliances [22]. For detailed information on the different aspects of disassembly, we refer the reader to a couple of recent books [23,24].

There is a vast amount of literature on the use of sensor-based technologies on after-sale product condition monitoring. Starting with the study of [25], different methods of data acquisition from products during product usage

were presented by the researchers [26-28]. In all of these studies, the main idea is the use of devices with memory to save monitoring data generated during the product usage. Although most of these studies focus on the development of SEP models, only few researchers presented a cost-benefit analysis. [29] analyzed the trade-off between the higher initial manufacturing cost caused by the use of an electronic data log in products and cost savings from the reuse of used motors. [30] improved the cost-benefit analysis of [29] by considering the limited life of a product design. They showed that, in that case, servicing provides more reusable components compared to EOL recovery of parts. [31] investigated the effectiveness of embedding sensors in computers by comparing several performance measures in the two scenarios-with embedded sensors and without embedded sensors. The performance measures considered include average life cycle cost, average maintenance cost, average disassembly cost, and average downtime of a computer. However, they do not provide a quantitative assessment of the impact of SEPs on these performance measures. Moreover, since only one component of a computer (hard disk) was considered, the disassembly setting does not represent the complexity of a disassembly line which is generally used to disassemble EOL computers. By extending [31], [32] analyzed the effect of SEPs on the performance of an EOL computer disassembly line which is used to disassemble three components from EOL computers, namely, memory, hard disk and motherboard. Due to relatively simple structure of an EOL computer, they did not consider the precedence relationships among the components. However, disassembly of a particular component is restricted by one or more components in some products. That is why, these products are disassembled according to a route determined based on the precedence relationships. In this study, we investigate the quantitative impact of SEPs on different performance measures of a disassembly system. The disassembly setting we consider is a disassembly line which is used to disassemble components from EOL dryers and dish washers. We also consider the precedence relationships among the components together with the routing of different EOL product types through the disassembly line.

3. Appliance Disassembly Process

EOL dryers and dish washers (DWs) are disassembled on a five-station disassembly line. Physical configuration of the stations in the disassembly line is given in Figure 1. Figure 2 presents the components disassembled at different stations of the disassembly line together with the disassembly sequence and routing of EOL dryers and dish washers. According to this figure, EOL dryers travel only in downstream direction since the precedence relationships among their components follow the sequencing of

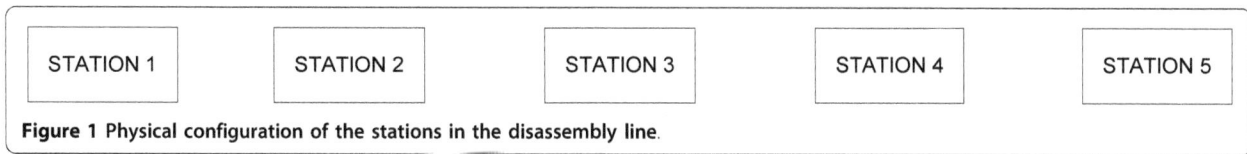

Figure 1 Physical configuration of the stations in the disassembly line.

disassembly process. However, EOL DWs can travel in both upstream and downstream directions depending on which component is to be disassembled next.

There are two common components shared by EOL dryers and dish washers, viz., metal cover and electric motor. Drum is only included in dryers while timer and circuit board are the components that can be disassembled only from EOL dish washers. All disassembled components are demanded except for the metal cover. Table 1 presents the precedence relationships among the components. Disassembly times at stations, demand inter-arrival times for components and EOL product inter-arrival times are all distributed exponentially.

Figures 3 and 4 present disassembly flow charts for conventional and sensor embedded appliance disassembly processes, respectively. Conventional appliances (ones with no sensors) visit all stations. Following the disassembly at each station, components are tested. The testing times are normally distributed with the means and

standard deviations presented in Table 1. Sensor embedded appliances visit only the stations which are responsible for the disassembly of functional components and their predecessor components. In addition, no testing is required for this case because of the sensor information available on the condition of the component.

Excess products, subassemblies and components are disposed of using a small truck with a load volume of 475 cubic feet. Whenever the total volume of the excess product, subassembly and component inventories become equal to the truck volume, the truck is sent to a recycling facility. Any product, subassembly or component inventory which is greater than *maximum inventory level* is assumed to be excess. Component volumes are given in Table 1. The volumes of EOL DWs and EOL dryers are taken as 20 cubic feet and 22 cubic feet, respectively. A multi kanban system (MKS) developed by [33] is used to control the disassembly line.

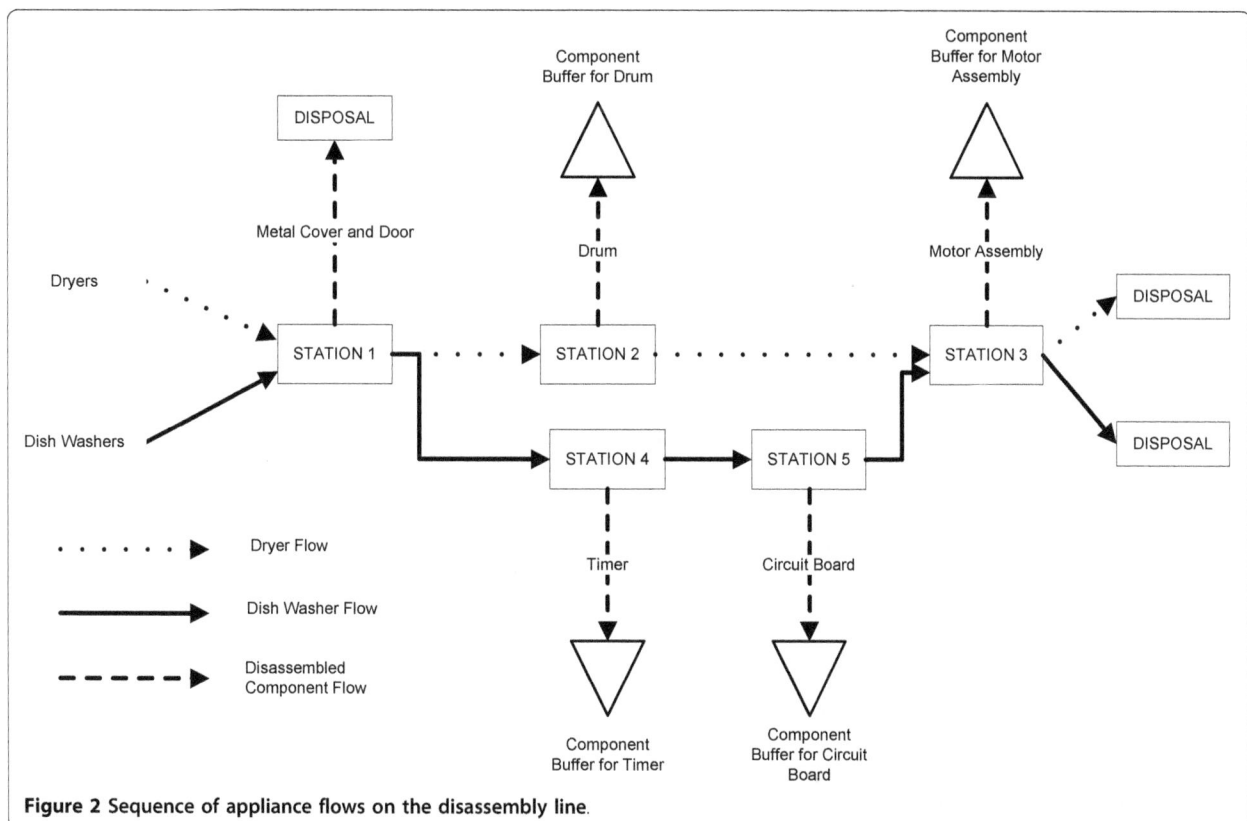

Figure 2 Sequence of appliance flows on the disassembly line.

Table 1 Specifications for DW and Dryer Components

Component Name	Code	Precedence Relationship		Testing Time (minutes)		Volume (cft)	Weight (lbs)
		DW	Dryer	Mean	Std. Dev.		
DW Metal Cover	A	-	-	-	-	0.720	*
Dryer Metal Cover	B	-	-	-	-	0.800	*
Drum	C	-	B	6	1.5	5.000	*
Motor Assembly	D	A, E, F	B, C	12	2	0.150	*
Timer	E	A	-	2.5	0.5	0.020	1
Circuit Board	F	A, E	-	6	1	0.030	1

*DW Metal Cover, Dryer Metal Cover, Drum and Motor Assembly weights are factors in the design of experiments study. For the weight ranges defined for these components, see Table 2.

4. Design of Experiments Study

In this section, we compare SEPs against CPs under different experimental conditions. The factors and factor levels considered in the experiments are given in Table 2. In this table, weights and prices of components have been estimated based on an online web search of various DW and dryer component sellers in USA. Further online web search was performed of various recyclers throughout the USA in order to estimate the steel scrap revenue per pound, disposal cost per pound, disposal cost increase factor for EOL products and scrap revenue decrease factor for EOL products. User and service manuals of various DW and dryer manufacturers were employed while estimating the mean disassembly and testing times of components together with small component weight factor. Maximum inventory level was estimated by making some trial simulation runs with different maximum inventory level values and investigating the changes in the number of products and components waiting in queues and various cost parameters. All the remaining parameter values (viz., non-functional and missing component probabilities, mean demand rates

for components, mean arrival rates of products, backorder cost rate, holding cost rate, testing cost per minute and disassembly cost per minute) were estimated based on the values used in the literature.

A full factorial design with 39 factors requires an extensive number of experiments (viz., 4.05E+18). Therefore, experiments were performed using orthogonal Arrays (OAs) [34] which allow for the determination of main effects by running a minimum number of experiments. Specifically, L_{81} OA was chosen since it requires 81 experiments while accommodating 40 factors with three levels [35]. DES models for both cases were developed using Arena 11 [36] to determine profit value together with various cost and revenue parameters for each experiment. Animations of the simulation models were built for verification purposes. In addition, models' output results were checked for reasonableness. Dynamic plots and counters providing dynamic visual feedback were used to validate the simulation models. The replication time for each DES model was 60480 minutes, the equivalent of six months with one eight hour shift per day. DES models were replicated 10 times for each OA experiment.

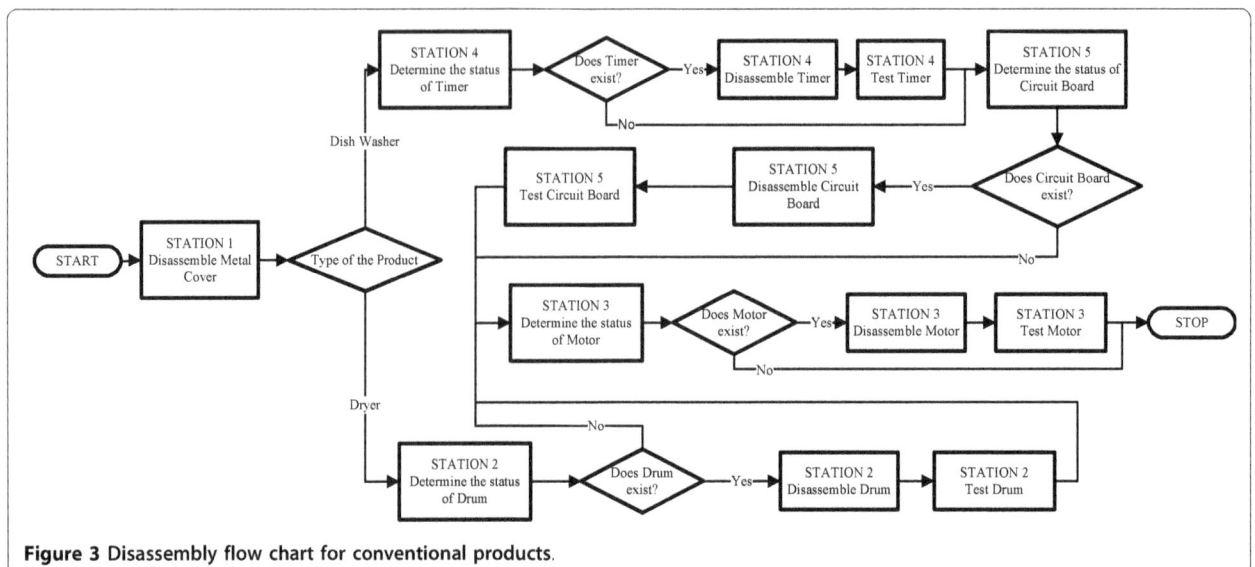

Figure 3 Disassembly flow chart for conventional products.

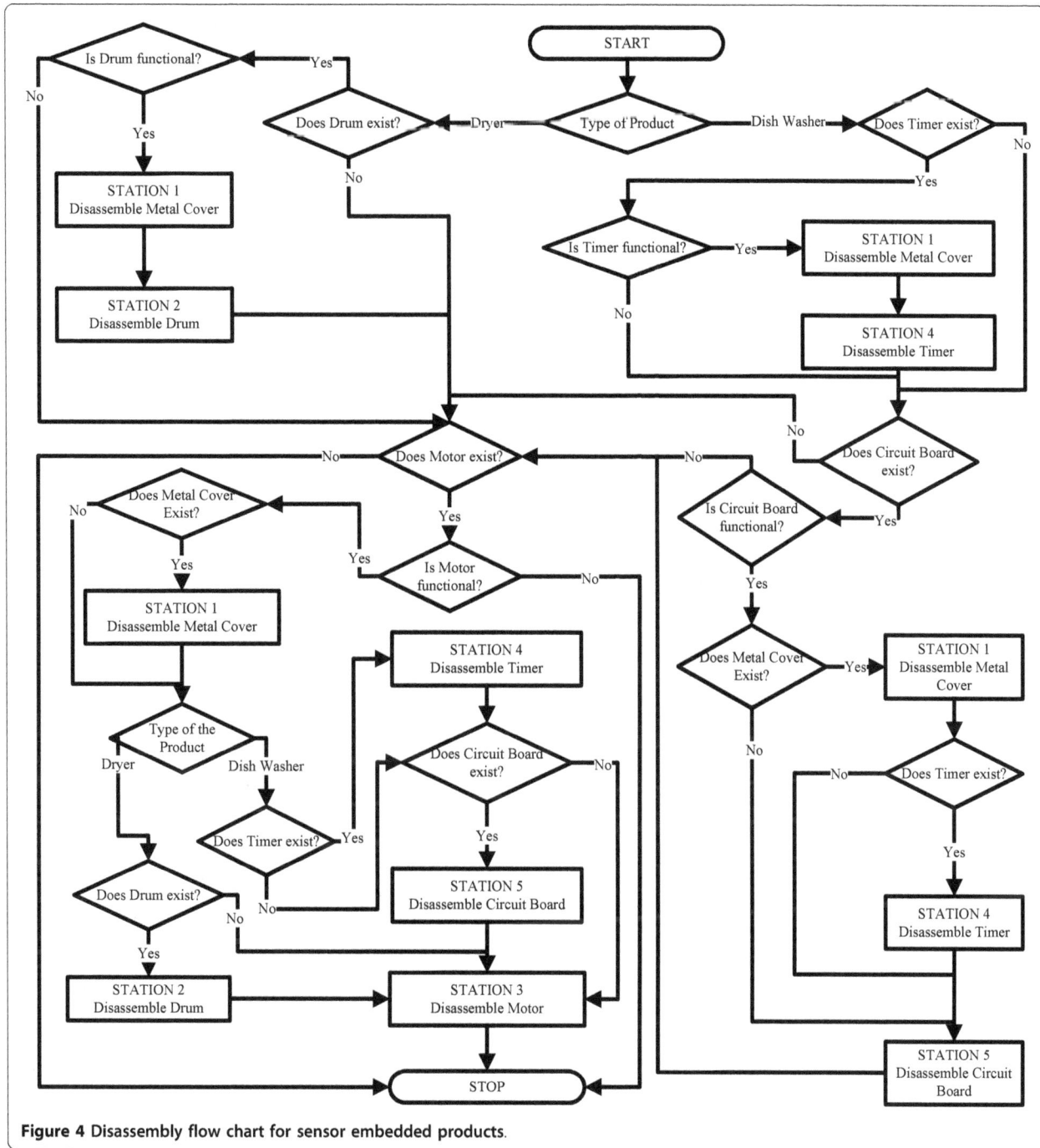

Figure 4 Disassembly flow chart for sensor embedded products.

Flow chart for the demand process is given in Figure 5. Figures 6 and 7 present the flow charts for the disassembly processes initiated by component kanbans for the CPs at the stations other than the last station and at the last station, respectively. Figures 8 and 9 present the flow charts of the disassembly processes initiated by component kanbans for the SEPs at the stations other than the last station and at the last station, respectively.

Flow charts for the disassembly processes initiated by subassembly kanbans for CPs and SEPs are depicted in Figures 10 and 11, respectively.

The following equation presents the formula used in the DES models for the calculation of profit value.

$$Profit = \overbrace{(SR + CR + SCR)}^{\text{Total Revenue}} - \overbrace{(HC + BC + DC + DPC + TC + TPC)}^{\text{Total Cost}} \quad (1)$$

Table 2 Factor levels

Number	Factor	Levels		
		1	2	3
1	Disposal cost increase factor for EOL products	0.06	0.12	0.18
2	Scrap revenue decrease factor for EOL products	0.06	0.12	0.18
3	Mean demand rate for Drum (components per hour)	8	12	16
4	Mean demand rate for Motor Assembly (components per hour)	8	12	16
5	Mean demand rate for Timer (components per hour)	8	12	16
6	Mean demand rate for Circuit Board (components per hour)	8	12	16
7	Mean arrival rate of EOL DWs (products per hour)	8	16	24
8	Mean arrival rate of EOL Dryers (products per hour)	8	16	24
9	Mean disassembly time for station 1 (minutes)	0.40	0.80	1.20
10	Mean disassembly time for station 2 (minutes)	0.75	1	1.25
11	Mean disassembly time for station 3 (minutes)	0.75	1	1.25
12	Mean disassembly time for station 4 (minutes)	0.75	1	1.25
13	Mean disassembly time for station 5 (minutes)	0.75	1	1.25
14	Backorder cost rate	0.40	0.60	0.80
15	Disassembly cost per minute ($)	0.75	1.5	2.25
16	Testing cost per minute ($)	0.50	0.60	0.70
17	Holding cost rate	0.20	0.30	0.40
18	Weight for Metal Cover of DW (pounds)	4	8	12
19	Weight for Metal Cover of Dryer (pounds)	5	10	15
20	Weight for Drum (pounds)	6	12	18
21	Weight for Motor Assembly (pounds)	5	10	15
22	Weight of other steel components of DW (pounds)	70	90	110
23	Weight of other steel components of Dryer (pounds)	80	100	120
24	Price for Drum ($)	30	50	70
25	Price for Motor Assembly ($)	40	60	80
26	Price for Timer ($)	20	40	60
27	Price for Circuit Board ($)	25	50	75
28	Disposal cost per pound ($)	0.40	0.50	0.60
29	Steel scrap revenue per pound ($)	0.20	0.25	0.30
30	Maximum inventory level	6	12	18
31	Small component weight factor	0.05	0.10	0.15
32	Probability of a non-functional Drum	0.12	0.24	0.36
33	Probability of a non-functional Motor Assembly	0.12	0.24	0.36
34	Probability of a non-functional Timer	0.12	0.24	0.36
35	Probability of a non-functional Circuit Board	0.12	0.24	0.36
36	Probability of a missing Drum	0.06	0.12	0.18
37	Probability of a missing Motor Assembly	0.06	0.12	0.18
38	Probability of a missing Timer	0.06	0.12	0.18
39	Probability of a missing Circuit Board	0.06	0.12	0.18

The different cost and revenue components used in the equation 1 can be defined as follows:

- SR : The total revenue generated by the component sales during the simulated time period (STP).
- CR : The total revenue generated by the collection of EOL products during the STP.
- SCR : The total revenue generated by selling scrap components during the STP.

- HC : The total holding cost of components, EOL products and subassemblies during the STP.
- BC : The total backorder cost of components during the STP.
- DC : The total disassembly cost during the STP.
- DPC : The total disposal cost of components, EOL products and subassemblies during the STP.
- TC : The total testing cost during the STP.
- TPC : The total transportation cost during the STP.

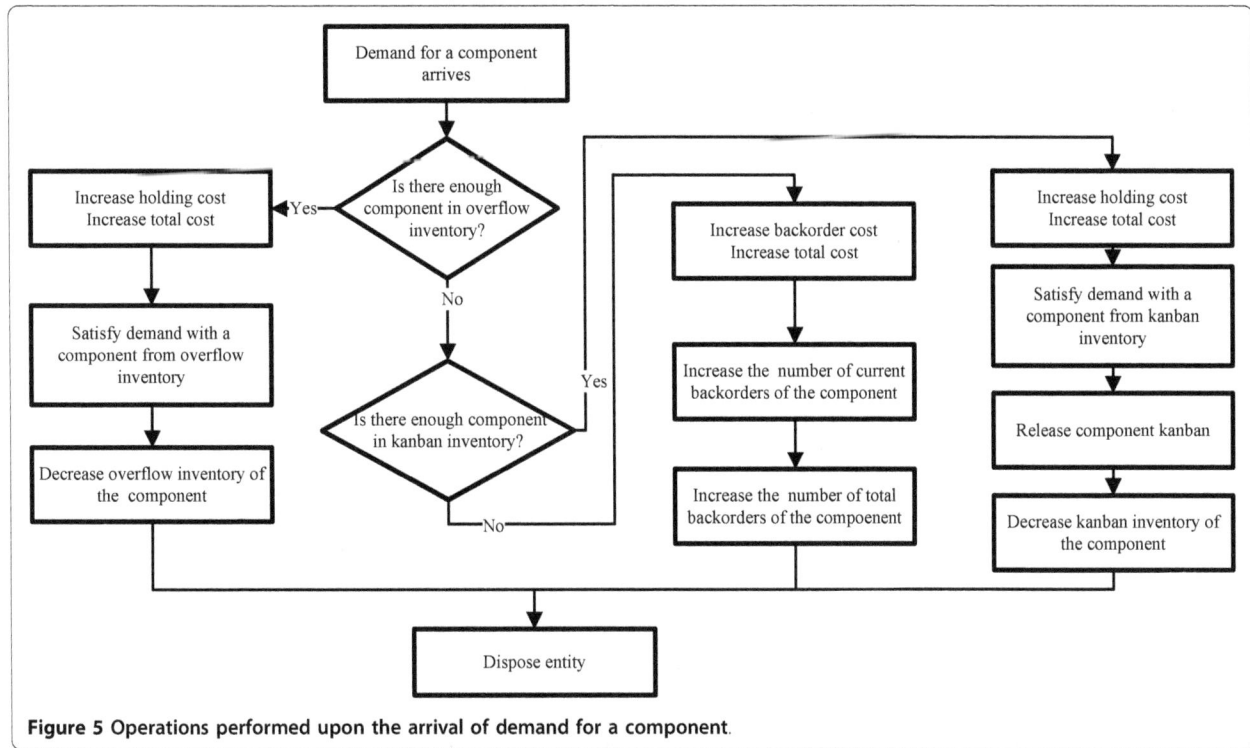

Figure 5 Operations performed upon the arrival of demand for a component.

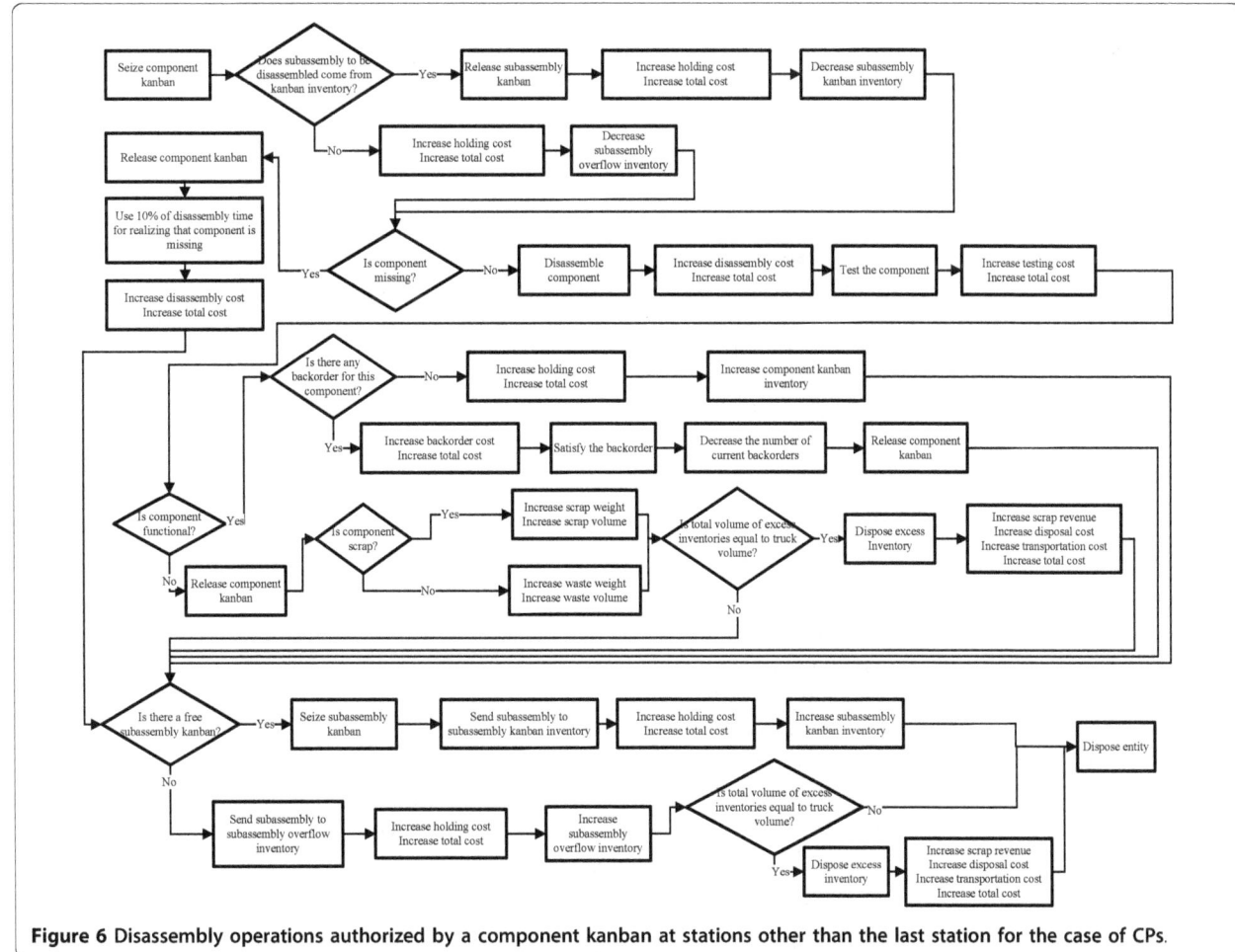

Figure 6 Disassembly operations authorized by a component kanban at stations other than the last station for the case of CPs.

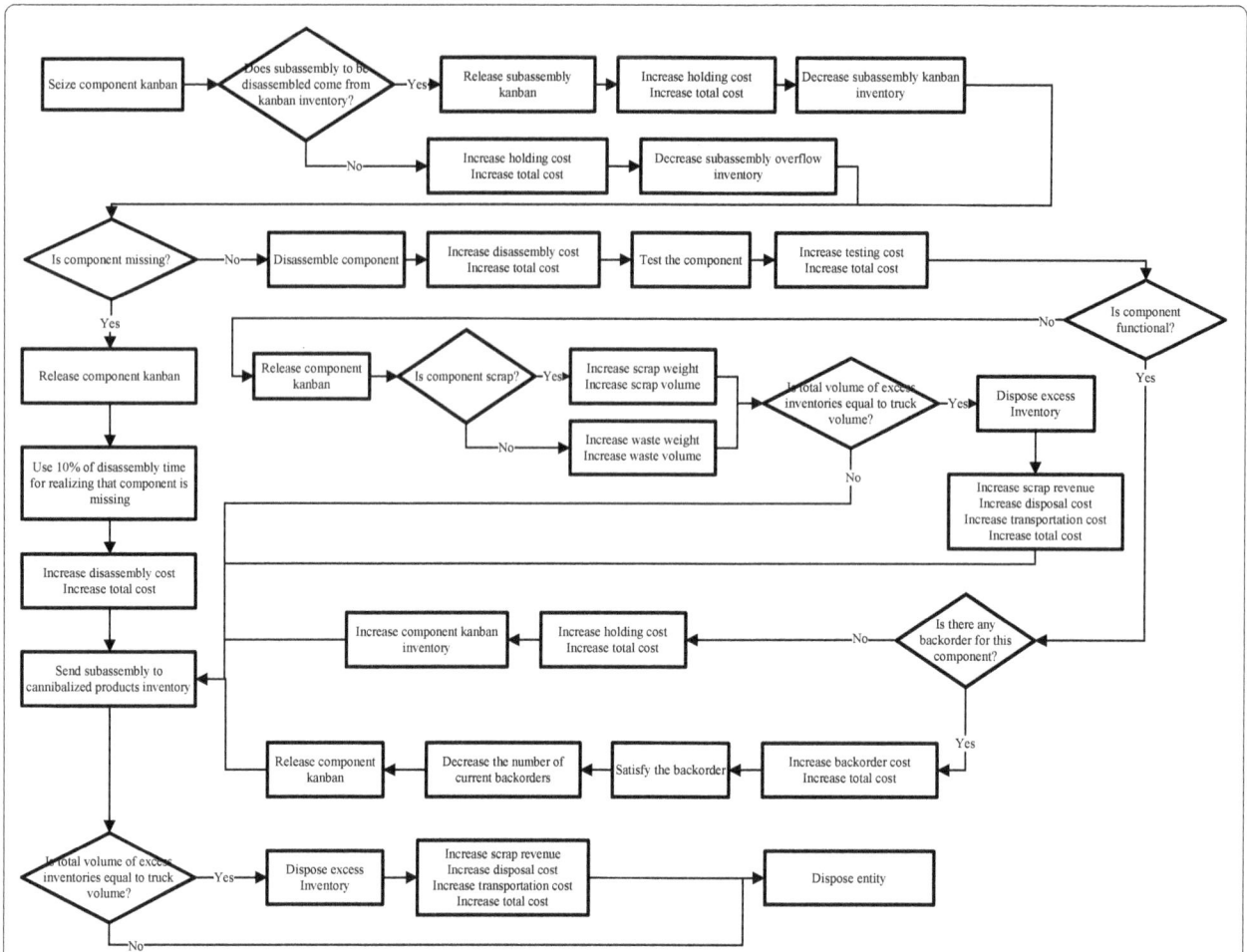

Figure 7 Disassembly operations authorized by a component kanban at the last station for the case of CPs.

In each DW, metal cover, door and other steel components (i.e., side and bottom steel plates) are sold as steel scrap. Metal cover, door, drum (if it is disposed due to excess inventory) and other steel components (i. e., side and bottom steel plates) are sold as steel scrap in each dryer. If the motor assembly of a dryer is disposed due to excess inventory, it is considered as a waste component. If timer, circuit board or motor assembly of a DW is disposed, it is considered as a waste component. In order to determine the total weight of small components such as screws, cables, total weight of the main components of a DW or a dryer is multiplied by a *small component weight factor*. These small components are considered as waste components.

It should be noted that there is no demand for metal cover and other steel components. That is why, there is no price determined for these components. Since holding cost is calculated based on the price of a component, holding cost for these components is not calculated. However, there is a demand and an associated price for drum. Consequently, the holding cost for drum is calculated based on its price.

Disposal cost of a waste component (D_c) is calculated using the following expression:

$$D_c = (W_c) * (dcp) \tag{2}$$

where W_c is the weight of the component in pounds and dcp is the disposal cost per pound. Disposal cost for subassemblies and products (D_s) are calculated as follows:

$$D_s = (W_s) * (dcp) * (dcif) \tag{3}$$

where W_s is the total weight of waste components in subassembly or product, dcp is the disposal cost per pound and $dcif$ is the disposal cost increase factor. This factor is employed in order to consider the fact that disposal of subassemblies and products create higher nuisance than components since they may involve multiple and/or hazardous materials.

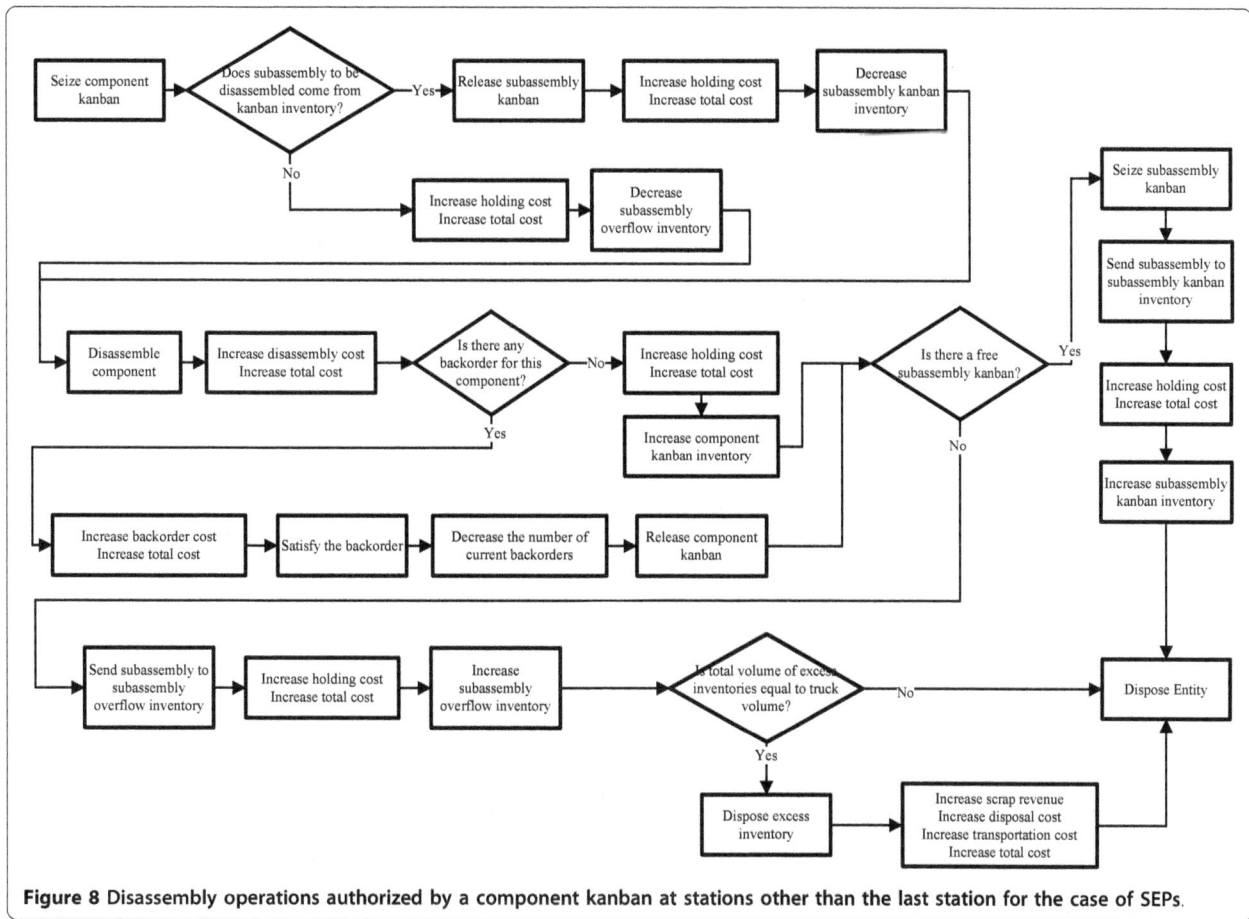

Figure 8 Disassembly operations authorized by a component kanban at stations other than the last station for the case of SEPs.

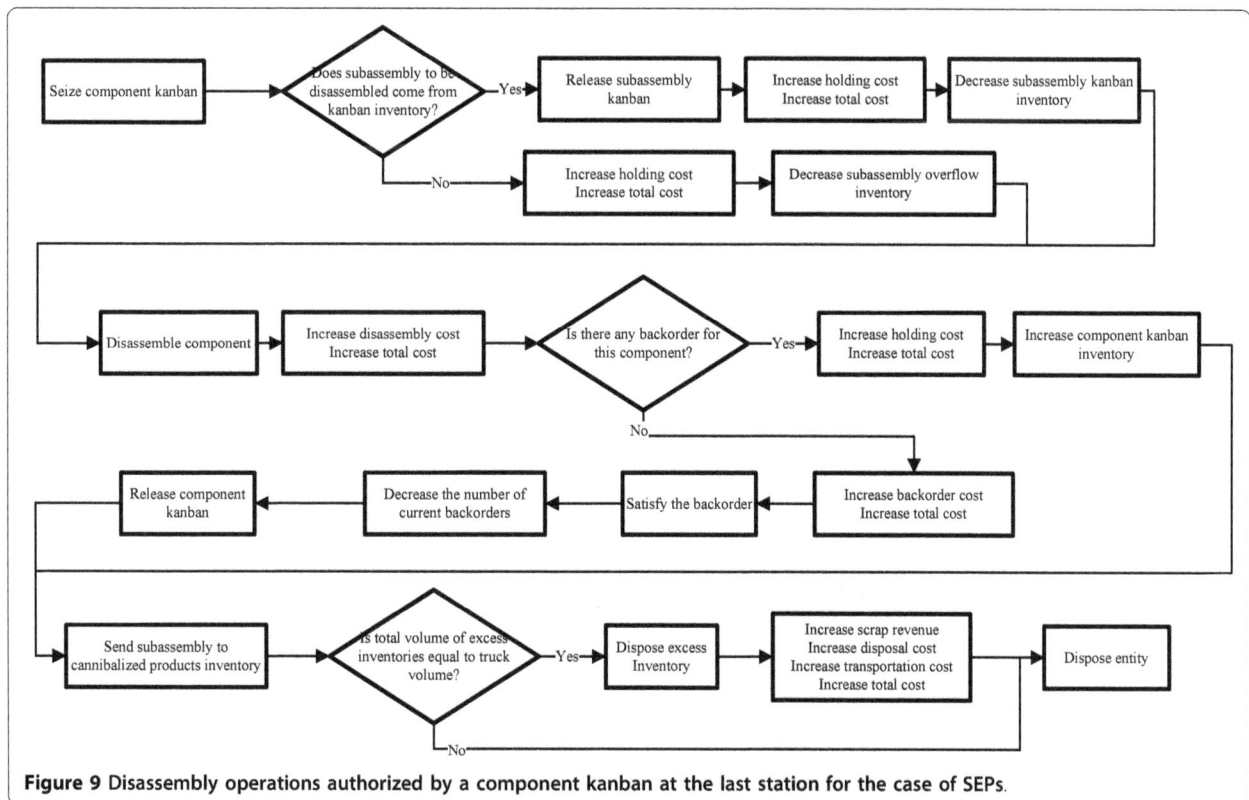

Figure 9 Disassembly operations authorized by a component kanban at the last station for the case of SEPs.

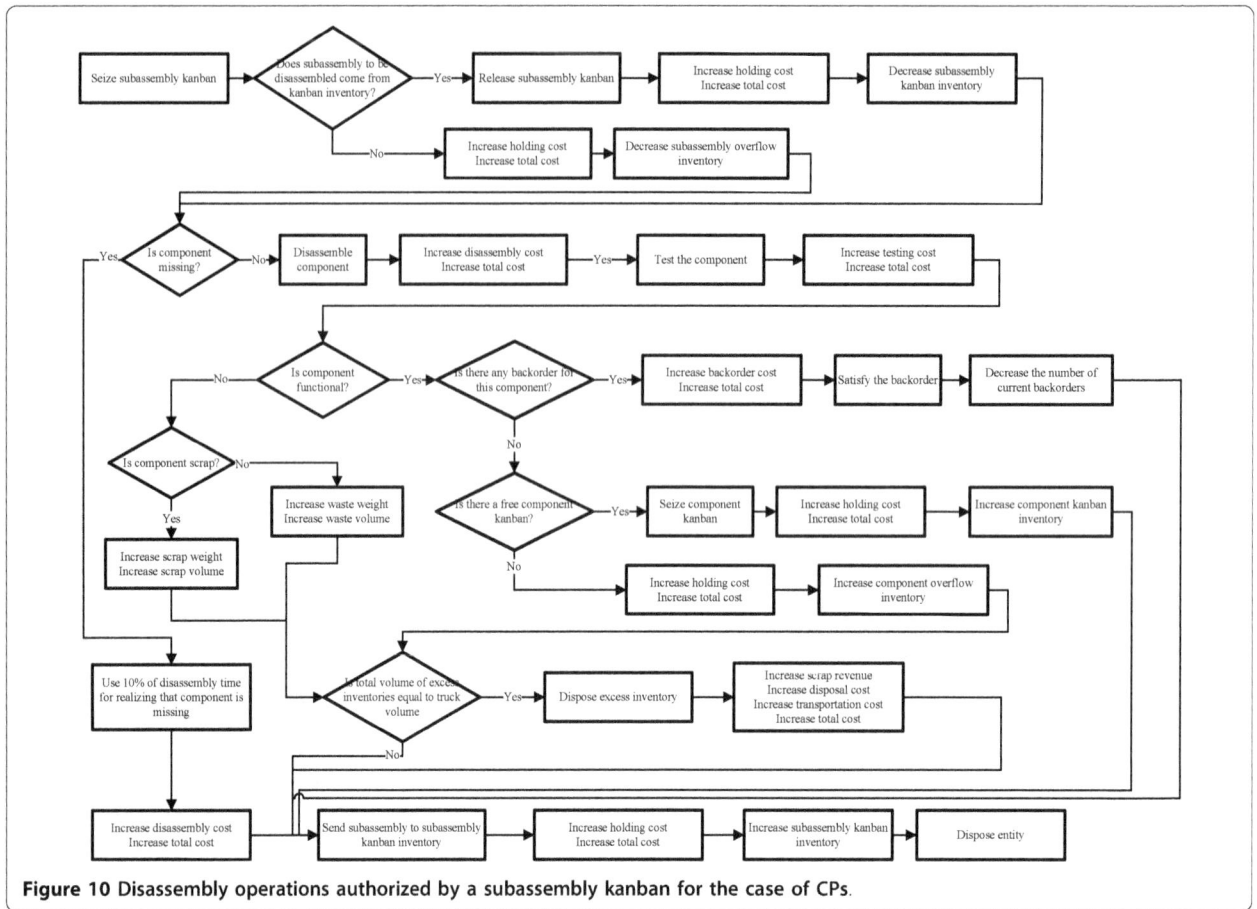

Figure 10 Disassembly operations authorized by a subassembly kanban for the case of CPs.

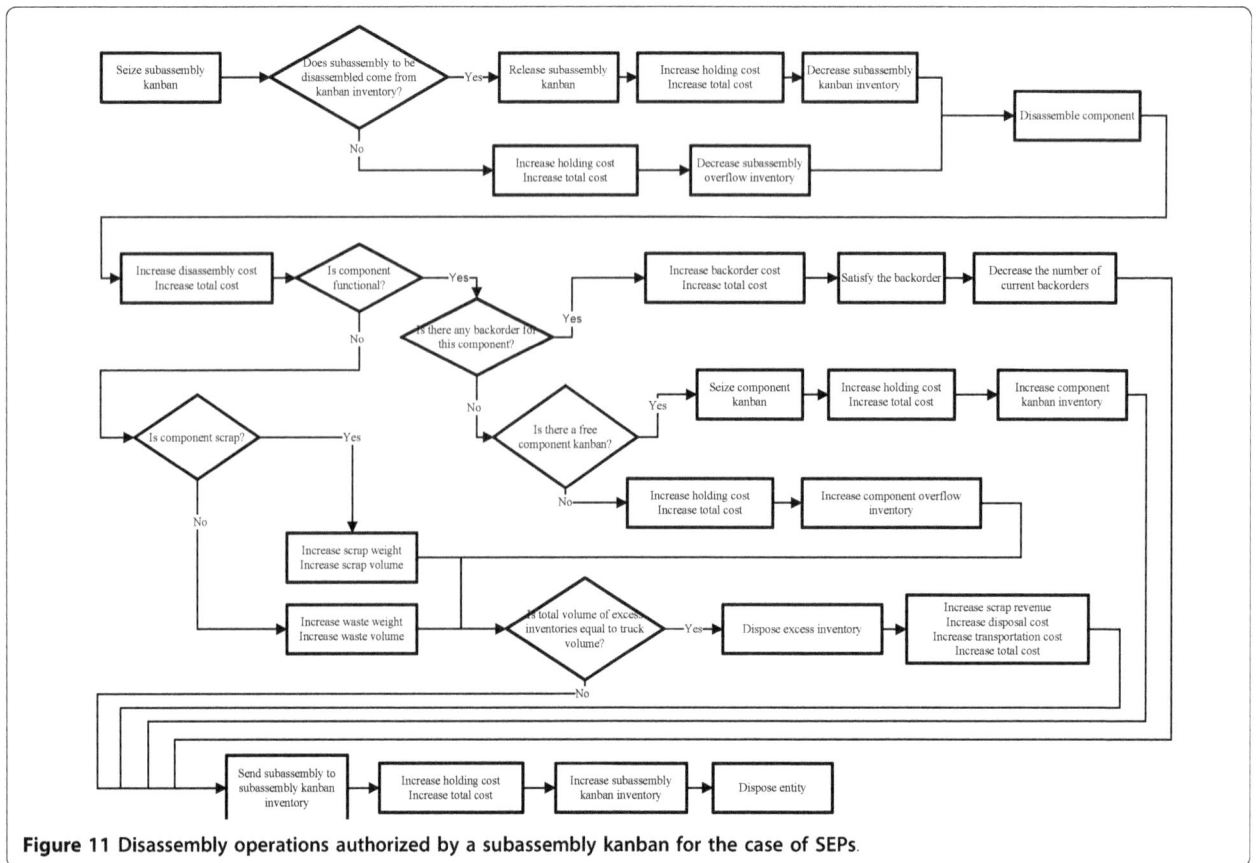

Figure 11 Disassembly operations authorized by a subassembly kanban for the case of SEPs.

Scrap revenue for a steel component (R_c) is calculated as follows:

$$R_c = (W_c) * (ssr) \qquad (4)$$

where W_c is the weight of the component in pounds and ssr is the steel scrap revenue per pound. Scrap revenue for subassemblies and products (R_s) are calculated as follows:

$$R_s = (W_s) * (srp) * (srdf) \qquad (5)$$

where W_s is the total weight of steel components in subassembly or product, srp is the steel scrap revenue per pound and $srdf$ is the scrap revenue decrease factor. This factor is employed in order to consider the additional costs associated with further material separation operations that might have to be performed on products or subassemblies before disposal.

While estimating the testing cost for SEPs, the time required to retrieve information from the sensors prior to disassembly is assumed to be 20 seconds and 15 seconds for DWs and dryers, respectively. In the calculation of transportation cost, the operating cost associated with each trip of the truck is assumed to be $55. The collection fee for EOL DWs and EOL dryers is $10.

5. Results

Three dimensional graphs given in Figures 12 and 13 present the values of four performance measures (viz., profit, disassembly cost, disposal cost, backorder cost) against the different levels of two factors (i.e., demand rate for motor and DW arrival rate) for SEPs and CPs, respectively. By visually comparing the graphs in Figures 12 and 13, we can easily see that SEPs result in higher profit values while having lower backorder, disposal and disassembly costs. However, there is a need for statistical comparison in order to have a quantitative assessment of the impact of SEPs on disassembly line performance measures.

That is why, design of experiments scheme presented in Section 4 was run for SEPs and CPs. Then, pair-wise t-tests were carried out for each performance measure. Table 3 presents the 95% confidence interval, t-value

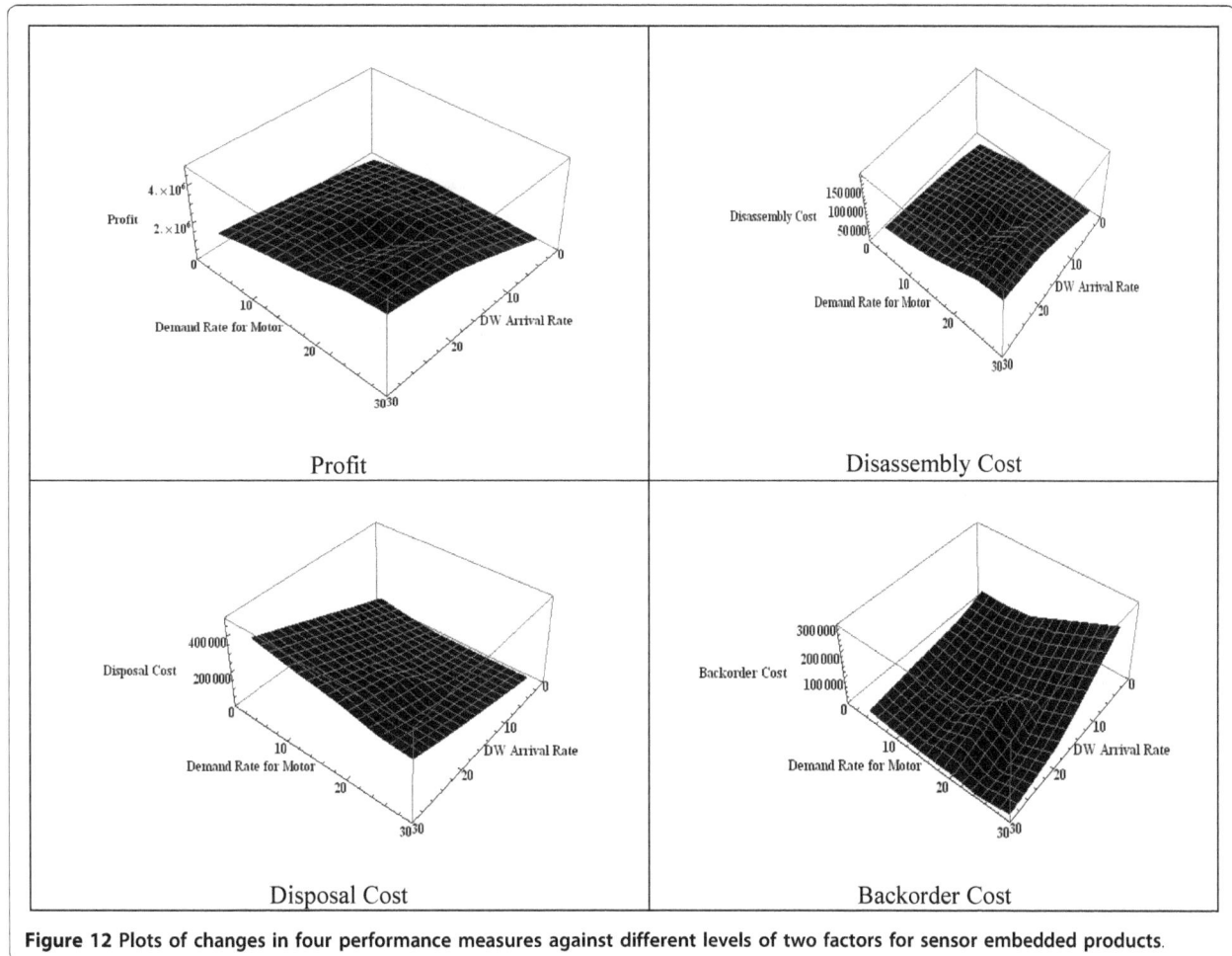

Figure 12 Plots of changes in four performance measures against different levels of two factors for sensor embedded products.

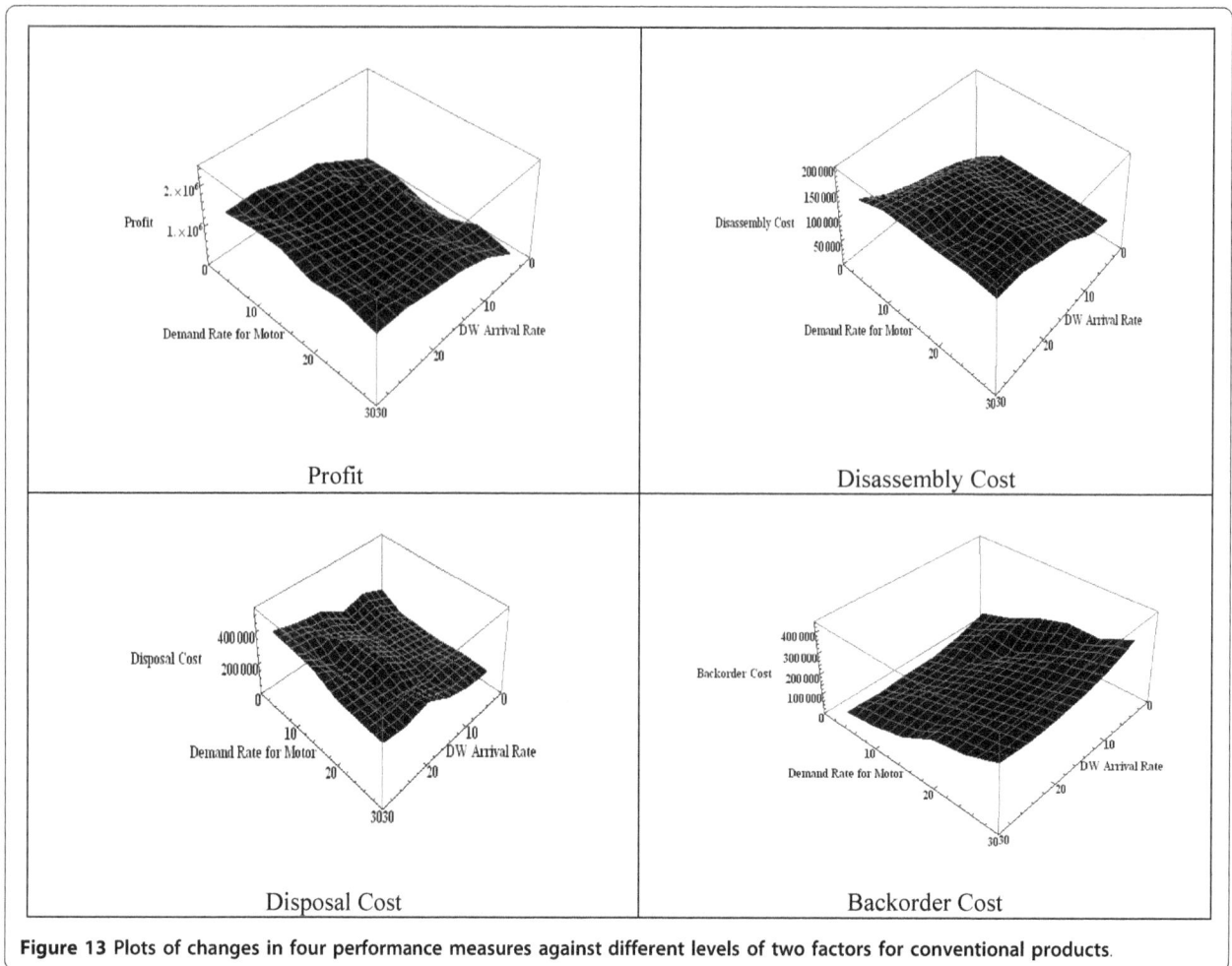

Figure 13 Plots of changes in four performance measures against different levels of two factors for conventional products.

and p-value for each test. According to this table, SEPs achieve statistically significant savings in holding, backorder, disassembly, disposal, testing and transportation costs. Moreover, there are statistically significant improvements in total revenue and profit for the case of SEPs.

In order to determine the average value provided by the sensors embedded in an EOL product, we first take the difference in profit values for SEPs and CPs for each experiment. By dividing this difference by the total number of EOL products collected, the value of sensors in an EOL product is determined for that experiment. Then, average value of sensors in an EOL product across all experiments is calculated by dividing the sum of individual experiment values by the total number of experiments. These calculation steps are presented in

Table 3 Pairwise t-test results for the comparison of SEPs against CPs.

Performance Measure	95% Confidence Interval on Mean Difference (Sensor -No Sensor)	t-value	p-value
Holding Cost	(-247.5, -180.7)	-12.76	0.000
Backorder Cost	(-94213, -71263)	-14.35	0.000
Disassembly Cost	(-45516, -35950)	-16.95	0.000
Disposal Cost	(-93485, -60931)	-9.44	0.000
Test Cost	(-118577, -108416)	-44.45	0.000
Transportation Cost	(-33331, -30669)	-47.85	0.000
Total Cost	(-370178, -322602)	-28.98	0.000
Total Revenue	(479013, 614665)	16.04	0.000
Profit	(810575, 975883)	21.51	0.000

Table 4 Use of experimental design study results to determine the average value of sensors

(1) Experiment	(2) Profit (SEPs)	(3) Profit (CPs)	(4) Difference in Profit ((2)-(3))	(5) Total Number of EOL Products Collected	(6) Value of Sensors in an EOL Product ((4)/(5))
1	892765.57	689984.74	202780.84	16161.60	12.55
2	1478995.73	1048313.37	430682.35	16173.60	26.63
3	1476935.85	1055570.37	421365.48	16185.60	26.03
4	2050911.70	1350521.63	700390.07	32327.60	21.67
5	3063884.18	2798715.87	265168.31	32339.60	8.20
.	.	.	.	.	.
.	.	.	.	.	.
.	.	.	.	.	.
80	1801372.32	1370260.27	431112.05	24271.20	17.76
81	1748333.24	1315398.01	432935.24	24247.20	17.86
Average	**2259563.26**	**1366333.99**	**893229.27**	**32320.83**	**28.64**

Table 4. According to Table 4, average value of sensors in an EOL product across all experiments is $28.64. This value can be useful in the determination of the cost associated with embedding sensors in products. In other words, as long as this cost is less than $28.64, embedding sensors in products is a profitable business decision. In this study, the value of sensors was determined by considering only EOL processing. It must be noted that if we had considered the additional benefits of sensors during the working lives of the products such as during maintenance, the value of sensors would have been further enhanced.

6. Conclusions

As a result of stricter environmental regulations, increasing public awareness toward environmental issues and economic reasons, remanufacturing has become a viable alternative to the traditional way of manufacturing products using new parts and/or components. In remanufacturing, used components and/or parts disassembled from EOL products as well as new parts/components are used during the manufacturing process. Due to missing and/or non-functional components, the number of parts that can be recovered from an EOL product is highly uncertain. In this study, we analyzed the use of sensors embedded in EOL products in determination of the condition of components prior to disassembly. First, separate design of experiments studies based on orthogonal arrays were carried out for CPs and SEPs. Then, pair-wise t-tests were conducted to compare the two cases based on different performance measures. According to the test results, SEPs not only decreased various costs (viz., disassembly, disposal, testing, backorder, transportation, holding) but also increased revenue and profit. The range of monetary resources that could be invested in SEPs was also

determined based on the improvements achieved by SEPs on profit for different experiments.

Acknowledgments
This research is supported in part by JSPS Grants-in-Aid for Scientific Research No. 23510193.

Author details
[1]Department of Industrial Engineering, Dokuz Eylul University, Buca 35160, Izmir, Turkey [2]Laboratory for Responsible Manufacturing, 334 SN Department of MIE, Northeastern University, Boston, MA 02115, USA [3]Department of Information and Creation, Kanagawa University, Yokohama, 221-8686, Japan

Authors' contributions
MAI reviewed the literature on the use of sensor-based technologies on after-sale product condition monitoring and developed the simulation models. He carried out the orthogonal arrays- based statistical experiments. SMG defined the research subject and directed the research from the start to the end. He provided important advices throughout the study and helped in editing the manuscript. KN provided information in the disassembly of dish washers and dryers and helped in writing and revising the manuscript and preparing the figures. All authors read and approved the final manuscript.

Competing interests
The authors declare that they have no competing interests.

References
1. Savaskan R, Bhattacharya S, Van Wassenhove L: Closed-loop supply chain models with product remanufacturing. *Management Science* 2004, 50:239-252.
2. Pochampally KK, Nukala S, Gupta SM: *Strategic planning models for reverse and closed-loop supply chains* Boca Raton, Florida: CRC Press; 2009.
3. Ijomah WL, Childe SJ: A model of the operations concerned in remanufacture. *International Journal of Production Research* 2007, 45:5857-5880.
4. Ijomah WL: A tool to improve training and operational effectiveness in remanufacturing. *International Journal of Computer Integrated Manufacturing* 2008, 21:676-701.
5. Ijomah WL, McMahon CA, Hammond GP, Newman ST: Development of robust design-for-remanufacturing guidelines to further the aims of sustainable development. *International Journal of Production Research* 2007, 45:4513-4536.

6. Ijomah WL, McMahon CA, Hammond GP, Newman ST: **Development of design for remanufacturing guidelines to support sustainable manufacturing.** *Robotics and Computer-Integrated Manufacturing* 2007, **23**:712-719.

7. Ijomah WL: **Addressing decision making for remanufacturing operations and design-for-remanufacture.** *International Journal of Sustainable Engineering* 2009, **2**:91-102.

8. Yamada T, Mizuhara N, Yamamoto H, Matsui M: **A performance evaluation of disassembly systems with reverse blocking.** *Computers & Industrial Engineering* 2009, **56**:1113-1125.

9. Gungor A, Gupta SM: **Issues in environmentally conscious manufacturing and product recovery: a survey.** *Computers & Industrial Engineering* 1999, **36**:811-853.

10. Ilgin MA, Gupta SM: **Environmentally conscious manufacturing and product recovery (ECMPRO): A review of the state of the art.** *Journal of Environmental Management* 2010, **91**:563-591.

11. Kongar E, Gupta SM: **Disassembly sequencing using genetic algorithm.** *The International Journal of Advanced Manufacturing Technology* 2006, **30**:497-506.

12. Tripathi M, Agrawal S, Pandey MK, Shankar R, Tiwari MK: **Real world disassembly modeling and sequencing problem: Optimization by Algorithm of Self-Guided Ants (ASGA).** *Robotics and Computer-Integrated Manufacturing* 2009, **25**:483-496.

13. Barba-Gutierrez Y, Adenso-Diaz B, Gupta SM: **Lot sizing in reverse MRP for scheduling disassembly.** *International Journal of Production Economics* 2008, **111**:741-751.

14. Gungor A, Gupta SM: **Disassembly line in product recovery.** *International Journal of Production Research* 2002, **40**:2569-2589.

15. Tang Y, Zhou MC: **A systematic approach to design and operation of disassembly lines.** *IEEE Transactions on Automation Science and Engineering* 2006, **3**:324-329.

16. Ding L-P, Feng Y-X, Tan J-R, Gao Y-C: **A new multi-objective ant colony algorithm for solving the disassembly line balancing problem.** *International Journal of Advanced Manufacturing Technology* 2010, **48**:761-771.

17. McGovern SM, Gupta SM: **A balancing method and genetic algorithm for disassembly line balancing.** *European Journal of Operational Research* 2007, **179**:692-708.

18. Kongar E, Gupta SM: **Disassembly to order system under uncertainty.** *Omega* 2006, **34**:550-561.

19. Giudice F, Kassem M: **End-of-life impact reduction through analysis and redistribution of disassembly depth: A case study in electronic device redesign.** *Computers & Industrial Engineering* 2009, **57**:677-690.

20. Kazmierczak K, Neumann WP, Winkel J: **A case study of serial-flow car disassembly: Ergonomics, productivity and potential system performance.** *Human Factors and Ergonomics in Manufacturing* 2007, **17**:331-351.

21. Rios PJ, Stuart JA: **Scheduling selective disassembly for plastics recovery in an electronics recycling center.** *IEEE Transactions on Electronics Packaging Manufacturing* 2004, **27**:187-197.

22. Kara S, Pornprasitpol P, Kaebernick H: **Selective disassembly sequencing: a methodology for the disassembly of end-of-life products.** *CIRP Annals - Manufacturing Technology* 2006, **55**:37-40.

23. Lambert AJD, Gupta SM: *Disassembly modeling for assembly, maintenance, reuse, and recycling* Boca Raton, FL: CRC Press; 2005.

24. McGovern SM, Gupta SM: *The Disassembly Line: Balancing and Modeling* New York: McGraw Hill; 2011.

25. Scheidt L, Shuqiang Z: **An approach to achieve reusability of electronic modules.** *Proceedings of the IEEE International Symposium on Electronics and the Environment; May 2-4 San Francisco, CA* 1994, 331-336.

26. Karlsson B: **A distributed data processing system for industrial recycling.** *Proceedings of 1997 IEEE Instrumentation and Measurement Technology Conference; May 19-21; Ottawa, Canada* 1997, 197-200.

27. Klausner M, Grimm WM, Hendrickson C, Horvath A: **Sensor-based data recording of use conditions for product takeback.** *Proceedings of the 1998 IEEE International Symposium on Electronics and the Environment; May 4-6 Chicago, IL* 1998, 138-143.

28. Petriu EM, Georganas ND, Petriu DC, Makrakis D, Groza VZ: **Sensor-based information appliances.** *IEEE Instrumentation & Measurement Magazine* 2000, **3**:31-35.

29. Klausner M, Grimm WM, Hendrickson C: **Reuse of electric motors in consumer products.** *Journal of Industrial Ecology* 1998, **2**:89-102.

30. Simon M, Bee G, Moore P, Pu J-S, Xie C: **Modelling of the life cycle of products with data acquisition features.** *Computers in Industry* 2001, **45**:111-122.

31. Vadde S, Kamarthi S, Gupta SM, Zeid I: **Product life cycle monitoring via embedded sensors.** In *Environment conscious manufacturing.* Edited by: Gupta SM, Lambert AJD. Boca Raton, FL: CRC Press; 2008:91-103.

32. Ilgin MA, Gupta SM: **Analysis of a Kanban Controlled Disassembly Line with Sensor Embedded Products.** *Proceedings of the 2009 Northeast Decision Sciences Institute Conference; April 1-3; Uncasville, Connecticut* 2009, 555-560.

33. Udomsawat G, Gupta SM: **Multikanban system for disassembly line.** In *Environment conscious manufacturing.* Edited by: Gupta SM, Lambert AJD. Boca Raton, FL: CRC Press; 2008:311-330.

34. Aksoy HK, Gupta SM: **Buffer allocation plan for a remanufacturing cell.** *Computers & Industrial Engineering* 2005, **48**:657-677.

35. Phadke MS: *Quality engineering using robust design* New Jersey: Prentice Hall; 1989.

36. Kelton DW, Sadowski RP, Sadowski DA: *Simulation with arena.* 4 edition. New York: McGraw-Hill; 2007.

8

An intelligent decision supporting system for international classification of functioning, disability, and health

Wei-Fen Hsieh[1], Lieu-Hen Chen[1*], Hao-Ming Hung[1], Eri Sato-Shimokawara[2], Yasufumi Takama[2], Toru Yamaguchi[2], Eric Hsiao-Kuang Wu[3] and Yu-Wei Chen[4]

Abstract

Background: In recent years, the population structure in Taiwan has changed so dramatically. Based on concerns of social welfare issues, Taiwanese government began to seek principles for assessment of disability. After seven years of carefully evaluation, the World Health Organization's International Classification of Functioning, Disability, and Health (Abbreviated to ICF) is officially adopted as Taiwan's assessment standard while most of the assessment procedures of ICF are sophisticated, and time consuming.
In this paper, we propose a sensor based decision supporting system for ICF. Our prototype system aims to reduce the burden of medical staffs, and to assist subjects to perform the assessments.

Methods: This paper integrate multiple devices including ASUS XtionTM, temperature/acceleration/gyro sensors on Arduino, and Zigbee to measure the mobility of limbs and joints. The subject's log of assessments is then recorded in the database so that the medical staffs can remote-monitor the co ndition of subjects immediately, and analyze the results later. Additionally, in our system, a user-friendly interface is implemented for the detection of dementia.

Results: In this paper, three experiments have been conducted for different purpose. The experiment was conducted to compare the variation between thermometer and our device. Moreover, we invited 20 elders aged for 65 to 80 to use our system and all of them gave positive feedback. Two elders were invited to perform full assessment for dementia and the results show that both of them didn't have sign of dementia. Also, the assessment of joint movement was performed by a 67 year-old elder and the result shows that the elder had well physical function and could take care of daily life.

Conclusions: The proposed system has potential for aiding users to perform the ICF testing better and provide benefits to medical staffs and society. With current technology, integration between sensor network systems and artificial intelligence approaches will more and more important. We develop a simple interface for user to manipulate and perform the ICF assessment. In addition, the early detection of dementia likely has the potential to provide patients with an increased level of precaution, which may improve quality of life.

Keywords: Social welfare; ICF; Arduino; Xtion Pro Live™; Sensoring techologies

Introduction

"People with Disabilities Rights Protection Act" was revised and promulgated in 2007, also new disability classification assessment method and applying disability certification process were scheduled to practice on July 11, 2012 (Directorate-General for the Information Society and Media 2010). People who are physically challenged can apply for the social welfare services according to the results of disabilities assessment.

At the same time, Taiwan is facing the impacts of "aging society with fewer children", including demographic imbalance, dependency burden, and lack of elderly nursing care services. Based on concerns about these social welfare issues, Taiwanese government began to seek principles for assessment of disability.

* Correspondence: lhchen@csie.ncnu.edu.tw
[1]Computer Science and Information Engineering, National Chi Nan University, Puli, Taiwan
Full list of author information is available at the end of the article

Figure 1 Xtion PRO live.

Figure 2 TMP36.

After seven years of carefully evaluation, as a result, the World Health Organization's International Classification of Functioning, Disability, and Health (Abbreviated to ICF) is officially adopted as Taiwan's assessment standard.

For this new international standard, WHO changed the classification from 16 items in the previous policies to 8 coding system (Directorate-General for the Information Society and Media 2010; World Health Organization 2001; World Health Organization 2007). The old assessment methods and service applying processes are also modified by WHO. One of the most important change in the new policies is the evaluation results are no more available for the entire lifetime. From now on, several aspects of lives have to be investigated at least once every five years.

With these changes, not only the methods that we used to apply social welfare service and attend disabilities assessment are transformed, but also the government is forced to establish new follow up social welfare service, evaluation index, evaluation tool, evaluation flow, etc. Though ICF provides a better system for evaluating disability in accordance with systematic regulations, it takes both people with disabilities and medical staffs more time to perform and complete the evaluation (Liao and Huang 2009; Shotton et al. 2011).

Due to the serious lack of human resource of medicine in Taiwan, in this paper, we propose a sensor based decision supporting system for ICF. At the moment, we focus on the part of assessment for disability and dementia. We developed the friendly interface which leads users to perform the actions required for ICF assessments according to Mobility of joint functions (ID b701). In order to detect users' movement, this project use Xtion PRO Live to measure and assess subject's actions and activities; combined with OPENI2 and NiTE2 to conduct the structure of skeleton. We also integrated multiple devices including ASUS Xtion™, Zigbee, and temperature/acceleration/gyro sensors on Arduino as wearable device to capture users' movement. There are many reports on using Kinect, Xtion PRO LIVE and other related devices for the purpose of motion capture (Mulvenna and Nugent 2010). There are also many software developing tools available for implementing interfaces using these devices. Subject can easily wear our multi-sensors monitoring device on the wrist like wearing a bracelet during the movement assessment test. This wearable device combines temperature, acceleration, and gyro sensors with Arduino. The temperature sensor measures subject's skin temperature on wrist vein so that inspector can monitor subject's health state during the test. Besides, the device includes gyroscope to detect the rotations of the wrist for testing items and it can assure that the subject raising their hands in correct orientation. We also implement accelerator to check whether the subject's movement is smooth or not. If the subject finishes the testing item but with uneven movement, the subject might have potential joint disease. The

Figure 3 L3G4200D.

Figure 4 MMA7455L.

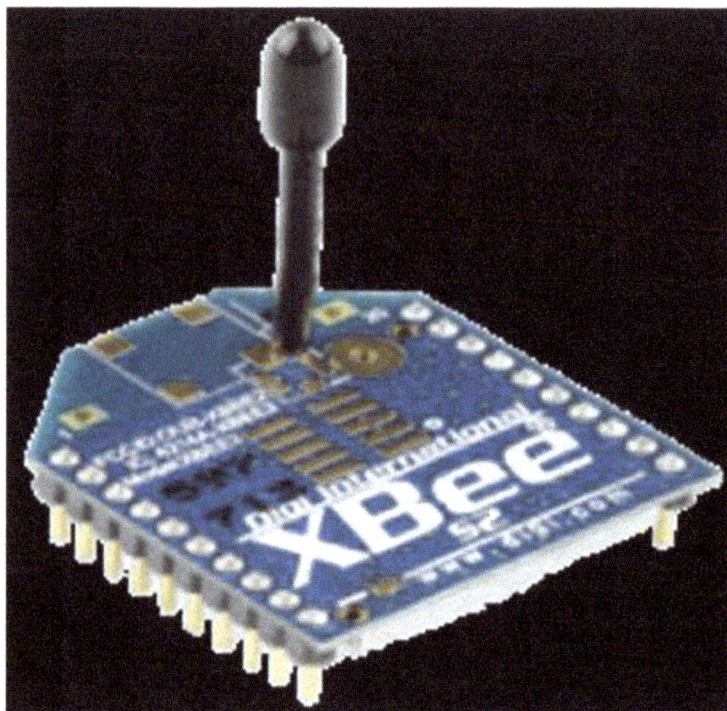

Figure 5 XBee - Series 2.

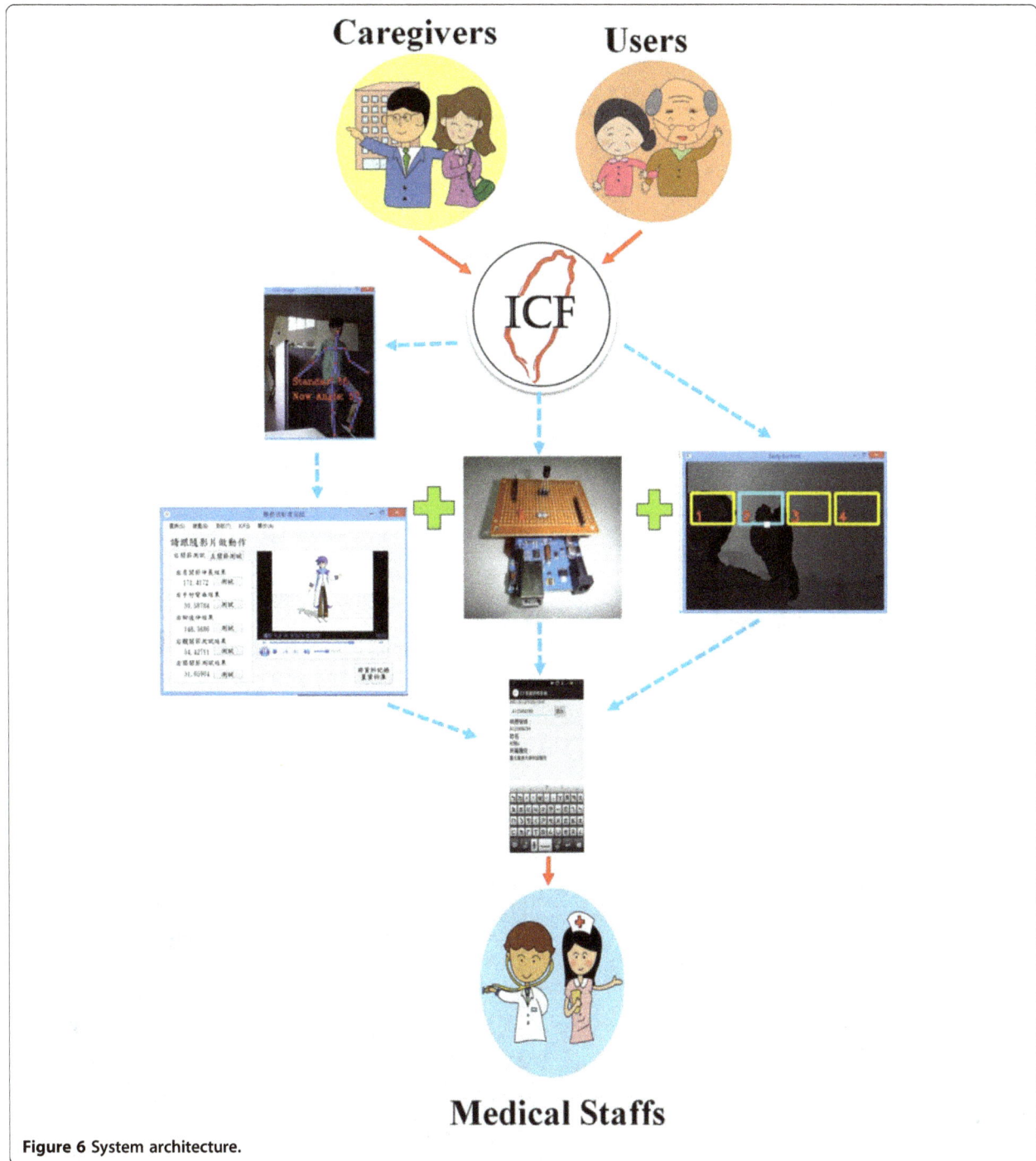

Figure 6 System architecture.

observation of movement can be used as meaningful information of health.

A dementia assessment interface with gesture recognizer on virtual button is developed to enable users to perform the test under a more convenient and comfortable condition. This project use Webcam combining with OpenCV to develop this interactive GUI interface. We designed the dementia questions according to the items referring to dementia according to

Orientation functions (ID b114), Intellectual functions (ID b117), Memory functions (ID b144), and Higher-level cognitive functions (ID 164) in ICF. In this project we assume that subjects all have the basic abilities to answer the question and to recognize words so that subjects can do dementia assessment test by themselves.

For the purpose of releasing the burden of medical staffs and increasing efficiency of evaluation work of ICF, our system collects the subjects' test results in a

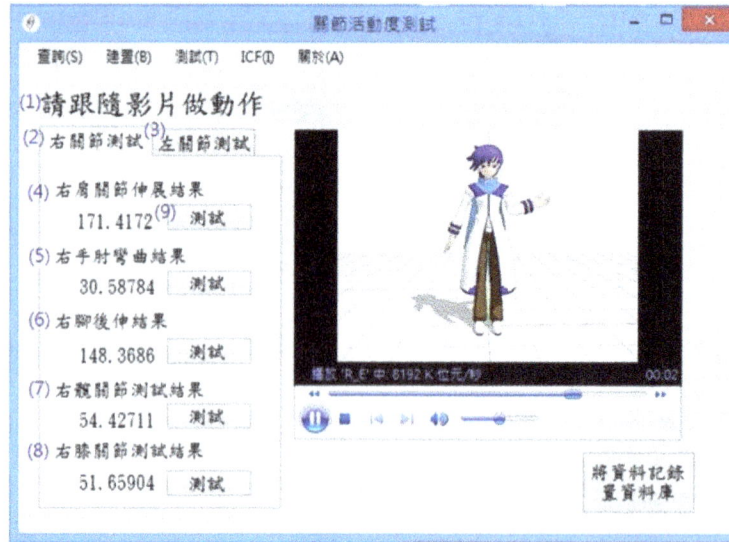

Figure 7 Joints mobility measurement interface. (Figure translation: (1) 請跟隨影片做動作: Please follow the video to perform the test movement, (2) 右關節測試: Assessment of Right Joints, (3)左關節測試: Assessment of Left Joints, (4) 右肩關節伸展結果: The result of right shoulder extension, (5) 右手軸彎曲結果: The result of right elbow extension, (6) 右□後跟伸結果: The result of right hip flexion, (7) 右髖關節測試結果: The result of right hip extension, (8) 右膝關節測試結果: The result of right knee extension).

database so that the medical staffs can keep up-to-date with subjects' condition through mobile phones or Internet (Eriksson et al. 2005; Morris 1993).

In the current version of prototype system, we still need a trained volunteer managing the movement assessment assistance interface on the computer to assist subjects to do the test.

Background

As a classification, ICF systematically groups different domains for a person in a given health condition within two parts and each part with two components.

Part one Functioning and Disability has the component (a) Body Functions and Structures, (b) Activities and Participation. Part two Contextual Factors has the

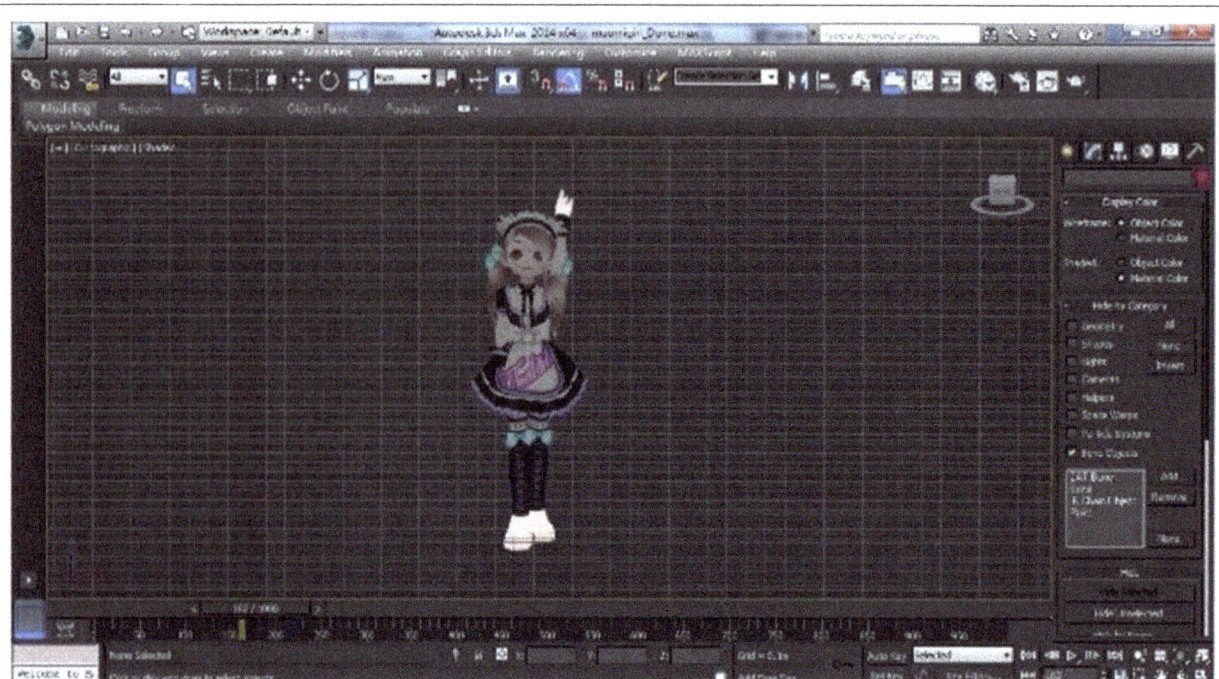

Figure 8 Recording demonstration.

Table 1 The joints mobility measurement result of 67 year-old subject

Age	Range of motion	Shoulder flexion	Elbow flexion	Hip flexion	Knee flexion
67	Right Limb	165.321	141.412	120.265	140.965
	Left Limb	165.242	142.732	119.253	139.961

component (c) Environmental Factors, (d) Personal Factors.

In this project we implement a method to calculate the degree of joint movement to measure the activity of body movement. As for questions in dementia assessment test are design by us according to part which don't need medical staffs ask subjects in person from Clinical Dementia Rating (abbreviate CDR).

Skeleton tracking

In this project, we use Xtion PRO LIVE as a sensor to track skeleton. Xtion PRO LIVE is made by ASUS uses infrared sensors, adaptive depth detection technology, color image sensing and audio stream to capture a users' real-time image, movement, and voice, making user tracking more precise (Figure 1). We use visual C++ to develop the joint detection.

OpenNI defined 24 joints for a person but the NITE2 offers tracking of 15 of them.

Dementia assessment test

The Clinical Dementia Rating or CDR was developed at the Memory and Aging Project at Washington University School of Medicine in 1979 for the evaluation of staging severity of dementia (Reed and Bufka 2006). The Clinical Dementia Rating is a five-point scale in which CDR-0 connotes no cognitive impairment, and then the remaining four points are for various stages of dementia:

- CDR-0.5=very mild dementia
- CDR-1=mild
- CDR-2=moderate
- CDR-3=severe

There are six aspects in Clinical Dementia Rating scale. They are memory, orientation, judgment and problem solving, home and hobbies and community affairs (Muilder and Stappers 2009). Moreover, Dementia assessment according to the items in ICF are: Orientation functions (ID b114), Intellectual functions (ID b117), Memory functions (ID b144), and Higher-level cognitive functions (ID 164). We designed the questions referring to above aspects except for home and hobbies and community affairs in Clinical Dementia Rating scale. Caregivers would deal with the questions consisting of the left aspects.

Temperature sensing

Normal human body temperature depends upon the place in the body at which the measurement is made, and the time of day and level of activity of the person. Temperatures cycle regularly up and down through the day.

TMP36 is a low voltage, precision centigrade temperature sensor (Figure 2). It provides a voltage output that is linearly proportional to the Celsius temperature. The output voltage can be converted to temperature easily using the scale factor of 10 mV/°C. Arduino is a single-board

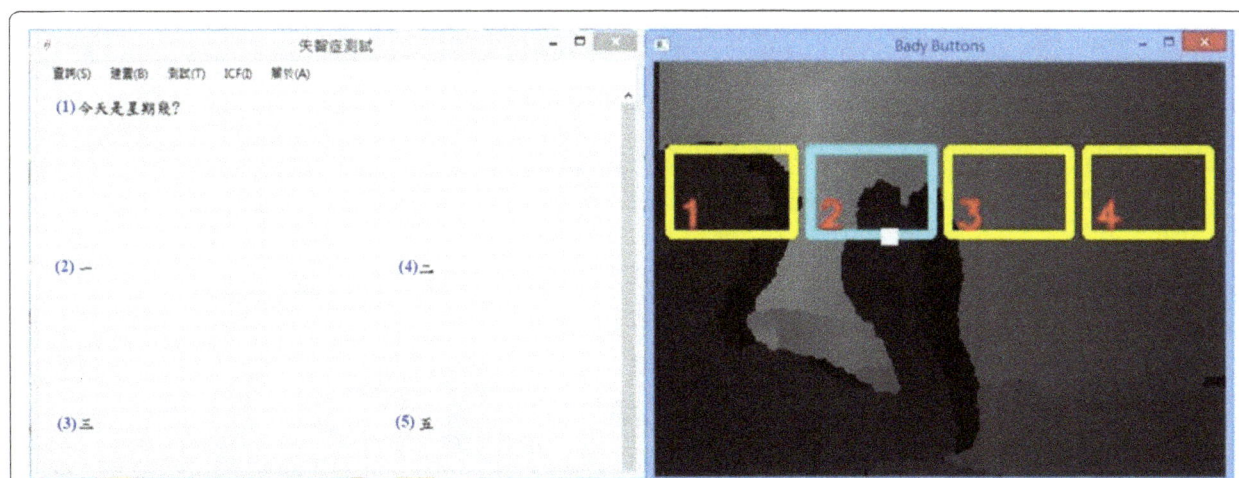

Figure 9 Dementia assessment interface. (Figure Translation: (1) 今天是星期幾? : What day is today? (2) 一: Monday, (3) 三: Wednesday, (4) 二: Tuesday, (5) 五: Friday).

Figure 10 The elder doing the dementia test.

microcontroller to make using electronics in multidisciplinary projects more accessible. In this project we develop this temperature sensing device by using open source Arduino software. We use TMP36 as the sensor and combine it with Arduino to sensor the temperature via wrist vein.

Gyroscope

In this paper we implement L3G4200D as the sensor to detect the rotation of wrist (Figure 3). It is a 3 Axis ultra-stable digital MEMS motion sensor. It gives stable sensitivity over temperature and time. L3G4200D module features an on board low drop out voltage regulator which takes input supply in the range of 3.6 V to 6 V DC. The L3G4200D has user selectable full scale of ±250, ±500, ±2000 degrees per second.

Accelerator

The MMA7455L is a Digital Output, low power, low profile capacitive micro-machined accelerometer featuring pulse detect for quick motion detection (Figure 4).

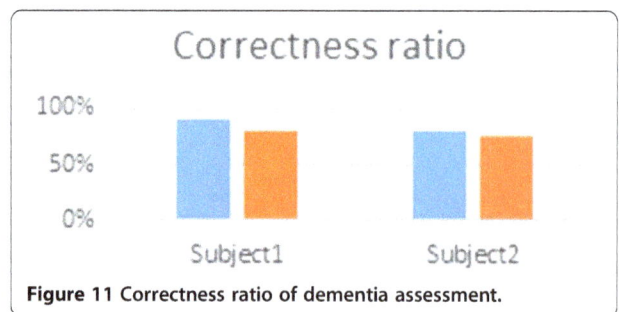

Figure 11 Correctness ratio of dementia assessment.

0 g offset and sensitivity are factory set and require no external devices. The 0 g offset can be customer calibrated using assigned 0 g registers and g-Select which allows for three kinds of acceleration selection ranges (2 g/4 g/8 g). It is suitable for handheld battery powered electronics because MMA7455L includes a Standby Mode.

Wireless technology

ZigBee is the latest wireless technology using high level communication protocol to create personal network area (Figure 5). Each ZigBee module can transmit and receive data from the other ZigBee module. These modules allow reliable communication between micro-controllers and computers. We use XCTU to manage and configure ZigBee module. It's a free and multiple platform application designed to enable developers to interact with Zigbee.

Methods

A user-friendly interface was developed for the ICF assessment in this paper. The system is based on multi-sensor technology which integrate multiple

devices including ASUS XtionTM, temperature/acceleration/gyro sensors. And we invited subjects to perform the assessment through our system. The subject's log of assessments is then recorded in the database. The log includes subject's personal information, assessment result and medical history. Our system is likely to extract useful information for medical staffs and support decision making. Written informed consent was obtained for the publication of this report and any accompanying images.

System architecture

As shown in Figure 6, we have implemented two main assessment functions in our prototype system. These functions include: joints mobility measurement, dementia progression assessment.

The first process of our system is joints mobility measurement. In the beginning of the joints mobility measurement, in order to let Xtion PRO LIVE detect the subject successfully, the subject needs to stand in the restricted position. After detecting successfully, the subject can start the measurement. In addition, during the assessment, the subject is asked to wear our sensing device on his/her wrist for detecting subject's body temperature. There are three sensors installed on this device, including a body temperature sensor, an accelerator sensor, and a gyroscope sensor. After the subject completes a designated posture, the inspector can keep going on next movement. After all testing movements are completed, the results are then recorded into the database.

As for dementia progression assessment, our system picks out 10 questions from 100 designed questions in the database randomly. Our system implement a multimodal interface for subjects to input their answers either by using a traditional keyboard-mouse combination, or by using gestures directly with virtual push-buttons on screen. The reason why we implemented two methods for users to manipulate the

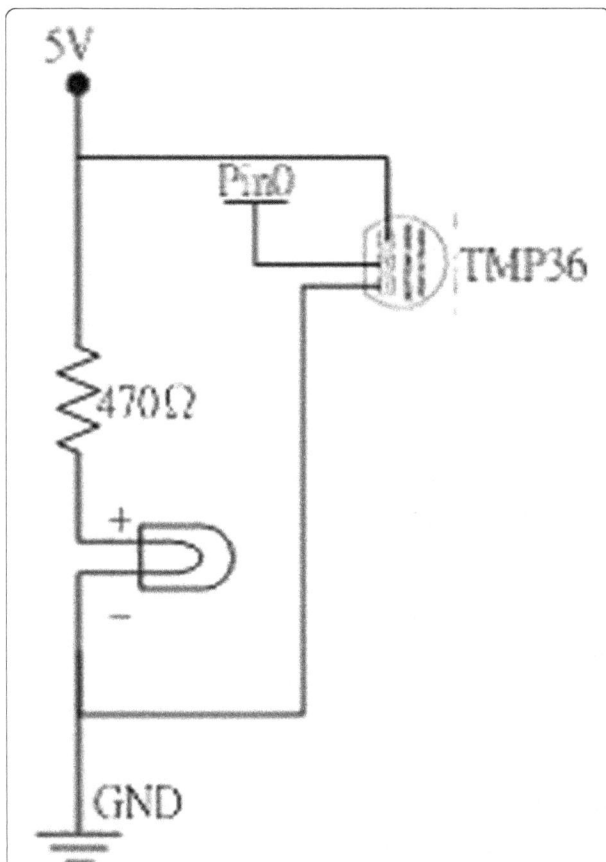

Figure 12 Temperature sensing circuit.

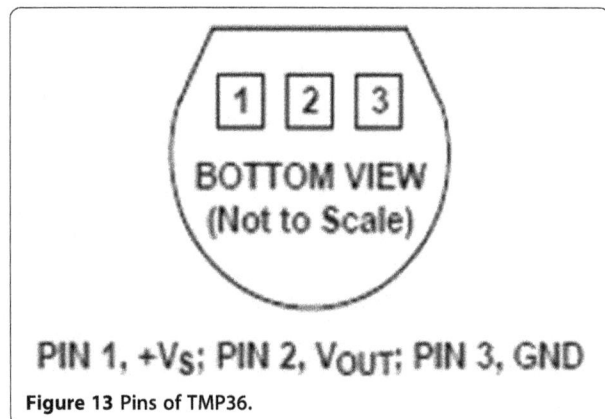

Figure 13 Pins of TMP36.

Figure 14 Temperature sensing device.

Figure 15 Wearable device.

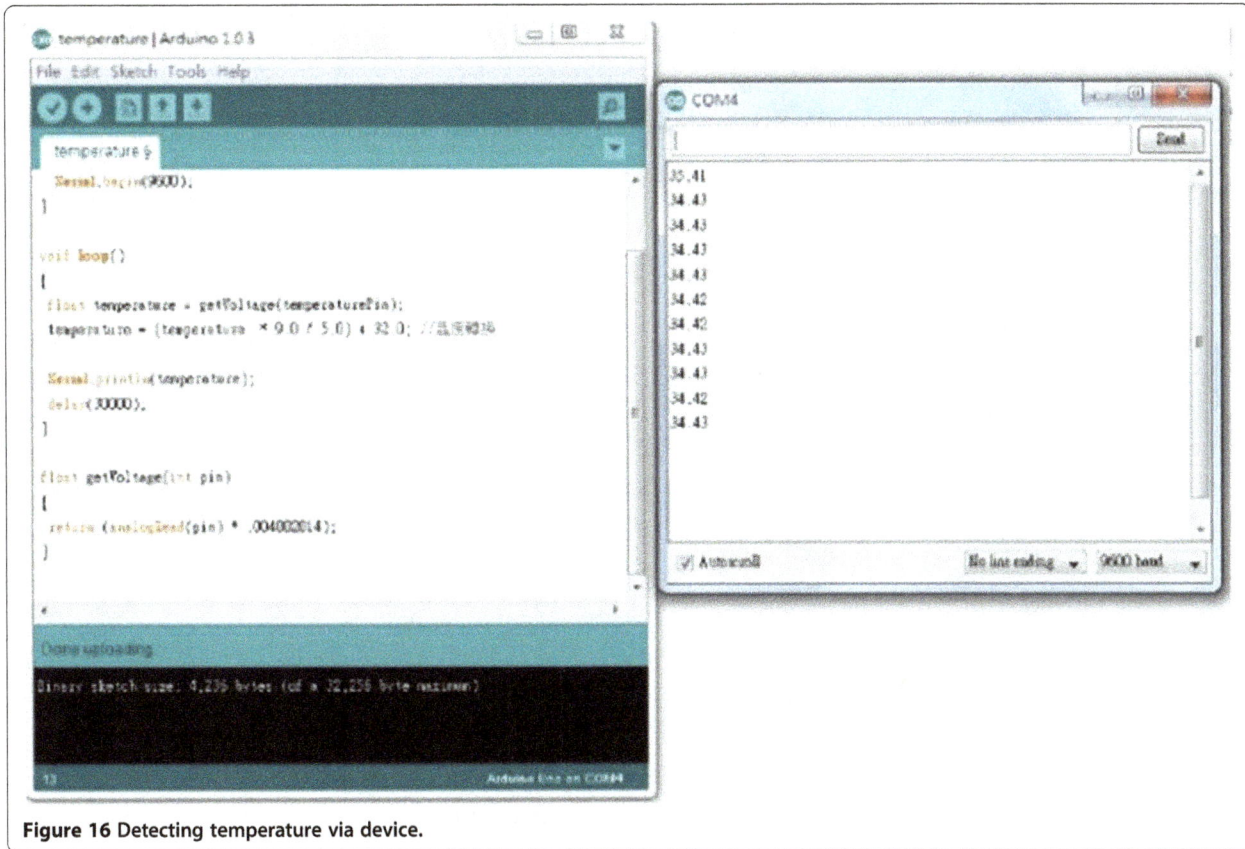

Figure 16 Detecting temperature via device.

system is because we are afraid that the elder are unable to control the mouse properly. Instead of mouse, we think manipulate by gesture is a more instinct way for the elder.

Finally, Medical staffs can query and check the assessment results through their mobile devices or computers by requesting the database.

Results

Joints mobility measurement

With the purpose of acknowledging subject's body's condition, the system will take subject's whole body picture through the webcam on Xtion PRO LIVE while beginning the measurement. Next, the system will show the video of the measurement item on

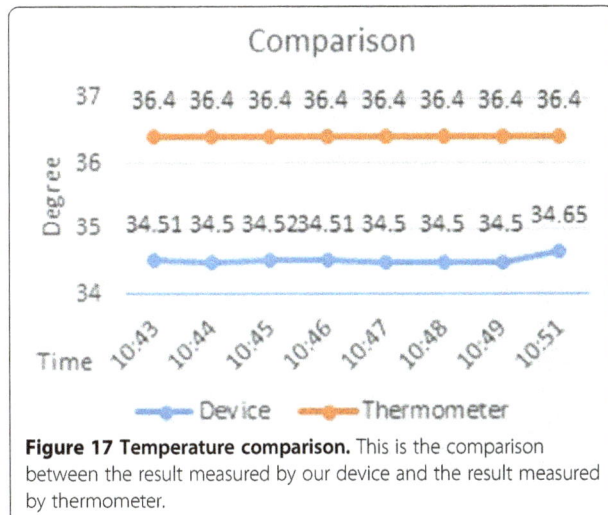

Figure 17 Temperature comparison. This is the comparison between the result measured by our device and the result measured by thermometer.

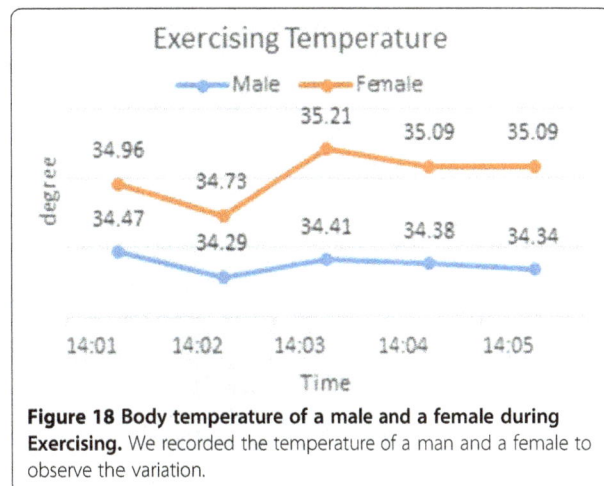

Figure 18 Body temperature of a male and a female during Exercising. We recorded the temperature of a man and a female to observe the variation.

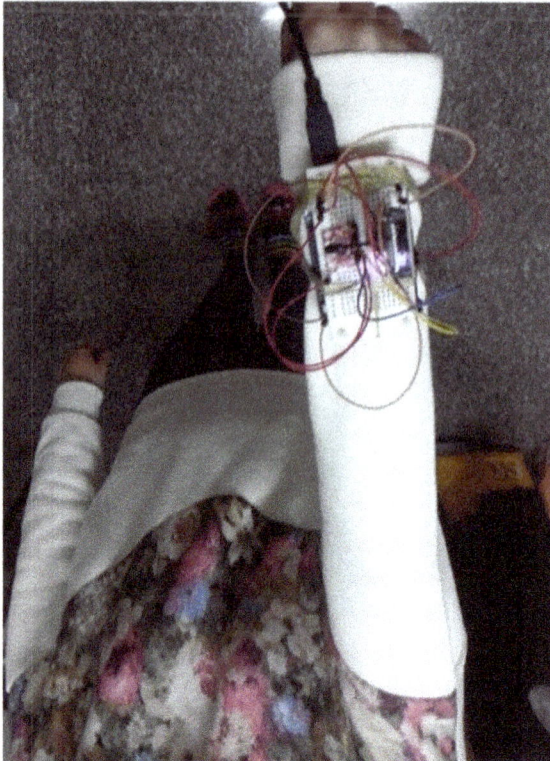

Figure 19 Wearable wrist detection device.

the interface of joints mobility measurement to instruct users what to do directly instead of words expression or oral direction (Figure 7). As for the instruction video, we previously recorded the demonstration animation so that the subject can imitate

the movement in the animation to finish the testing items (Figure 8). We use Xtion PRO LIVE for joint skeleton tracking then calculate whether user's joint movement reach the standard degree or not. If users can't reach the standard degree within 60 seconds, the system would record the latest detected extension and continue to next measurement item.

Joints mobility measurement comprises two parts as following: upper extremity and lower extremity.

According to the items from Mobility of joint functions (ID b701) in ICF, we pick out the relative measurement testing items as below:

Upper extremity

- Mobility of shoulder
  Range of motion at the shoulder extension
  (Standard degree: 180)
  Range of motion at the shoulder flexion
  (Standard degree: 60)
  Joint function (standard degree: 240).
- Mobility of elbow
  Range of motion at the elbow extension
  (Standard degree: 145)
  Range of motion at the elbow flexion
  (Standard degree: 0)
  Joint function (standard degree: 145)
- Mobility of wrist
  Range of motion at the wrist extension
  (Standard degree: 80)
  Range of motion at the wrist flexion
  (Standard degree: 70)
  Joint function (standard degree: 150)

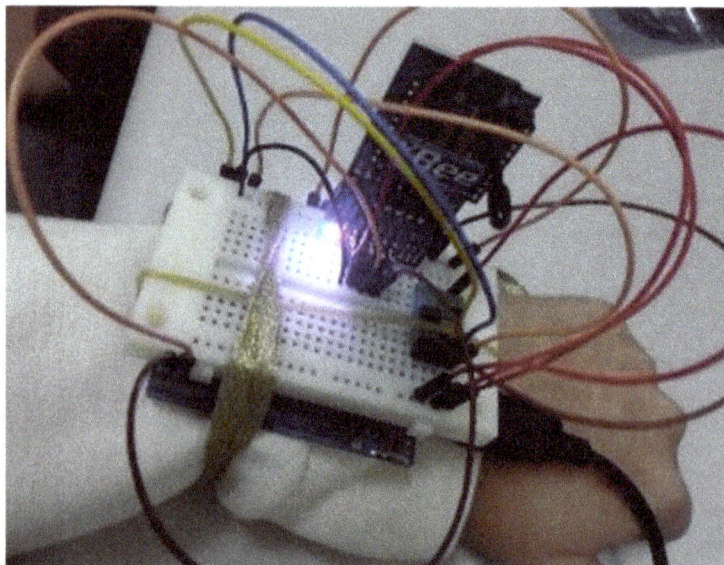

Figure 20 Wearable rate detection device.

Lower extremity

- Mobility of hip
  Range of motion at the hip extension
  (Standard degree: 125)
  Range of motion at the hip flexion
  (Standard degree: 10)
  Joint function (standard degree: 135)
- Mobility of knee
  Range of motion at the knee extension
  (Standard degree: 145)
  Range of motion at the knee flexion
  (Standard degree: 0)
  Joint function (standard degree: 140)

In this paper, our system was performed by a 67 year-old elder. Through our system, the subject's mobility of joints could be measured and quantified as shown in Table 1. According to this result, this elder had well physical function and could take care of daily life.

Dementia assessment interface

The dementia assessment interface is based on gesture recognition (Figure 9). Users can choose the answers through pressing the virtual button on the screen or just using the mouse. It provides two simple and easy methods for users to manipulate the system. We design the questions and options according to the different aspect in Clinical Dementia Rating scale respectively and the referring items from ICF. The system pick out 10 questions from 100 designed questions in the database randomly. The questions are all multiple choice which subjects choose one answer from four. The sequence of questions and options are randomly to prevent subjects from memorizing the answer from doing the test last time. Our system was tested by 20 elders aged from 65 to 80 under the help of volunteers (Figure 10). All of them feel that using our system is not only interesting but also fascinating. Their reactions show that our system arouses their interest and increase their willingness to do dementia test.

In order to improve the accuracy of our system, full examination was performed by two elders aged 67 and 77 respectively. And their family, as caregiver, was asked to answer the AD8 questionnaire which is the version for caregiver. All the questions are based on daily life, for example, the color judgment, shape recognition, and simple calculation. We couldn't determine if the subject have dementia by merely one test. The assessment factor should consist of assessment test, the questionnaire for caregiver and the judgment by medical staffs. The main purpose of our system is preliminary assessment and it's clearer for medical staff to understand subject's situation. The standard of high risk population is that if answer's correctness percentage is lower 80% and the

questionnaire answered by caregiver got low score, the subject need to be kept under observation.

From Dementia Assessment, we could learn the subject's ability in different aspects. During the process, we found that it was helpful for the subject to keep calm and focus on the assessment when their family was sitting around. According to the result and the questionnaire answered by their family, both of subjects didn't have the sign of Dementia (Figure 11).

Multisensor device

As previously mentioned, we integrated temperature/acceleration/gyro sensors on Arduino which we developed the device with open source Arduino software. We implemented Zigbee to detect users' body temperature and sophisticated movement during the test. The detail

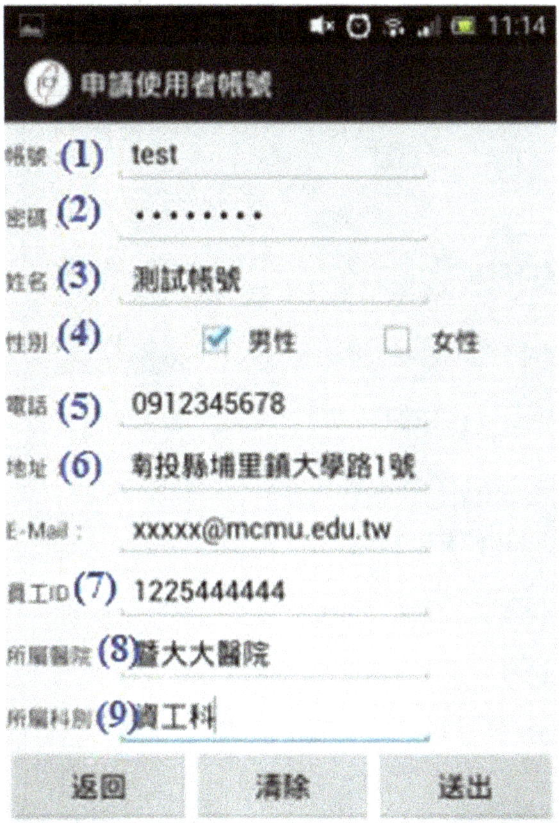

Figure 21 Create a medical staff's account. (Figure Translation: (1) 帳號: Account, (2) 密碼: Password, (3) 姓名: Name, (4) 性別: Gender, (5) 電話: Telephone, (6) 地址: Address, (7) 員工ID: Staff ID, (8) 所屬醫院: Institution, (9) 所屬科別: Hospital Department).

prototype would be described in this session respectively. For the convenience of measurement, these devices were integrated into a sensing module for users to wear on the wrist so that won't cause their uneasiness.

Temperature sensing device

The temperature sensing circuit is based on TMP36 and Arduino UNO (Figure 12). Owing to the device without switch, the LED on Arduino board can make sure whether the device is on or not. The resistance prevents electric current from passing through the LED and damaging the LED.

The system would read the voltage on the Vout pin shown in Figure 13, the output voltage can be converted

to temperature easily using the scale factor of 10 mV/°C (Eq.1]).

$$\text{Temperature in } °C = [(\text{Voutin mV}) - 500]/10 \qquad (1)$$

In order to show the result on the system, we connect Arduino and computer via USB and use the 5 V power on Arduino as power. We connect Vout on TMP36 to Analog IN Pin0 on Arduino UNO so that the device does not need other external power supply (Figure 14).

Subject can wear this device to record the temperature variation and which testing item will cause the variation most during the joints mobility measurement (Figure 15). If the temperature changes higher than 0.5 degree the system will record the time and temperature into the

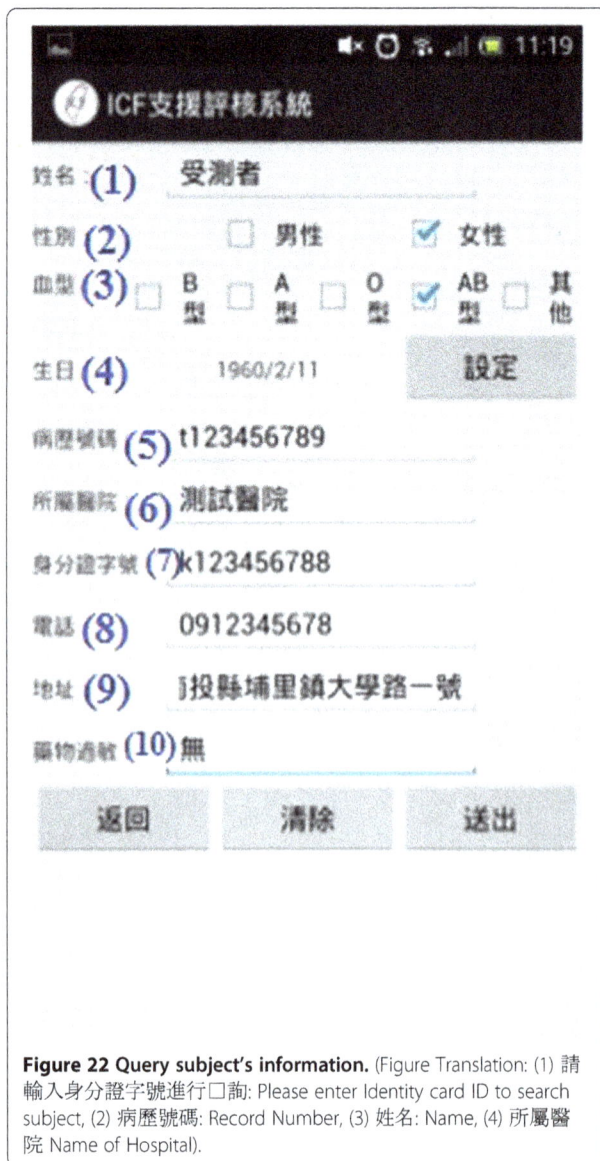

Figure 22 Query subject's information. (Figure Translation: (1) 請輸入身分證字號進行□詢: Please enter Identity card ID to search subject, (2) 病歷號碼: Record Number, (3) 姓名: Name, (4) 所屬醫院 Name of Hospital).

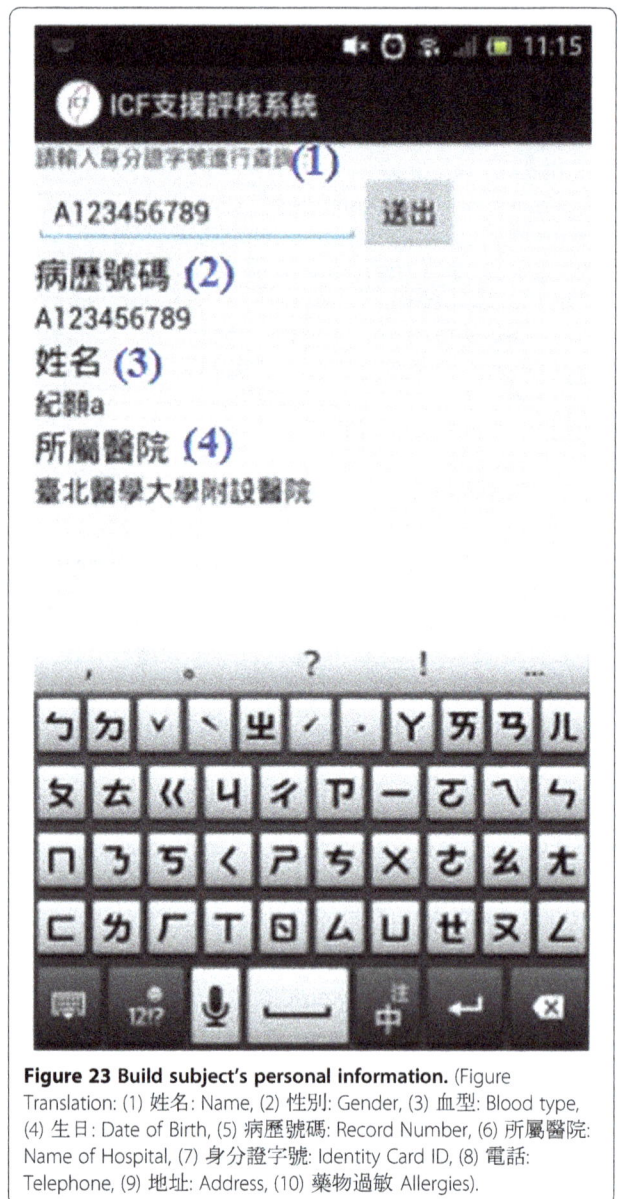

Figure 23 Build subject's personal information. (Figure Translation: (1) 姓名: Name, (2) 性別: Gender, (3) 血型: Blood type, (4) 生日: Date of Birth, (5) 病歷號碼: Record Number, (6) 所屬醫院: Name of Hospital, (7) 身分證字號: Identity Card ID, (8) 電話: Telephone, (9) 地址: Address, (10) 藥物過敏 Allergies).

database so that the medical staffs can acknowledge subject's health state during the measurement (Figure 16).

In this project, we focus more on understanding the variation of subject's body temperature rather than the accuracy of degree. In order to prove that thermometer can be substituted by our device, we compare the result through our device with the result form thermometer when the indoor temperature was 26°C. The comparison result shows that the device's variation is consists with thermometer's variation with the stable average difference of 1.88°C (Figure 17). Thus, Subject can wear our device instead of using the thermometer.

Another experiment is designed to compare the differences between a male and a female user while they were doing indoor bicycling exercise at temperature 26°C. According to the result, we acknowledged that female's body temperature was normally higher under the same environmental condition and body temperature roses gradually (Figure 18). Through observing the variation of body temperature, we could know subject's body condition and the medical staffs can analyze this information to know more about subject's health condition.

In this paper, in order to observe the sophisticated movement of wrist, we use L3G4200D to detect the rotation of wrist. And moreover, to observe if the subject's movement is fluently or not, our system use MMA7455L as the sensor to detect the rate of subject's movement.

We let user wear it with the temperature sensor so that the device can also detect the wrist motion regarding the test item of wrist extension and flexion (Figure 19).

During the test, the purpose of doing these tests is not only finishing the test but also observing the health condition of the elder. The elder might have potential joints problems if they reach the standard movement with uneven movement (Figure 20).

Query interface on the mobile and on the computer

We developed the query interface on the mobile and on the computer which can connect to the database and access the assessment result so that medical staffs can query the results immediately. To enter the system, medical staffs need to create an account. After entering the system, if subjects' information has not been established yet, medical staffs need to create subjects' basic information for the sake of convenience querying. Then the assessment can start after establishing this information.

Functions on the mobile and computer interface mentioned above are listed as follows:

- Connect to the database
- Medical staffs can register accounts (Figure 21)
- Build subject's basic information (Figure 22)
- Query subject's information by identify ID (Figure 23)
- Record the result of the test
- Query the results of test (Figures 24 and 25).

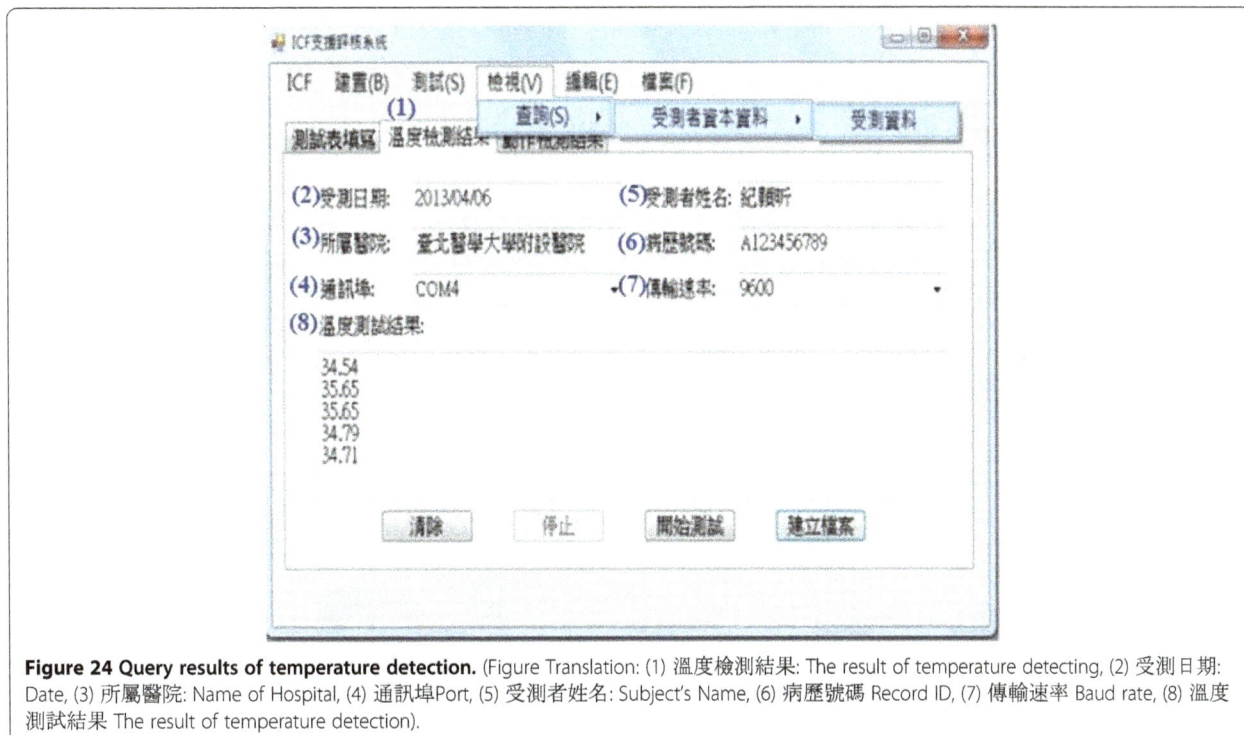

Figure 24 Query results of temperature detection. (Figure Translation: (1) 溫度檢測結果: The result of temperature detecting, (2) 受測日期: Date, (3) 所屬醫院: Name of Hospital, (4) 通訊埠Port, (5) 受測者姓名: Subject's Name, (6) 病歷號碼 Record ID, (7) 傳輸速率 Baud rate, (8) 溫度測試結果 The result of temperature detection).

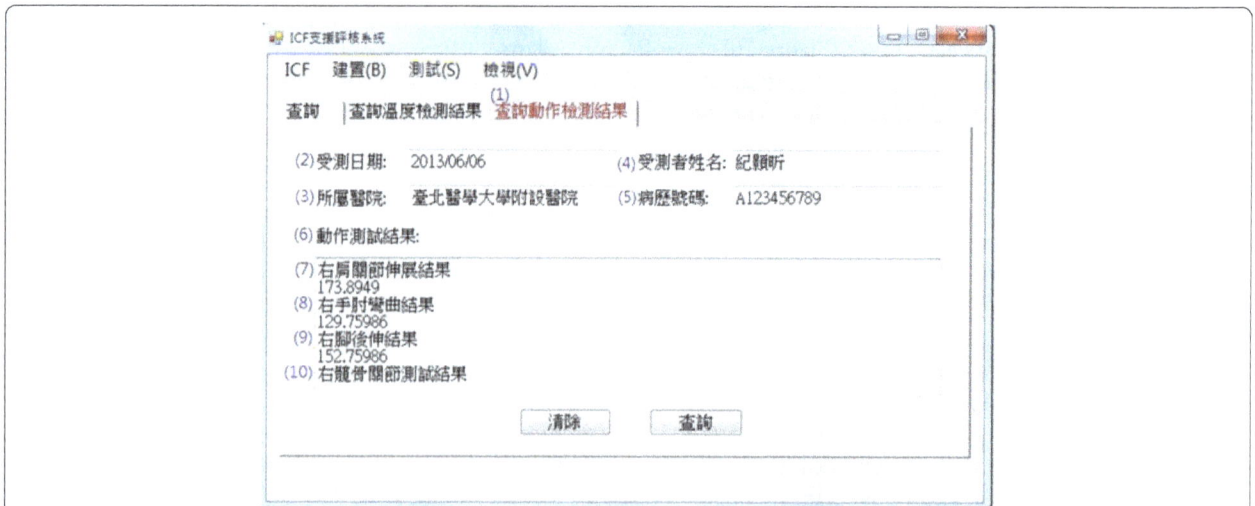

Figure 25 Query result of joints mobility assessment. (Figure Translation: (1) □詢動作檢測結果: Query the result of joints mobility assessment, (2) 受測日期: Date, (3) 所屬醫院: Name of Hospital, (4) 受測者姓名: Subject's Name, (5) 病歷號碼 Record Number, (6) 動作測試結果: The result of joints mobility assessment, (7) 右肩關節伸展結果: The result of right shoulder extension, (8) 右手軸彎曲結果: The result of right elbow extension, (9) 右□後跟伸結果: The result of right hip flexion, (10) 右髖關節測試結果: The result of right hip extension.

Discussion

In the joint mobility measurement, subject was able to perform the measurement according to the demonstration animation and the degree of joint mobility was recorded in database. Through the subject's moment we could observe not only if he/she has the ability to take care of themselves but also the inconvenience might occur in everyday life. The experiment results show that our system has potential for enabling the ICF assessment to be performed by users themselves or with assistance from family, and therefore releasing medical staffs' burden.

In the experiment, we found that the elder could recollect their memories by asking question about family affairs. During dementia test, when the subjects were asked about their date of birth, they both incidentally mentioned funny stories in their childhood. Subject 1 is a traditional farmer living with family. Subject 2 is subject 1's wife who is a housekeeper Table 2. Both of them were confused when they were asked about "What date is today?", though they could remember "What date is today?" From the results, we could tell that subject1 was good at calculating because he could count the price correctly for harvest. In contrast, Subject 2's ability of calculating was not as good as Subject 1. Though subject 2 performed badly at calculation we figured out that it was caused by less use of calculating in life. The questionnaire answered by their family/caregiver is also an important information for decision supporting. Through our system, early detection of dementia might be found and possibly provides patients with benefit from better treatment option and may improve patients and their family's life quality.

Conclusion

Our research results contain the following three parts: First, our system can successfully track user's movement. And consequently, most of the measurement results, which was originally performed by subject with the assistant of medical staffs, can now be recorded automatically under the help of volunteer. Therefore our system is able to release medical staffs' burden. Second, our system can measure and visualize user's body temperature automatically. In addition, through our graphical interface, helpers and medical staffs can

Table 2 Subject's information

	Gender	Age	Note
Subject1	Male	77	Farmer
Subject2	Female	67	Housekeeper

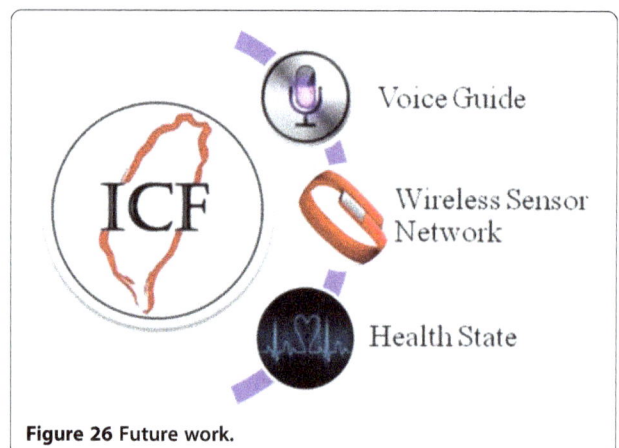

Figure 26 Future work.

acknowledge auxiliary information of subject's body condition more easily than the conventional way. Third, our system provides a user-friendly multimodal interface for elders, such as the virtual buttons with webcam-based gesture recognizer, etc.

For the current version, we still have lots of work to do. In the part of joint mobility measurement, we are adding a voice guide to encourage users while performing assessment. Also, we are also implementing a wireless sensor network so that we can overcome some measuring problems for sophisticated joints, such as ankle and wrist.

Moreover, understanding the massage presented by the computer is not always easy, especially for the elderly. Therefore, we would like to develop a more flexible and customizable interface which can provide necessary services according to the user's need by introducing more AI approaches (Figure 26).

Comparison of the joint mobility measurement results of the system and those done by an expert as gold standard is suggested for validation. We may need a lot of subjects to have the statistics of P-values. The questions about dementia are even more complicated. We may need comparison between the answers by subjects using computer and the answers by the expert. We are planning to test our system on more patients with the help of medical staffs in the near future. Finally, being encouraged by the positive feedbacks from both medical staffs and elders, we are also planning to further develop our system as a more general platform for visualizing medical metadata.

In this paper, we described an ongoing project of a sensor-based decision supporting system for ICF. Many parts of this project are still at its early stage. However, the first goal which aims to develop a conceptual prototype by utilizing the advices from medical staffs, is reported in this paper. And the requirements for the research of this paper in the medical field are higher and higher.

Competing interests
The authors declare that they have no competing interests.

Authors' contributions
LHC, WEH and HMH participated in the sequence alignment and drafted the manuscript. YWC gave the professional suggestion as medical staff. EHW gave conceptual advice of medical mobile device. ESS and TY gave the conceptual advice of Quality of life. YT gave the suggestion for visualization in user's daily log. All authors read and approved the final manuscript.

Author details
[1]Computer Science and Information Engineering, National Chi Nan University, Puli, Taiwan. [2]Faculty of System Design Tokyo Metropolitan University, Hino, Tokyo, Japan. [3]Computer Science and information Engineering, National Central University, Taoyuan, Taiwan. [4]Department of Internal Medicine, Landseed Hospital, Taoyuan, Taiwan.

References
Directorate-General for the Information Society and Media. (2010). *Advancing and applying Living Lab methodologies: an update on Living Labs for user-driven open innovation in the ICT domain*. European Union: European Network of Living Labs publication.

Eriksson, M, Niitamo, V-P, & Kulkki, S. (2005). State-of-the-art in utilizing Living Labs approach to user-centric ICT innovation–a European approach. *Technology 13, 1,* 1–13.

Liao, H-F, & Huang, A-W. (2009). Introduction to International Classification of Functioning, Disability and Health (ICF) and recommendations for its application in disability evaluations in Taiwan. *Formosan Journal of Physical Therapy, 34*(5), 310–318.

Morris, JC. (1993). The Clinical Dementia Rating (CDR): current version and scoring rules. *Neurology, 43,* 2412–2414.

Muilder, I, & Stappers, PJ. (2009). *Co-creating in practice: result and challenges, collaborative innovation: emerging technologies, environments and communities* (pp. 22–24).

Mulvenna, MD, & Nugent, CD. (2010). *Supporting people with dementia using pervasive health technologies*. Berlin: Springer.

Reed, G, Bufka, L. (2006). Using the ICF in Clinical Practice, Annual North American Collaborating Center Conference on ICF; 12, [Online] http://www.wcpt.org/sites/wcpt.org/files/files/KN-ICF-Clinical_practice.pdf.

Shotton, J, Fitzgibbon, A, Cook, M, Sharp, T, Finocchio, M, & Moore, R. (2011). Real-time human pose recognition in parts from single depth images. *Computer Vision and Pattern Recognition (CVPR), 2,* 1297–1304.

World Health Organization. (2001). *International classification of functioning, disability and health: ICF-CY*. Geneva: World Health Organization.

World Health Organization. (2007). *International classification of functioning, disability and health: children and youth version: ICF-CY*. World Health Organization.

9

Reverse Engineering Technologies for Remanufacturing of Automotive Systems Communicating via CAN Bus

Stefan Freiberger[*], Matthias Albrecht and Josef Käufl

Abstract

Nowadays, as mechatronic and electronic systems have found their way into vehicles, the technological knowledgebase of traditional remanufacturing companies erodes rapidly and even the industrial principle of remanufacturing is at risk. Due to the fact that modern cars incorporate up to 80 of these mechatronic and electronic systems that are communicating with each other e.g. via the vehicle controller area network (CAN), remanufacturing of these automotive systems requires innovative reverse engineering knowhow, methodological innovations and new technologies, especially focusing on the tasks testing and diagnostics of systems and their subassemblies. The European research project "CAN REMAN", conducted by Bayreuth University in cooperation with two other universities and eight industrial partners, focuses on these needs in order to enable companies to remanufacture modern automotive mechatronics and electronics with innovative reverse engineering skills as well as to develop appropriate and affordable testing and diagnostics technologies.

In order to operate and test the mechatronic device with CAN interface outside the vehicle environment, an appropriate simulation of the vehicle network and all connected sensors of the device under test (DUT) is essential. This implies an electrical analysis of the connectors of the DUT, a content-related analysis of the CAN-bus, a sensor hardware simulation and a CAN-bus simulation.

All electrical measurements and results were taken using conventional multimeters or oscilloscopes. The CAN-bus analysis and simulations were conducted using the Vector Informatics software tool "CANoe" (Version 7.1) and a suitable CAN-bus hardware, e.g. the CANcardXL and the IOcab8444opto. All hardware simulations were executed with a conventional wave form generator or a microcontroller evaluation board (Olimex AVR-CAN) and an appropriate electric setup.

In order to initially readout the failure memory and to investigate the diagnostic communication of the DUT, garage testers such as "Bosch KTS 650" or "Rosstech VAG-COM" were used.

The results of the project are application-orientated methods, test benches and skills for remanufacturing companies to find out the working principles of the CAN-bus communication between automotive mechatronic and electronic systems within vehicles.

The knowhow presented in this article enables remanufacturing companies to remanufacture modern automotive mechatronic and electronic systems which are communicating via the CAN-bus and similar communication types.

Keywords: Remanufacturing, Mechatronics, Electronics, CAN-bus, Reverse Engineering, Testing, Diagnosis, Vehicle Network Topology

* Correspondence: stefan.freiberger@uni-bayreuth.de
Chair of Manufacturing and Remanufacturing Technology, Bayreuth
University, Universitaetsstrasse 30, 95447 Bayreuth, Germany

1. Introduction

Raising requirements on occupant safety and comfort on the one hand and the introduction of new emission regulations on the other hand, forces the automotive manufacturers to enhance their products continuously. In order to achieve these improvements, electronic systems, based on microcontrollers, have found their way into modern cars and they contributed considerably to many new advantages in terms of safety and comfort such as Electronic Stability Program (ESP), Anti-lock Brake System (ABS), Parking Assist System (PAS), Electro Hydraulic Power Steering (EHPS) or Electro Assisted Steering (EAS). Nevertheless, the new trend of modernization has an immense impact on the remanufacturing business. It can be seen that new branches in electronic remanufacturing arise. In contrast to that, the knowhow of traditional remanufacturing companies has eroded rapidly and even the industrial principle of remanufacturing is at risk [1]. Due to the fact that modern cars incorporate up to 80 of these mechatronic and electronic systems that are communicating with each other e.g. via the CAN-bus, remanufacturing of these automotive systems requires innovative reverse engineering knowhow, methodological innovations and new technologies especially focussing on the tasks testing and diagnostics of systems and their subassemblies. Since, traditional remanufacturing companies do not have much capacity to build up the appropriate knowhow, the Chair of Manufacturing and Remanufacturing Technologies at Bayreuth University assists these companies in reverse engineering, as well as finding new methodologies and technologies for remanufacturing [2,3].

In the following chapters, reverse engineering methodologies, technologies and results for automotive components will be presented on the example of an EHPS pump. The results have been obtained within the European research project "CAN REMAN" which is conducted by Bayreuth University, Linköping University (Sweden), the University of Applied Sciences Coburg, Fraunhofer Project Group Process Innovation and eight industrial partners. The target of this project is to enable independent aftermarket (IAM) companies to remanufacture modern automotive mechatronics and electronics with innovative reverse engineering skills as well as to develop appropriate and affordable testing and diagnostics technologies [4]. The described, close to industry results, will contribute to the remanufacturing research theory by the upcoming PhD-thesis of engineers of the Chair of Manufacturing and Remanufacturing Technology.

2. Automotive Mechatronics Change Today's Remanufacturing

The term "mechatronics" was formulated in 1969 in Japan and it is an artifice that describes a system which combines mechanics, electronics and information technologies. A typical mechatronic system gathers data, processes the information and outputs signals that are for instance converted into forces or movements [5].

2.1. Technological Change of Vehicles

Automotive parts should not longer be seen as isolated standalone applications with few mechanical and electrical inputs and outputs. Now, they have the capability to communicate to each other and to share the same information. Subsequently, the communication of the different automotive subsystems helps the original equipment manufacturers (OEMs) to reduce weight and cost by sharing the same sensors and reducing cable doubling (cable length) in modern vehicles. For the driver the network and communication within the car remains invisible and he feels the car behaving like ten years ago despite of some additional comfort functions. Figure 1 demonstrates the radical shift in the automotive technological development.

But if we take a closer look, nowadays modern vehicles resemble more or less a distributed system. Several embedded computers - often referred to electronic control units (ECUs) - communicate, share information and verify each other over the vehicle network. One of the commonly used communication networks in vehicles is the CAN-bus. Within this network structure, each control unit has at least one unique identifier (ID) on which it broadcasts messages that again incorporate different signals and information [6]. Easily speaking, in case of a missing or faulty participant in the network, all other controllers will notice the participant as they have a lack of information. The lack of information or errors on the CAN-bus can force other systems to operate in a "safe mode" or cause that these systems never start their operation. In reverse, a controller not connected to the specific vehicle network will not start its regular operation patterns.

2.2. Difficulties for Remanufacturers

As stated before, the introduction of electronic networks into modern cars entails enormous problems for remanufacturers. Modern electronic and mechatronic vehicle components cannot be tested as easily as traditional electrical and mechanical ones [7-9]. While it was usually sufficient to link electrical systems to the power supply (battery), modern mechatronic and electronic systems gather a lot of information from the vehicle environment and driving conditions using plenty of sensors and the CAN-bus network of the vehicle. As a consequence, connecting all sensors and the power plug to the DUT is insufficient unless the device is connected to the network of a real car or an adequate simulation of the communication in the vehicle.

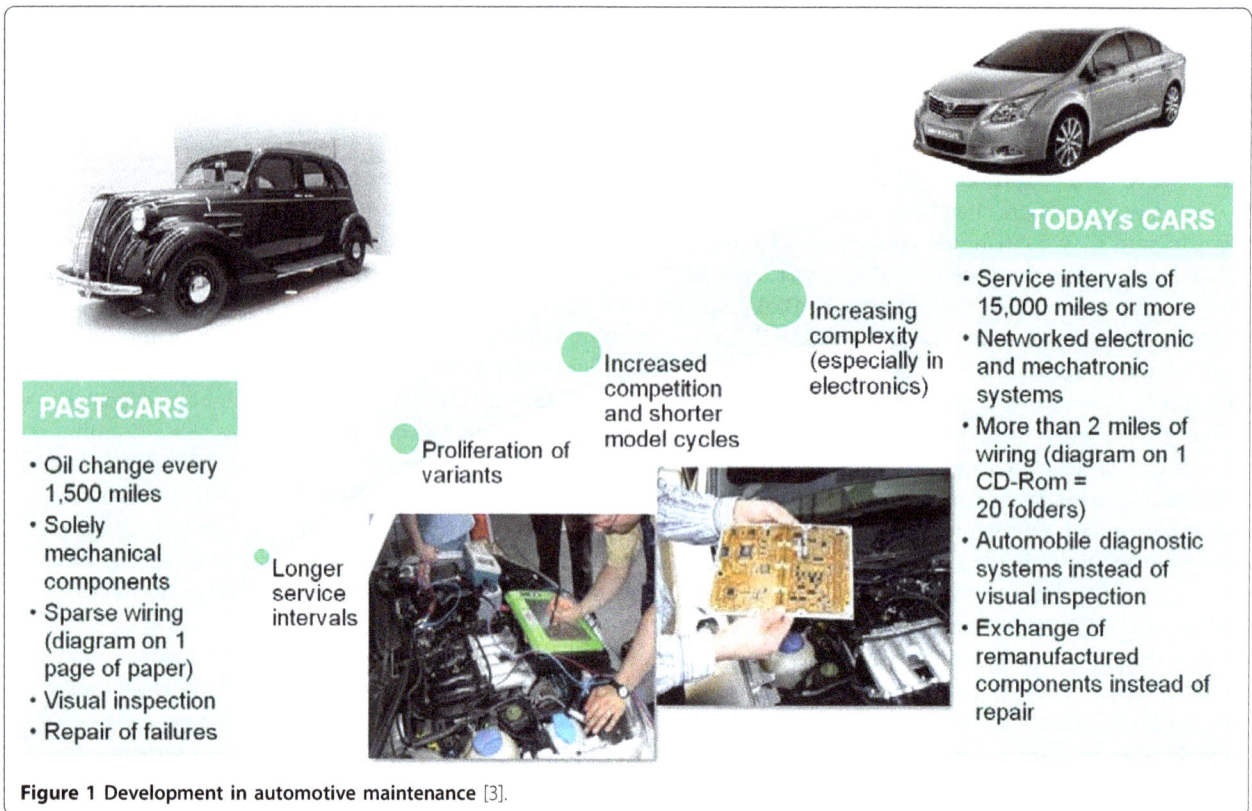

Figure 1 Development in automotive maintenance [3].

Following these statements, the key for successful remanufacturing and testing of a certain automotive system lies in the simulation of the complete network communication in the vehicle. In each case, the car matrix (CAN database) of the specific vehicle model is required to build a simulation of the CAN communication in a vehicle. However, the OEMs will not release any information on the communication parameters to non-OEs and therefore they will not support the independent remanufacturing business. As a consequence, the independent remanufactures - onto which this paper focuses - have to do a lot of reverse engineering themselves or in cooperation with others in order to design their remanufacturing process chain and to come up with test solutions to ensure the quality of their products [10-12]. These reverse engineering activities focus on the system, its components, the system behavior in the vehicle and the vehicle CAN-bus communication.

2.3. The Remanufacturing Process Chain for Automotive Mechatronics

Following the previous aspects, the state-of-the-art process chain for remanufacturing, needs to be reconsidered when it comes to mechatronics, as shown in Figure 2. Regarding the process steps, disassembly, cleaning and reassembly, great progress has been made on mechanic systems, as it can be found in the literature

[13-16]. This progress on the mechanical systems can also be transferred to the mechanic components inside of mechatronic systems. However, the diagnostics and testing differs to a certain extent from the traditional (final) testing of mechanics, as it has already been discussed before. In addition to this, it was found that a lot of failures of parts and its subassemblies can only be detected or isolated with a test of the completely assembled mechatronic system [2,17], e.g. by utilization of the onboard-diagnostics and readout of the fault memory inside a mechatronic system. This means that the process chain for remanufacturing of mechatronic systems should be extended by an additional first step as it is shown in Figure 2.

In the initial entrance diagnostics of the system to be remanufactured all communication patterns have to be reverse engineered in order to simulate the vehicle network and get access to the fault memory of the system. An appropriate vehicle network simulation will also prevent new fault memory records to be stored in the DUT.

3. Reverse Engineering an Automotive Mechatronic System

The term *"reverse engineering"* has its origin in the mechanical engineering and describes in its original meaning the analysis of hardware by somebody else

Quality Assurance

1. Initial Diagnosis of the System

2. Complete Disassembly of the Product

3. Thorough Cleaning of all Parts

4. Inspection and Sorting of all Parts

5. Reconditioning of Parts and/or Replenishment by new Parts

6. Product Reassembly

Final Testing

Figure 2 Adopted remanufacturing process chain for mechatronics [20].

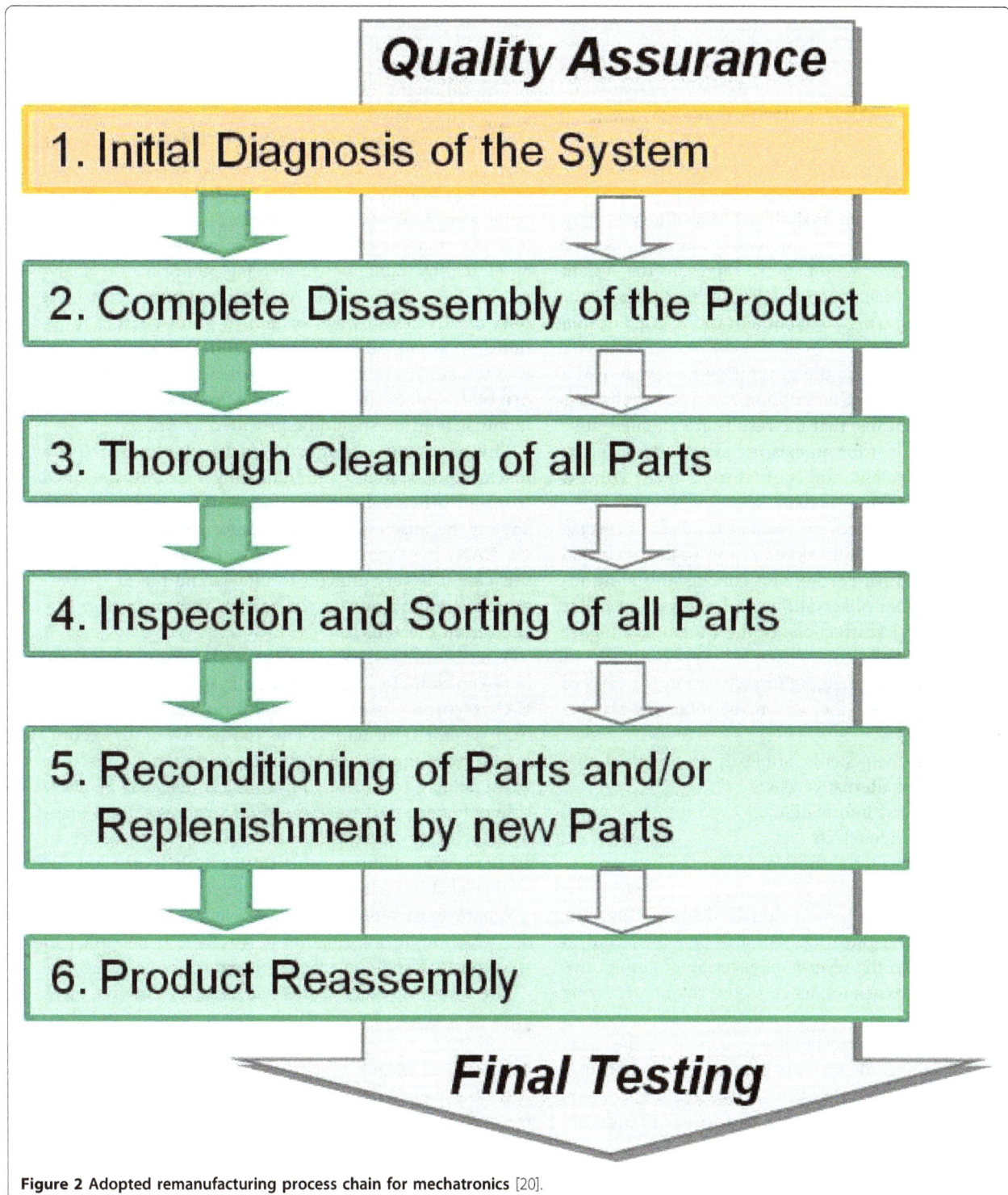

than the developer of a certain product and without the benefit of the original documentation or drawings. However, reverse engineering was usually applied to enhance your own products or to analyze the competitor's products [18]. According to Cifuentes and Fitzgerald (2000), an analog term is *"reengineering"* (of software) which does not refer to the process of analyzing software only, but which also intends to translate software into a new form, either at the same or a higher level of abstraction. In addition to this, the two authors summarize the different types of software reverse engineering. It can be differentiated between black and white box reverse engineering. While black box reverse engineering only looks at the behavior of a program and its documentation (if it's available) without examination of the internals of the program, white box reverse engineering involves looking at the internals of a program so that its working can be understood [19].

Chikovsky and Cross (1990) describe reverse engineering in the context with software development and the software life cycle as an analysis process of a system, in order to identify the system (sub-) components, to investigate their interaction and to represent the system at a higher level of abstraction [18]. In this context, they also clarifie the terms "redocumentation" and "design recovery".

"Redocumentation" is the generation or revision of a semantically equivalent description at the same abstraction level. This means, that the results are an alternative representation form for an existing system description. However, redocumentation is often used in the context of recovering "lost" information [18].

The term "design recovery" defines a subset of reverse engineering that includes domain knowledge, external information (of third parties) and conclusions additionally to the original observations and analyses in order derive meaningful abstractions of the system at a higher level [18].

Overall, reverse engineering of software in the field of software development focuses on the following six targets [18-20]:
  - Coping with the system complexity
  - Generation of alternative views
  - Recovery of lost information
  - Detection of side effects
  - Synthesis of higher abstractions
  - Facilitation of reuse

These targets, that have originally been defined for software reverse engineering, can also be transferred to a certain extent to the reverse engineering of automotive mechatronic systems and hence to the remanufacturing of these systems.

First, remanufacturers will have to cope with the complexity of mechatronic systems as stated before. "Cope" means in this context, that it must be possible to operate an automotive mechatronic system independently from its original environment (the vehicle).

Second, universal taxonomies have to be detected in order to transfer the gained knowledge to similar mechatronic systems or to other variants of the system. Especially the high degree of variation of similarly looking mechatronic systems and control units makes it difficult for the remanufacturers to manage the complexity of automotive components that usually differ by a slight detail [21].

Third, recovery of missing, rather than lost, information will be one of the most important aspects for the remanufacturing.

The following chapter demonstrates how a reverse engineering analysis can be conducted for an automotive mechatronic system.

4. Analyzing an Automotive System in Five Steps

After a reference system (a reference system in this case is a commonly used automotive subsystem; for example an electro-hydraulic power steering pump) for the analysis has been chosen it is necessary to procure at least one, ideally brand-new, system to grant correct functionality, for all following investigations. In order to analyze the system in its normal working environment, the original vehicle, in which the reference system commonly is built in, should be procured as well.

This investment might be unavoidable, because a mechatronic system communicating via CAN, detached from all other vehicle communication will not work anyway, because essential input information, transmitted via CAN, is missing otherwise (refer to chapter 2). In this case it is very difficult to understand the ECU communication and put up the system into operation isolated from the vehicle.

A cheaper way to investigate the communication between vehicle and reference system is to create a CAN recording using a software tool such as "CANoe" from Vector Informatics. This tool allows easily recording of the complete vehicle communication for instance while doing a test drive with a vehicle that may be available only once. But this procedure requires careful planning prior the test drive is carried out, in order to record every driving condition which is needed for further analyses without having the vehicle available.

Whatever strategy is chosen, it is essential to figure out which input information (CAN data) is necessary to start, operate and control the system.

The following subsections will describe the five most important steps of the analysis process more in detail.

4.1. Electrical Wiring

After having obtained a reference system, it is essential to know the pinout of all connectors of the system. Therefore, the very first step is to find out which pin belongs to which wire and signal.

First of all, the power connector (ground and positive terminal), including ignition, must be identified. One opportunity to obtain this information is the utilization of wiring diagrams or similar credentials. If such documents are not available, for example a visual inspection of the connectors and wire harness in the vehicle or continuity measurements can be beneficial.

Afterwards, it is indispensible to identify the CAN connection pins. These can easily be recognized by inspecting the cable harness (in most cases two twisted wires, but single wire CAN connection is possible, too) or by measuring a terminating impedance of 60 Ω between to cables.

Finally, all connectors for sensors and actuators (auxiliary power and sensor/actuator signal) must be known as well to go further in the analysis process.

4.2. Vehicle Network Topology

The investigation of the structure of all bus systems in the vehicle is placed in front of the proper CAN-bus analysis step. It is necessary to determine how many (CAN-bus) networks are established and in which network the reference system is located. Additionally, the network speed, the presence of a separate diagnosis network (e.g. K-Line), and all ECUs of the specific networks must be found out, especially those ECUs that provide essential input as mentioned before. Furthermore, possible gateway ECUs, which are linking different networks, should be identified.

A feasible solution to gain this information can be for example a web inquiry, documents from the manufacturer of the vehicle or the system, third party documents or technical journals (e.g. ATZ, MTZ ...).

4.3. CAN Bus Communication

In order to understand the vehicle communication more in detail, all ECUs and its associated CAN message IDs must be determined. For this purpose CANoe can be used. First of all, a physical connection to access the CAN-bus using CANoe has to be installed in the vehicle, ideally nearby the reference system ECU. With the "trace functionality" of CANoe the bus communication and all CAN messages of all ECUs can be displayed easily (Figure 3). Beside of the CAN IDs, the cycle time and the length of each message can be analyzed. This information is relevant later on for a rest bus simulation of all participating ECUs to ensure correct functionality of the reference system.

The assignment of CAN ID and the associated ECU is more difficult. In the following, two options are described in detail.

One possibility to gather this information is to record the CAN communication initially with all ECUs connected to the bus using CANoe. Afterwards, each ECU is disconnected from the bus one after another and a CAN trace is stored again. Next, all recordings are compared to each other. Those IDs that are missing in the recording can be assigned to the disconnected ECU.

Another appropriate and more sophisticated way is to locate all ECUs which provide relevant data on the CAN bus and to separate the CAN wires out of the cable harness. Each end of the CAN wires in the vehicle must be connected to a computer via CAN hardware. Afterwards, a kind of software gateway (Figure 4) is installed in between the DUT and the other ECUs using CANoe and a simple CAPL (CAN Access Programming Language) program.

By this means, it is now possible to detect the messages on the bus as well as the transmit direction - receive or transmit. This step is repeated for each ECU which provides relevant input data for the reference system. Obviously, the time exposure for this kind of CAN-bus analysis is much higher due to fact that the gateway has to be placed in between every ECU which is connected to the CAN network. The higher the complexity level of the reference system (more inputs), the more time is needed to identify all ECU messages which transmit relevant data via CAN.

The second way is more satisfying, although it may be more time-consuming than the first one. The first option offers a good overview of all CAN messages and its original ECU, but it may be fault-prone and incomplete. No matter which way is chosen, the result is a complete CAN message structure. Both ways are targeting.

But not all identified messages are relevant for the DUT. Some are not recognized by the DUT and can be disregarded for further investigations. By adding filters for single messages in the gateway CAPL program or simply disconnecting whole ECUs from the network, an empty fault memory of the DUT will reveal unnecessary messages/ECUs and hence reduce data complexity. Hereby, an external garage tester can be used in most cases in order to readout the fault memory and in order to determine whether a failure was caused by removing certain data information.

After having identified the relevant CAN messages, it is inevitable to examine the message data bytes in detail to determine the physical signals. This can be achieved by generating physical inputs manually (e.g. open the throttle, drive, break ...) and observe the particular CAN messages as well as its bytes in parallel. After that, a correlation between a CAN message, its CAN data and a physical input value can be established.

Having performed the steps above, it is possible to setup the desired restbus simulation for the reference system.

4.4. Sensors

Besides the CAN data, analog inputs of sensors and analog outputs of actuators are important in order to ensure correct functionality of the reference system. Therefore, each sensor and nearly each actuator has to be analyzed and simulated, too.

Time	Diff time	ID /	DLC	Data
25.446944	0.020036	50	4	00 02 10 12
25.447194	0.007198	1A0	8	00 41 00 00 FE FE 00 03
25.442988	0.010980	280	8	48 26 40 39 27 2F 2C 00
25.443222	0.010844	288	8	D0 9C 10 00 00 49 D6 14
25.425994	0.023989	320	8	05 02 15 00 00 00 00 00
25.443468	0.010688	380	8	00 66 2F 00 00 00 00 1B
25.433546	0.019980	3D0	2	05 00
25.301656	0.200174	420	8	81 00 00 6E 9C 00 FF 00
25.421938	0.048553	470	5	00 11 00 FF 00
25.443710	0.020430	480	8	EC 00 F3 34 00 00 00 2B
25.443940	0.010680	488	8	6D 27 26 7A A6 00 00 B0
25.440248	0.013756	4A0	8	00 00 00 00 00 00 00 00
25.301278	0.200038	520	8	40 42 02 C0 0B FB EE 01
25.373787	0.100495	570	4	87 A0 B6 00
25.044520	1.000000	580	8	8E 00 00 1E 0B 2F 38 8C
25.444182	0.020430	588	8	08 5D 7A 82 00 00 00 00
25.440490	0.013756	5A0	8	7F 00 00 77 7E 00 10 2B
25.372967	0.100273	5D0	6	80 02 40 80 00 41
25.374279	0.101339	5D8	8	55 0C 00 00 00 00 00 00
25.314309	0.199940	5DE	5	00 00 00 05 00

Figure 3 CANoe trace with all CAN IDs (messages) that are sent by the different ECUs within the VW Polo (the first column shows the current time stamp in seconds, the second column shows the cycle time in seconds, the third one displays the IDs and the last column contains the data bytes of each message).

The sensors can be analyzed using an oscilloscope and a multimeter in order to characterize current consumption, supply voltage and signal transmission. Typically, sensor output signals are analog to:
- Current/voltage, amplitude
- Frequency/cycle time
- Pulse width/duty cycle

Or they are discrete in the following forms:
- Binary
- Multi-staged (different scaled)
- Multi-staged (equidistant) → digital

For the simulation, the measured values must be interpreted and emulated. For example, the internal resistance of a sensor (load) can be calculated from the sensor current consumption. Afterwards, the presence of the sensor can be simulated using a (simple) resistor.

The simulation of the sensor signal can be realized using a waveform generator, an analog circuit, a microcontroller or a combination of them, depending on the signal characteristics.

4.5. Diagnostics

Finally, to test the reference system completely detached from the vehicle, it is necessary to know how the diagnosis communication works in order to check the fault memory and to read internal sensor information of the ECU (e.g. for temperature).

First, the applied protocols for transport and application layer must be identified. Often, standardized communication protocols for ECU diagnostics are used (e.g. ISO TP, KWP2000 or UDS). In some cases OEMs use proprietary self-developed keyword protocols (e.g. KWP1281). Thus, it is more difficult to establish a

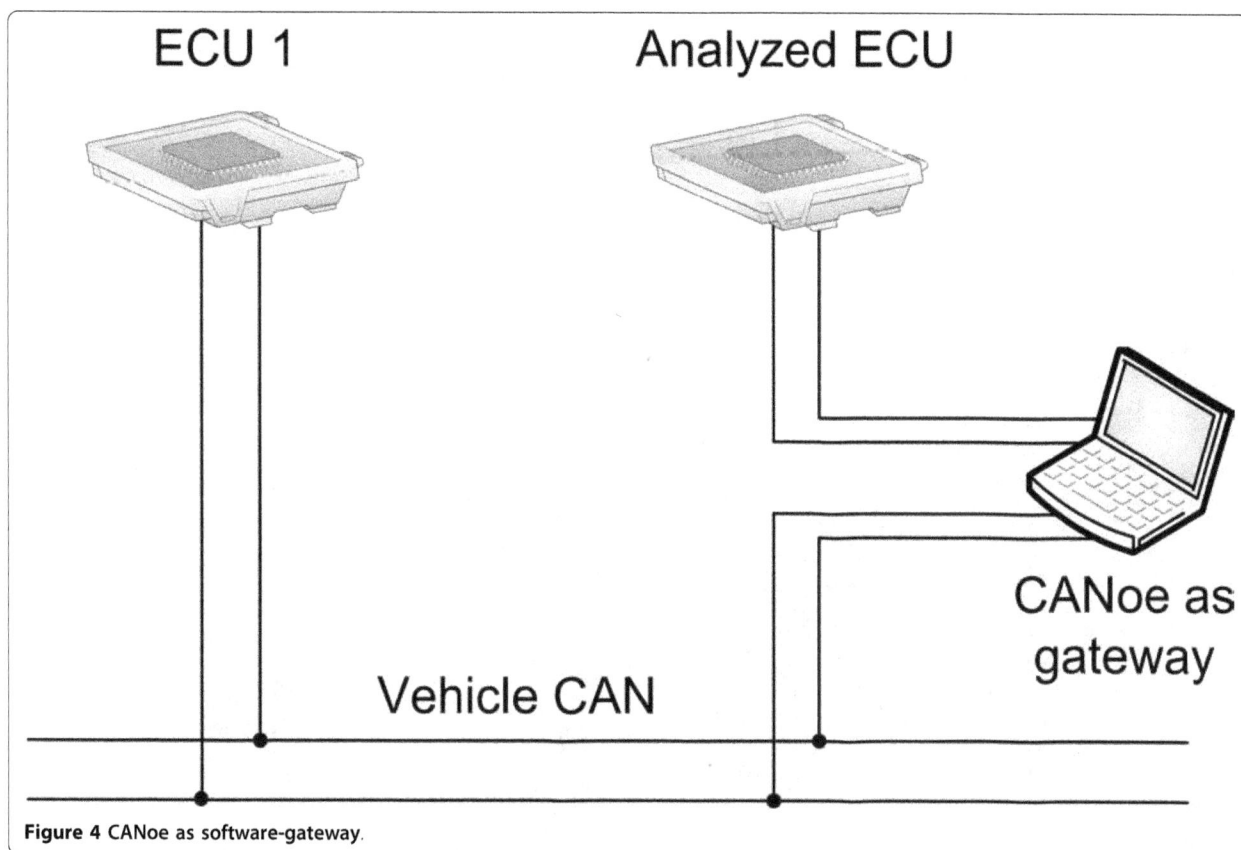

Figure 4 CANoe as software-gateway.

diagnosis connection to the reference system because the protocol specification is unknown to the remanufacturer. Hence, a detailed analysis of the CAN or K-Line communication during a diagnosis session is essential. Sophisticated reverse engineering capabilities are necessary in order to analyze, understand and recreate such a diagnosis communication. The message IDs, used for the communication, must be investigated independently by observing the diagnosis communication with CANoe. If the CAN IDs and protocols are known, the diagnose communication can be reproduced for example in CANoe using the CAPL environment.

After a remanufacturing company has accomplished all mentioned steps for the reference system, it is able to operate this system detached from all analog (sensor signals) or digital (CAN) inputs.

Finally, a test bench can be developed for entrance and final testing in series production scale.

5. Example: Remanufacturing of an Electro Hydraulic Power Steering (EHPS) Pump

An electro hydraulic power steering pump is a rotating oil pump driven by an electro motor. The pump converts electric power to hydraulic power. The hydraulic power is used to reduce the force the driver needs to steer the car. Most steering assistance is needed at low

driving speeds, maybe for parking, which makes it necessary that the EHPS pump has information about the actual driving speed. That information is communicated via CAN-Bus.

The following six steps describe the reverse engineering process on the basis of an EHPS pump that is used in a VW Polo which is seen in Figure 5.

5.1. Physical Analysis and Electrical Wiring of the EHPS

At the beginning, the EHPS has to be perceived as a black box with inputs and outputs. Because of the mechanical design and the general function of a hydraulic power steering, the output can be determined as the flow rate of the fluid [20]. The inputs are composed of an information about the internal combustion engine state (running or not running) and direct or indirect information about the necessary oil flow rate.

To get a first overview about the electrical connections of the device, a reference system (in this case the EHPS of the VW Polo - see Figure 6)) was completely disassembled. Large connector pins were good indicators for the general power supply by reason that the power consumption of the electric motor is supposed to be high. The ground pin of this connector was found by searching for a direct linkage between those pins and the ground plate of the circuit board. The other cable

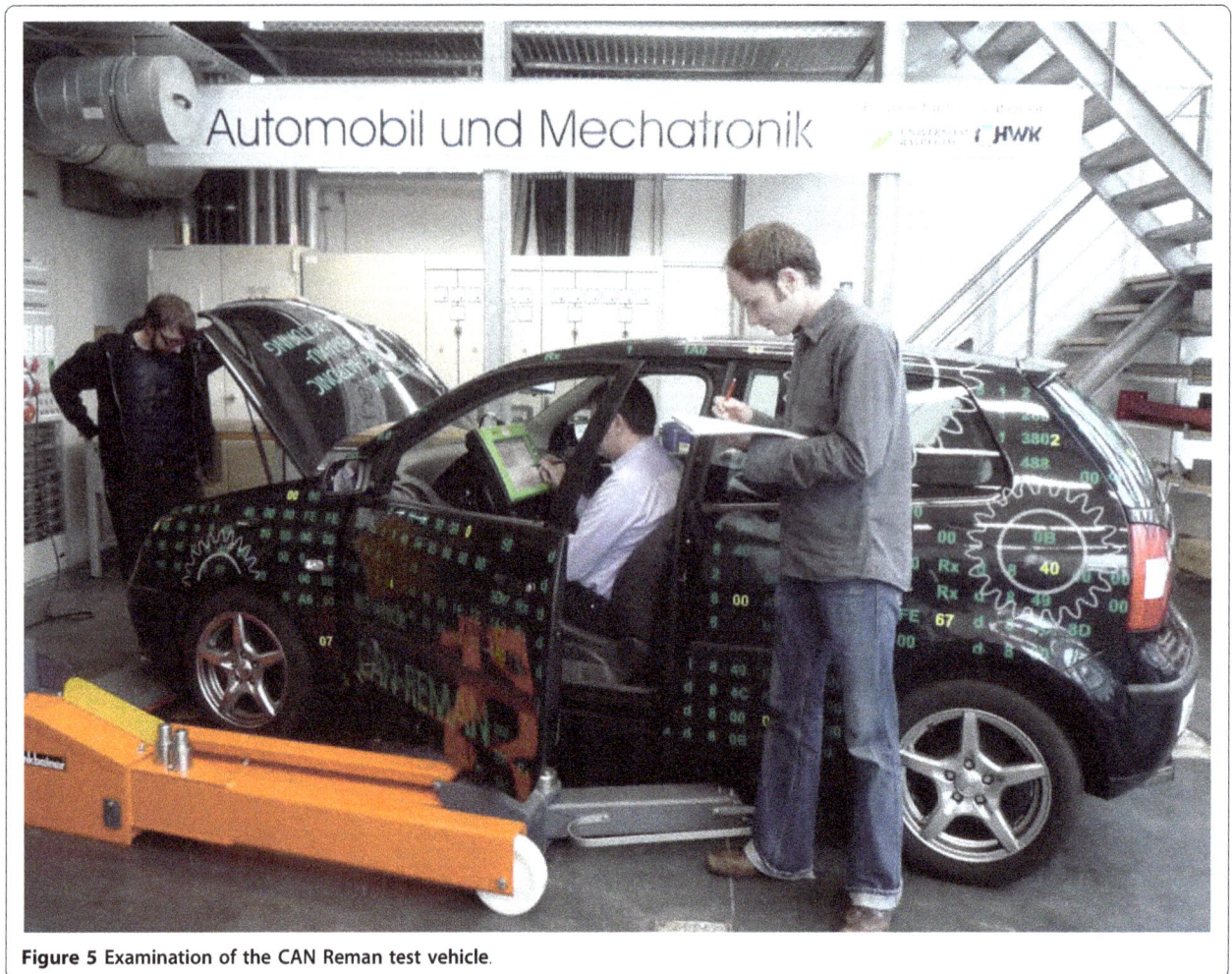

Figure 5 Examination of the CAN Reman test vehicle.

on the connector is the positive power supply. At this point, the connection of the steering angle sensor, which is directly mounted on the steering shaft, was disregarded. The third connector contained three cables. Two of them were twisted in the following cable harness. That was a perfect indication for CAN cables. The CAN-high cable rises from 2.5 V to 3.5 V and the CAN-low cable falls from 2.5 V to 1.5 V during active communication. When operating the vehicle, the last cable was on 12 V level and therefore it was assumed to be the signal for "ignition on". At this point the electrical analysis of the device was completed.

5.2. Vehicle Network Topology

On the example of the VW Polo EHPS, all relevant ECUs for operating the DUT have to be in the same CAN-bus network (Figure 7). Unfortunately, the CAN bus is not linked to the on-board diagnosis (OBD) connector of the test vehicle, whereas usually selected CAN bus data is also accessible through this connection. Therefore, the CAN wires in between of the EHPS and

the rest of the vehicle were separated in order to get access to this network for further investigations.

Assigning single messages/IDs to ECUs has simply been done by disconnecting single ECUs and locating missing messages/IDs. By a parallel readout of the internal fault memory of the DUT, relevant ECUs or single messages have been found.

5.3. CAN Bus Communication Investigations

This step can always be split into two parts. The first is the analysis of the communication in order to filter out and understand the relevant messages for the EHPS sent by other ECUs. The second is the simulation of the necessary CAN communication, which is called "restbus" in the following.

First, the start signal, transmitted to the EHPS via CAN bus, must be discovered as described in step 2. Therefore, a recording of the in-car CAN communication was made at a stationary test with well defined and reproducible conditions. After that, the recording was replayed to the test device outside the car and it started

Figure 6 Pins of the EHPS of the VW Polo.

its operation. Next, CAN messages were successively filtered out until the motor of the test device stopped. Hence, the last filtered message contained some kind of a start signal. Having performed in depth analyses, this signal was identified to be the RPM signal of the internal combustion engine. In order to eliminate or to find other input parameters, the same study was carried out using a recording of a real-road test. It was found that the vehicle speed is another input parameter for the EHPS.

Second, required input parameters were simulated with CANoe. Using a third party diagnosis garage tester (Bosch KTS 650), it was discovered that the fault memory of the external EHPS can only be erased when at least the presence of the missing messages of the in-car communication is simulated, too. This simulation of messages with and without data content is called restbus.

At this point the EHPS can completely be operated outside the car, but with a real steering angle sensor.

5.4. Simulation of Sensors
In order to operate the EHPS in a completely simulated environment, the angular velocity sensor had to be simulated.

Analog to step one, VCC and GND were identified on the sensor terminal using a multimeter. The third cable transfers the information about the angular velocity of the steering wheel. This signal was analyzed using an oscilloscope (Figure 8) and identified as a pulse width modulated signal. This signal was simulated by a

waveform generator. Furthermore, the sensor presence had to be emulated by a simple 600 Ω resistor matching the power consumption of the original sensor.

5.5. Diagnostic Functions of the Device
Most devices, including the present EHPS, can be diagnosed over CAN-bus with an external diagnosis garage tester. This tester can, as mentioned above, directly communicate with ECUs using a transport and a keyword protocol. The protocols are only partially defined and the communication differs from brand to brand tremendously. Therefore, the most efficient way to understand how e.g. the fault memory can be erased is to erase the fault memory with one of those testers and to try projecting the sequence onto known standards. In the present case, it were the standards KWP1281 and TP1.6. Even though the understanding of the diagnosis communication was very time-consuming, it was possible to erase and read the fault memory, to read the internal sensor data or duty cycles, to parameterize the device for different car models or even to completely reprogram the software.

Finally, all functions were implemented in CANoe using CAPL which can be controlled by a graphical user interface (GUI).

5.6. Operation Range
At last, the correlations between input and output values were determined in detail. For this reason, the

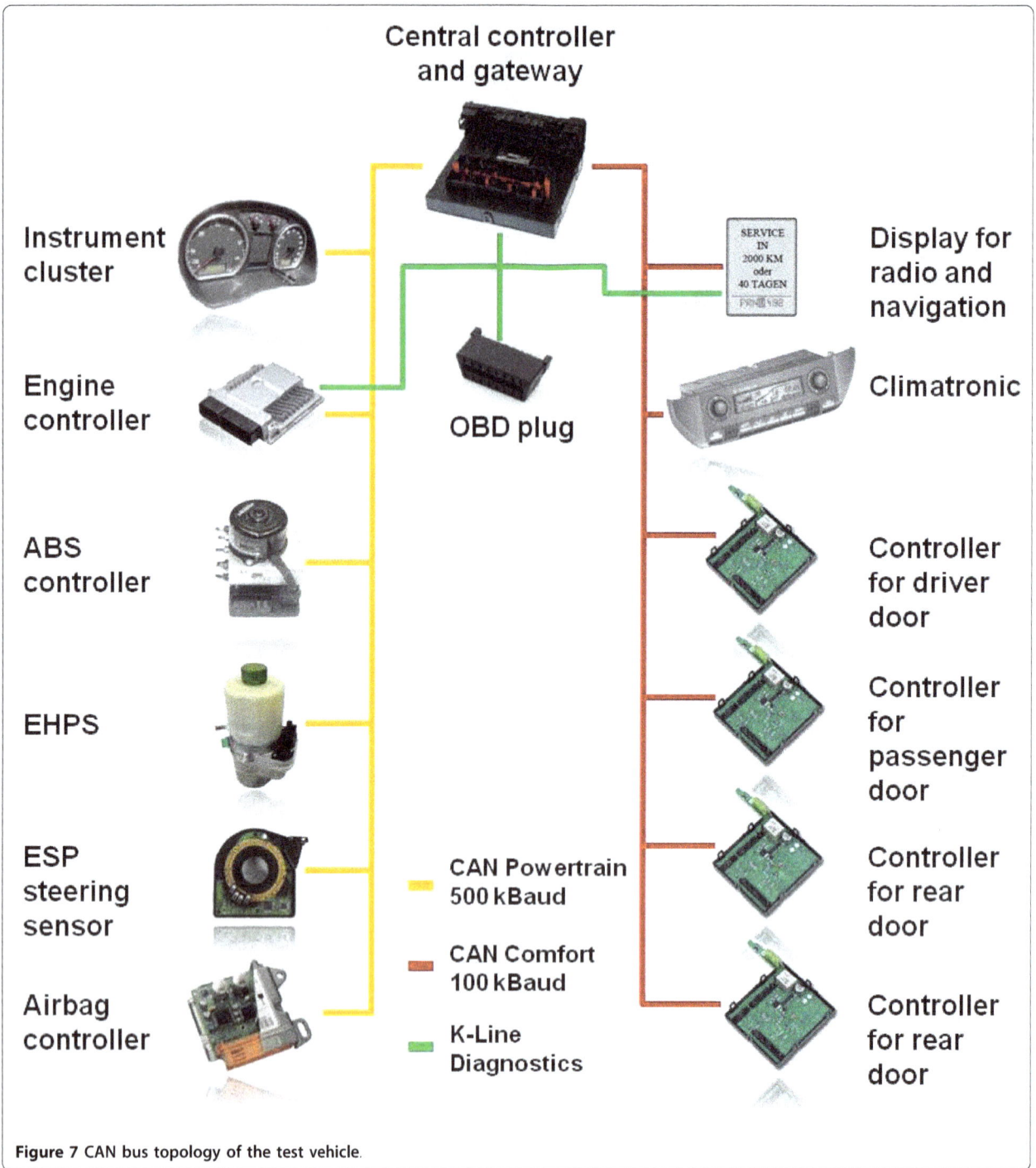

Figure 7 CAN bus topology of the test vehicle.

input parameters angular velocity, vehicle speed, RPM and the outputted oil flow rate were recorded simultaneously.

In this case the RPM signal only started the EHPS and was disregarded for the measurement. The vehicle speed was found in a particular message on the CAN bus as figured out in step 3. The angular velocity value is part of the sensor data provided by the EHPS in a diagnosis communication session as mentioned in step 5. The resulting oil flow rate was measured by installing an oil flowmeter to the low pressure side of the EHPS in the test vehicle. This flowmeter generates a frequency modulated signal which was converted to a CAN message by a microcontroller and broadcasted to the local in-car CAN network in a separate CAN message. Finally, all necessary input and output values were

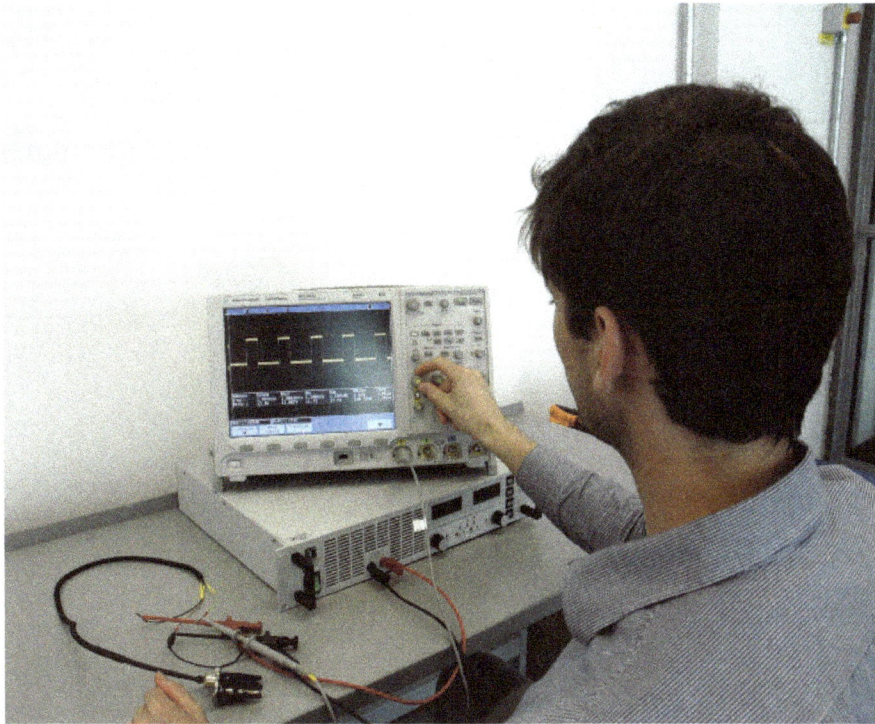

Figure 8 Measuring the sensor signal.

recorded from the CAN network time simultaneously using CANoe. Figure 9 depicts the flow rate of the steering oil as a result of vehicle speed and angular velocity, measured in a real-road test.

5.7. Practicability of the results

For further mechatronic systems an analysis time of 2 days to 2 month is required, depending on the system complexity. To give some examples for the time required: 2 days for example for another EHPS pump in another VW (each model needs new analyses), 5 days for another EHPS pump in different brand, 1 month for an absolutely new mechatronic system with medium complexity and 2 month or longer for a very complex mechatronic system like an automatic gear box. The costs for the analyses are splitted in the fix costs for the hard- and software of about 40.000 Euro and the costs for the employees for the days they work on. The reliability and safety of a remanufactured mechatronic system is in the same level compared with a new system. A mandatory regulation about standardization of the signals would decrease the costs significantly.

6. Conclusion

A still increasing number of mechatronic and electronic systems is built into today's vehicles. In the future, even more of these systems will be introduced to the cars as a result of increasing demand for comfort, safety and reduced fuel consumption. Remanufacturing of failing mechatronic systems offers a great opportunity for all, the OEMs and OEs which can safe resources and provide spare parts over a long period of time without the demand of long time warehousing; the remanufacturing companies as they can make a growing new business with these systems; and the customers that are benefitting from cheaper, but as good as new, spare parts.

Progress is not possible without its challenges, but it is achievable. The increasing complexity and variety of mechatronic end electronic devices cannot be handled with traditional methodologies. Therefore, remanufacturing companies have to build up new reverse engineering knowhow, find methodological innovations and they need to develop new technologies, especially focusing on the tasks testing and diagnostics of automotive systems and their subassemblies. After having met these challenges, new remanufacturing steps, such as the initial test, can be established and increase the productivity of the remanufacturing businesses e.g. in terms of an automated identification of systems or automated electronic test.

The paper outlines challenges, possible solutions and technological progress for the reverse engineering process of mechatronic automotive systems that are communicating via CAN-bus. In addition to this, the reverse

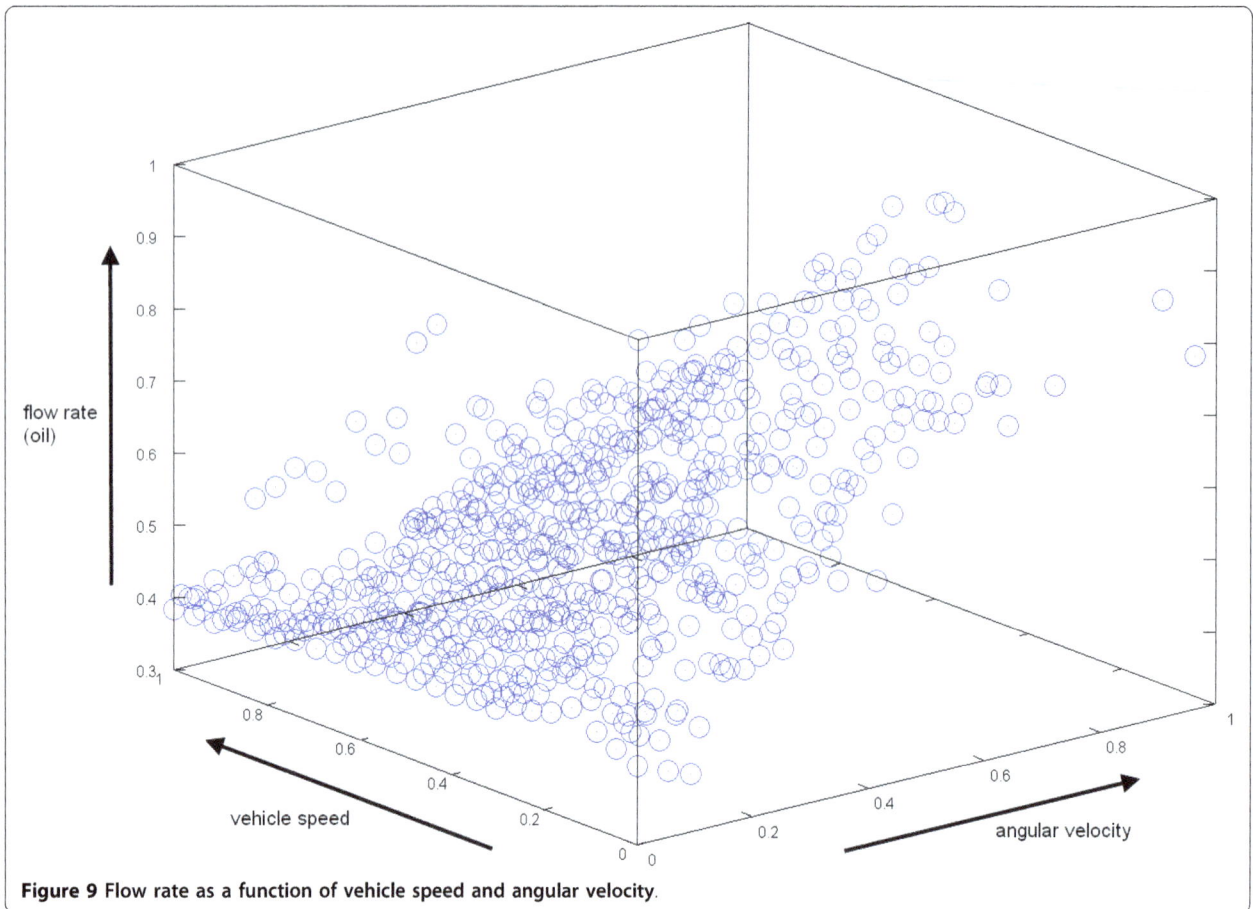

Figure 9 Flow rate as a function of vehicle speed and angular velocity.

engineering process is demonstrated on the example of an EHPS which is used in a VW Polo. Obviously, it was possible to completely understand the steering system of the VW Polo by reverse engineering. Now, it is possible to run and test the mechatronic system outside the car as well as to adopt the results for remanufacturing the system in series production scale. The same principle can also be applied to further automotive systems, so that everyone wins, regardless of perspective.

List of abbreviations

ABS: Anti-lock Breaking System; ATZ: Automobiltechnische Zeitschrift (title of an automotive and technical journal); CAN: Controller Area Network; CAPL: CAN Access Programming Language; CRV: Current Replacement Value; DUT: Device Under Test; EAS: Electro Assisted Steering; ECU: Electronic Control Unit; EHPS: Electro Hydraulic Power Steering; ESP: Electronic Stability Program; GND: Ground Connection; GUI: Graphical User Interface; PAS: Parking Assist System; IAM: Independent Aftermarket; ID: (CAN-bus) Identifier; MTZ: Motortechnische Zeitschrift (title of an engine related journal); OEM: Original Equipment Manufacturer; OPI: OEM Product-Service Institute; VCC: Positive Power Supply; VW: Volkswagen.

Acknowledgements

The research project "CAN REMAN" and the activities described in this paper have been financed by the German Federal Government Department for Education and Research (support code 16INE014). Nobody, beyond the mentioned author's, contributed materials essential for the study.

Authors' contributions

AB carried out the molecular genetic studies, participated in the sequence alignment and drafted the manuscript. JY carried out the immunoassays. MT participated in the sequence alignment. ES participated in the design of the study and performed the statistical analysis. FG conceived of the study, and participated in its design and coordination. All authors read and approved the final manuscript.

Authors' information
Dr.-Ing. Stefan Freiberger

Managing Engineer at:
Bayreuth University
Chair Manufacturing and Remanufacturing Technology and Fraunhofer Project Group Process Innovation
Chairman of the Mechatronics and Electronics Division of APRA (Automotive Parts Remanufacturers Association)
Consultant in the fields of: Remanufacturing, Material- and Energy Efficiency, Process Innovation in Production, Lean Management and Six Sigma
PhD thesis about: "Prüf- und Diagnosetechnologien zur Refabrikation von mechatronischen Systemen aus Fahrzeugen"; „Test and Diagnosis Technologies for Remanufacturing Automotive Mechatronic Systems".
M.Sc., Dipl.-Ing. (FH) Matthias Albrecht

Engineer and research assistant at
Bayreuth University
Chair Manufacturing and Remanufacturing Technology and Fraunhofer Project Group Process Innovation
Field of Activity: Research for remanufacturing of mechanical, electronic and mechatronic components.
Development of test and diagnosis methods for coupled mechatronic and electronic systems with CAN-bus. Transfer of developed technologies and test equipment to different systems and implementation of these systems in

remanufacturing companies. Development of industrial test equipment for mechatronic and electronic automotive components that is application-orientated and easy-to-use.

Dipl.-Ing. (FH) Josef Käufl
Research Engineer at:
Bayreuth University
Chair Manufacturing and Remanufacturing Technology and Fraunhofer Project Group Process Innovation
Field of activity: Technologies for remanufacturing of mechanical, electronic and mechatronic components. Development of test and diagnosis methods for coupled mechatronic and electronic systems with CAN-bus. Transfer of developed technologies and test equipment to different systems and implementation of these systems in remanufacturing companies.

Competing interests

The authors declare that they have no competing interests.

References

1. Freiberger S: **Finding profitable products for Remanufacturing.** *APRA Global Connection, Ausgabe Nr,6 Chantilly* 2010.
2. Steinhilper R, Rosemann B, Freiberger S: **Product and Process Assessment for Remanufacturing of Computer Controlled Automotive Concepts.** *13th CIRP International Conference on Life Cycle Engineering, Leuven, Belgium, May 31st - June 2nd* 2006.
3. Steinhilper R: **Automotive Service Engineering and Remanufacturing: New Technologies and Opportunities.** *15th CIRP International Conference on Life Cycle Engineering, Sidney, Australia, March 17th - 19th* 2008.
4. Freiberger S: **European Research Project "Major European reman project given the green light" Starts Now.** *ReMaTecNews 2/2009, RAI Langfords B. V./RAI Publishing House, Amsterdam* 2009.
5. Roddeck W: **Einführung in die Mechatronik.** B.G. Teubner Verlag/GWV Fachverlage GmbH, Wiesbaden, Germany;, 3 2006.
6. Zimmermann W, Schmidgall R: **Bussysteme in der Fahrzeugtechnik.** Vieweg + Teubner/GWV Fachverlage GmbH, Wiesbaden, Germany;, 3 2008.
7. Freiberger S, Steinhilper R, Heinrich A, Brüggemann D: **Failure Detection and Isolation through Infrared Thermal Imaging.** In *ReMaTecNews - Automotive Remanufacturing International. Volume 6.* RAI Publishing House, Amsterdam, Dezember; 2006.
8. Freiberger S, Steinhilper R, Stöber R, Fischerauer G: **How to remanufacture partially documented mechatronic systems.** *APRA Global Connection, Ausgabe 16, Chantilly* 2006.
9. Freiberger S: **Remanufacturing of Mechatronics and Electronics.** *APRA Mechatronics and Electronics Division, Harrisburg* 2006 [http://www.apra-europe.org].
10. Freiberger S, Landenberger D, Wrobel S: **FMEA in der Refabrikationsindustrie - Erfassen, bewerten, vermeiden.** *Quality Engineering, Ausgabe 04/2006* Konradin Verlag, Leinfelden-Echterdingen; 2006.
11. Freiberger S, Rosemann B, Steinhilper R: **Design for Recycling and Remanufacturing of Fuel Cells.** *Proceedings Eco Design 2005: 4th International Symposium on Environmentally Conscious Design and Inverse Manufacturing, Tokyo 12. bis 14* 2005.
12. Freiberger S, Rosemann B: **State of the Art Application and End-of-Life of Fuel Cell Systems.** *Proceedings 9th International Congress for Battery Recycling, Como, 2. bis 4* 2004.
13. Johnson MR, Wang MH: **Economical evaluation of disassembly operations for recycling, remanufacturing and reuse.** *International Journal of Production Research* 1998, **36(12)**:3227-3252.
14. Seliger G, Hentschel C, Wagner M: **Disassembly Factories for Recovery of Resources in Product and Material Cycles, pp 56 - 67.** In *Life-Cycle Modeling for innovative Products and Processes, Proceedings on life-cycle modeling for innovative products and processes, Berlin, Germany, November/December 1995.* Edited by: Jansen H, Krause F-L. Chapman 1995:.
15. Seliger G, Grudzien W, Zaidi H: **New Methods of Product Data Provision for a simplified Disassembly.** *Proceedings of the Life Cycle Design 99, Kingston, Kanada* 1999.
16. Westkämper E, Alting Arndt: **Life Cycle Management and Assessment: Approaches and Visions Towards Sustainable Manufacturing.** *CIRP Annals - Manufacturing Technology* 2000, **49(2)**:501-526.
17. Freiberger S: **Selected and Applied Test and Diagnosis Methods for Remanufacturing Automotive Mechatronics and Electronics.** In *Remanufacturing Automotive Mechatronics and Electronics* Edited by: Fernand J. Weiland, Germany; 2008:.
18. Chikofsky EJ, Cross JH II: **Reverse Engineering and Design Recovery: A Taxonomy.** *IEEE Software, IEEE Computer Society* 1990, 13-17.
19. Cifuentes C, Fitzgerald A: **The legal status of reverse engineering of computer software.** In *Anals of Software Engineering. Volume 9.* Springer Netherlands; 2000:(1):337-351.
20. Freiberger S: **Prüf- und Diagnosetechnologien zur Refabrikation von mechatronischen Systemen aus Fahrzeugen.** *Dissertation, Reihe: Fortschritt in Konstruktion und Produktion, Band 6, Shaker Verlag, Aachen, März* 2007.
21. Haumann M, Köhler DCF: **Coping with complexity in remanufacturing.** *Rematec News* **9(3)**:32-33.

10

Reverse logistics challenges in remanufacturing of automotive mechatronic devices

Erik Sundin[*] and Otto Dunbäck

Abstract

The remanufacturing industry as a whole and the automotive sector in particular have, over the years, proven to be beneficial to the environment and economically lucrative to the companies involved as well as to their customers. However, remanufacturing is associated with complicating characteristics, not least to mention the process of core acquisition.

The automotive industry is one of the earliest adapters of remanufacturing. Parts like engines, brake calipers and servo pumps are common targets for remanufacturing. Modern cars also have several embedded computers, often referred to as electronic control units that communicate, share information and verify each other over a Controller Area Network (CAN) bus. Due to their high value and an increasing trend in the amount of CAN bus mechatronic devices, interest in their remanufacture is growing.

Previous research has shown that it is preferable that the remanufacturer is an original equipment manufacturer (OEM), or has a close relation to the OEM, in order to achieve a well-performing remanufacturing business. In the automotive industry, there are many small and medium-sized enterprises (SMEs) that perform remanufacturing; for these enterprises, the challenges to have a profitable business are even harder. This is because the OEMs will not release any information on the communication parameters and therefore will not support the independent remanufacturing business. As a consequence, the independent remanufacturers, often SMEs, have to perform substantial reverse engineering.

This paper presents a qualitative research study, based on interviews at SMEs regarding challenges linked to the reverse logistics of SMEs remanufacturing and trading used automotive mechatronic devices, to identify specific challenges concerning the collection phase of automotive mechatronic remanufacturing. Challenges previously identified by researchers are confirmed, additional challenges within the collection phase are recognized, and challenges expected to arise when remanufacturing and trading automotive electronic CAN bus mechatronic devices are identified. The major concern for the involved companies when commencing future challenges is the handling, transportation and storing of cores. Even though the cores today mainly consist of mechanical devices, these challenges are still present; they are expected, however, to become even more crucial when cores contain a higher degree of mechatronic devices.

Keywords: Reverse logistics, Remanufacturing, Mechatronics, Electronics, CAN bus, Automotive, CAN REMAN, SME

Background

Remanufacturing is considered the ultimate form of recycling [1] and is sometimes referred to as a 'win-win-win' situation compared to traditional manufacturing since the customer pays less, the remanufacturing companies earn more and the environment benefits from less usage of raw materials and energy [2]. The benefits of remanufacturing have been put into figures by Giutini and Gaudette [3], who found that remanufactured products cost 40% to 65% less to produce than new products, are typically 30% to 40% cheaper for the customer to buy, and save globally the energy equivalent of 16 million barrels of crude oil annually. In a study by Sundin and Lee [4], it was noted that 11 of 12 environmental research studies found remanufacturing as a preferable option, at least in comparison to new manufacturing.

The remanufacturing industry has grown recently, and to date, close to 4,000 establishments are confirmed in

* Correspondence: erik.sundin@liu.se
Division of Manufacturing Engineering, Department of Management and
Engineering, Linköping University, Linköping SE-58183, Sweden

the USA alone, with more than 110 known product areas [5]. The automotive part remanufacturing industry is roughly estimated to be a $85 to $100 billion dollar industry worldwide, where the value of the remanufactured parts was estimated to be $40 billion in the USA in 2009 [6].

A fairly new and interesting area within remanufacturing in the automotive industry concerns mechatronics (e.g. power steering systems, central locking systems and anti-lock braking systems) and electronic systems (e.g. engine control units and distance control units) communicating via a Controller Area Network (CAN) bus. Developed by Bosch in 1983, CAN is a serial communication bus designed to provide robust, simple and efficient communication for in-vehicle networks [7]. The rapid growth in complexity of automotive electronics in the following decades made traditional point-to-point wiring increasingly expensive to manufacture, install and maintain; hence, CAN soon became adopted by all car manufacturers, resulting in a sales growth of CAN nodes from merely 50 million in 1999 to more than 340 million in 2003 [7]. Given the many CAN nodes in a modern vehicle and the high costs associated with replacing a malfunctioning device, ranging between 200 and 3,000 €, there is an economic incitement for remanufacturing such devices and an opportunity not yet exploited [8].

This research was part of a research project called 'CAN REMAN', with the target to develop innovative diagnosis methods and technologies for automotive mechatronics and electronic remanufacturing. The project, funded by the European Union, was conducted by Bayreuth University (Germany), Linköping University (Sweden), the University of Applied Sciences Coburg (Germany), Fraunhofer Project Group Process Innovation (Germany) and eight European small and medium-sized enterprises (SMEs). The aim of this paper is to identify previously unknown, and verify known, reverse logistics challenges experienced by SMEs that are about to remanufacture or trade automotive mechatronics and electronic systems communicating through the CAN bus system.

Reverse logistics challenges - in theory

Remanufacturing differs from traditional one-way manufacturing in several ways. These differences are also associated with manufacturing challenges and are necessities to realize a successful remanufacturing system (i.e. core acquisition, remanufacturing process and redistribution [9]). These challenges have been recognized by researchers, e.g. [9-12], but also have been summarized by Lundmark et al. [13], who also categorized them according to where and when in the remanufacturing system they occur.

In the literature about challenges within the collection phase of the remanufacturing system, there is a lot of emphasis on a lack of control regarding quantity, quality and timing of the returned products. This lack of control is recognized by, for example, [10-12,14-17] and is caused by:

- Reflection of the uncertain life of a product [11]
- Product life cycle stage and the rate of technological change [11,16]
- The dispose behaviour, which results in a stochastic return pattern [14,16]

The lack of control regarding quantity, quality and timing of the returned products (cores) is described as the major difference between a traditional production-distribution network and a product recovery network [18]. The handling of these control issues is stated as the key for creating profitable remanufacturing by Guide and Van Wassenhove [19]. In addition, there is also uncertainty regarding the demand of the remanufactured products. This uncertainty is caused by the following:

- The rate of technical development. The demand for a product might suddenly drop due to the technical development [16].
- Detailed forecasting is not possible to perform due to uncertainties regarding timing and quantities of the eturned products [16].

In order to maximize profit, a remanufacturer must be able to balance the return of cores with the demand from customers for remanufactured products. If not, the remanufacturer faces the risk of building up excessive amounts of inventory (when returns exceed demand) or low levels of customer service (when demand exceeds supply) [11]. The uncertainties in supply and demand make it hard for many remanufacturing companies to balance supply and demand [11]. All companies do not try to balance the supply with demand since the uncertainties in supply and demand makes inventory management and control functions more complicated [11]. The kind of motivation for returns could also affect the situation for the remanufacturing company since a take-back obligation might give the remanufacturing company an abundance of used products [12]. A survey conducted with 48 remanufacturing companies by Guide [11] showed that more than half of the companies had no control over the timing or the quantity of the returns. The remanufacturing companies that do not try to balance supply with demand instead dispose excess used products on a regular basis [11]. Excess used products might cost a lot of money, and the disposal cost might be high [11,15]. The storage area needed to store the excess used products is also often expensive [15]. Another challenge is that a remanufacturing firm typically has a large number of sources which means that a remanufacturing firm has to bring together a large number of small volume flows which increases the complexity [14].

According to Lundmark et al. [13], the uncertainties regarding quantity, quality and timing of the returned products are the main challenges for the collection phase of the remanufacturing system. The uncertainty in timing and quantity of the returned products also make the remanufacturing process less predictable than an ordinary manufacturing process [19]. This uncertainty makes production planning more difficult [12,15]. The uncertainty in quality adds challenges to the remanufacturing process in two different ways. Two returned products (cores) that are identical might yield a very different set of remanufacturable parts which makes inventory planning and control and purchasing more difficult [11].

Reverse logistics challenges - in CAN REMAN
Reverse logistics in CAN REMAN

Reverse supply chains comprise the activities, routes, intermediaries, etc. when transporting products in the opposite way compared to forward supply chains, i.e. from the customer or end user, via possible intermediaries to the remanufacturer, as can be seen in Figure 1. Intermediaries can, for instance, be retailers and repair shops in the forward supply chain and core brokers and scrap yards in the reverse supply chain. How the reverse supply chains are designed, i.e. how used products are brought back to the remanufacturer, plays an essential part in the remanufacturing system as a whole since it is of most importance to receive the right cores in the right quantities at the right time in order to be able to perform successful and profitable remanufacturing.

The designing of reverse supply chains is a delicate procedure with many variables and aspects to take into consideration, and there exists no general optimal supply chain. The number and type of cores (material, value, size, weight, etc.) that are to be transported highly influence the design of the reverse supply chain. In addition, types of core suppliers, types of core acquisition and

relationship between remanufacturer and supplier/customer have a great impact.

Core acquisition

A remanufacturer of automotive components normally has the possibility to choose from several core suppliers, as can be seen in the next section. The choice of supplier does not, however, automatically set the guidelines and rules on how the core acquisition is made but may enforce a certain acquisition type or open up for multiple choices. It is common that remanufacturers' ways to acquire cores vary between different types of core suppliers, but even the acquisitions within the same type may differ:

- Direct-order: in this situation, the supplier who also is the customer gives an order for the remanufacturing of a used product. The supplier/customer sends the core to the remanufacturer, which, after being remanufactured, is sent back to the supplier/customer. Within the scope of this research, there is a tendency that this type of acquisition is common when remanufacturing relatively complex products such as engines and more likely towards end users. In addition, it is common that the customer/supplier is responsible for the transportation to the remanufacturer.
- Reman-contract: this type of transaction is somewhat similar to direct-order since the supplier, which also is the customer, gives an order for remanufacturing. Also, the ownership of the core and the remanufactured product remains at the customer. However, this type is guided by a contract and spans over a longer time, with closer collaboration between remanufacturer and customer/suppler and also involves greater quantities. While direct-order is common towards end users, reman-contracts are more commonly used in collaborations with original equipment manufacturers (OEMs).

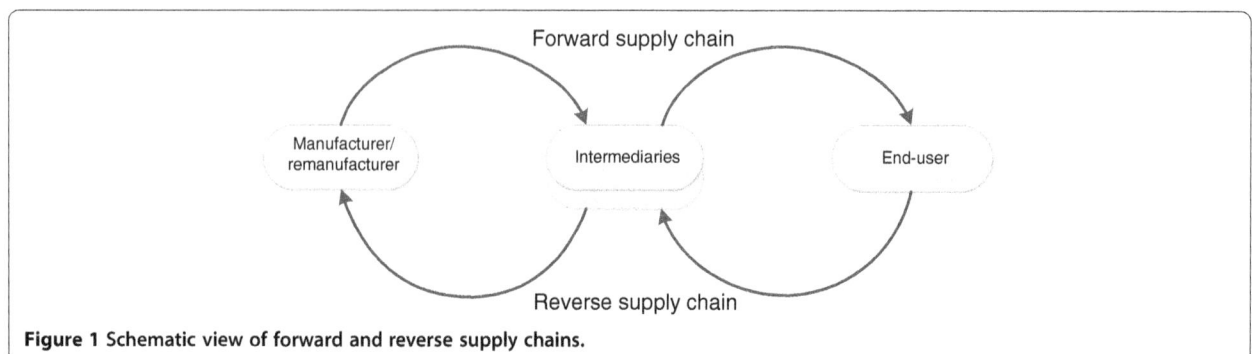

Figure 1 Schematic view of forward and reverse supply chains.

- Deposit-based: this means that when the customer buys a remanufactured product, the customer is obligated to return a similar used product. This type of transaction is frequent within automotive remanufacturing and in particular concerning components that are cheap and often exchanged at services (e.g. brake calipers).
- Credit-based: the customer receives credits for returning a core, which can be used as a discount when buying a remanufactured product. The supplier is also a customer in this case.
- Buy-back: the remanufacturer buys the core. One could say that the characteristic of this type of acquisition is the lack of relationship since neither the supplier nor the buyer has any further obligations after the transaction is made. It is common that remanufacturers buy cores from core brokers or scrap yards but could possibly be end users as well. The supplier is, in this case, seldom the customer.

Types of core suppliers

This section describes the different core supplier types that have been identified within this research. It is a somewhat simplified picture given since a classification is necessary in order to get a uniform view. It is noteworthy to observe that the reverse supply chains depicted are sources of used automotive components (cores), not virgin spare parts, etc. In a sense, end users normally supply all cores, but in this report, the categorisation and the type of supplier derive from the supplier closest down in the supply chain. The identified core suppliers are end users, scrap yards, core brokers, OEMs and independent aftermarket distributors (IAMDs).

Reverse supply chains in the CAN REMAN project

This paper presents a qualitative study of six SMEs, of which their primary, if not solely, business segments are the remanufacturing of, or trading with, automotive mechatronic devices. Four of the companies studied are German remanufacturers (companies A to D). The remaining two are Swedish (companies E and F), where the latter is a core broker and hence not conducting any remanufacturing but is still an important factor in the reverse supply chain. An overview of the participating companies can be seen in Table 1. In a parallel CAN REMAN research study on inter-organisational relationships, companies A to E are studied further with the same notation [20,21].

The following sections present reverse logistics challenges identified and verified during interviews within the CAN REMAN project. An overview of the challenges experienced by the interviewed companies can be seen in Table 2. An empty space in the table means that the company does not consider the specific subject as a challenge or that it has chosen not to answer.

Controlling quantity of cores

Company A is sometimes forced to buy larger quantities than needed by its customers since core suppliers require minimum order quantities. Strong competition among core acquisition for new article numbers forces company B to buy cores without having a concrete demand. Company D experiences difficulties acquiring a sufficient amount of cores for its fast-moving product variants. Although the lack of control regarding quality is considered a challenge, both Swedish companies (E and F) accept it as a part of their business.

Table 1 General characteristics of the participating companies

Company	A	B	C	D	E	F
Products	Automotive, marine, heavy-duty and industrial engines and parts	Alternators, starters and marine and heavy-duty engines	Engines, gearboxes and differential gears	Car engines and diesel injection parts	Brake calipers, DPFs, hydraulic servo pumps, water pumps, EGR valves	Wide range of automotive parts
Company size (employees)	Medium (100)	Small (24)	Medium (80)	Medium (80)	Medium (50 + 110)	Small (6)
Annual turnover (million €)	14	1.6	4	12 to 13	7	0.8
Number of variants	650 cylinder heads, hundreds of engines	2,500 internal variants	Hundreds of variants per product	200 variants of engines, uncountable variants of diesel injectors	2,800 variants of brake calipers, 3,000 in total	10,000 to 20,000
Core suppliers	End users, IAMDs, OEMs, scrap yards, core brokers	End users, IAMDs, OEMs, core brokers	IAMDs, OEMs, end users	OEMs, scrap yards, core brokers, end users	OEMs, core brokers, scrap yards, IAMDs	Scrap yards, core brokers, IAMDs
Relation to OEM	Independent, contracted	Independent, contracted	Independent, contracted	Contracted, independent	Contracted, independent	Independent

Table 2 Overview of reverse logistics challenges identified within CAN REMAN

	Controlling quantity of cores	Controlling quality of cores	Controlling timing of cores	Balancing supply and demand	Additional challenges	Expected CAN bus challenges
Company A	X					
Company B	X	X	X	X	X	X
Company C		X	X			
Company D	X				X	X
Company E	X	X	X	X	X	X
Company F	X	X	X	X	X	X

Controlling quality of cores

According to company E, the challenge of controlling the quality of the returned cores differs between types of products. It claims to have total control of the quality of the returned brake calipers. There are two reasons for that. Firstly, brake calipers have a simple design and are made out of a robust material but also contain few parts, which make units worn beyond their remanufacturability infrequent. Secondly, because of their design, it is easy to visually determine whether a unit is remanufacturable or not. Regarding servo pumps, the situation is different since one cannot determine a core's condition without disassembling it. Internal parts may be broken beyond remanufacturing, and it is not feasible to have every single core opened before they arrive at the facility.

Another quality-related challenge concerns the handling of the cores. Company E estimates that 60% to 70% of the initially sound diesel particulate filters (DPFs) from one of its core suppliers are being damaged when dismounted from the cars or when transported to company E's remanufacturing facility. The damages are assumed to be caused by the mechanics at the OEM dealers as well as by the companies responsible for the transportation of the cores - they do not recognize the value of the cores and thereby handle them in an incautious way. This is especially critical for DPFs since reckless handling and transportation can lead to damage not visible to the naked eye, hence first noticed in the remanufacturing process. The effects are increasing scrap rates and that cores originally suited for remanufacturing can no longer be remanufactured. This issue is also being referred to as a challenge by company B.

Company C uses check sheets that private customers and car dealerships/repair shops must fill out in advance to avoid receiving non-remanufacturable cores and cores too costly to remanufacture. This improves quality control, but does not eliminate the need for quality inspections at arrival.

Company F considers the lack of control of the cores' quality as a challenge, but in contrast to the remanufacturers, 90% of the cores are visually inspected before they are bought, making the degree of control higher in this case.

Controlling timing of cores

The lack of control regarding the timing of the returned cores is considered a challenge by companies E and F, but only if there is a lack of the specific core in the market. Company B controls the timing of deliveries for all supplier types except for private customers, who are harder to control. Company C lacks control of the core deliveries from OEMs, but the use of check sheets (see previous section) for private customers and car dealerships/repair shops facilitates influence over the timing of the incoming cores from those suppliers in a positive way.

Balancing supply and demand

Companies E and F experience a lack of control regarding balancing supply and demand, but their situation differs from each other. Company E usually gets a 12-month time horizon on the estimated demand from the contracted OEMs, which are core suppliers/customers that company E is contracted to perform remanufacturing for. From these forecasts, additional cores are bought if needed from other suppliers, e.g. scrap yards and core brokers. Statistics from previous years are also kept, thereby helping to distinguish trends in demand. The prerequisites are similar for company B. Company F has much shorter time horizons, usually only a week or a month ahead.

A different aspect on the issue of balancing supply and demand is how company F buys certain cores without having a concrete demand but believes it will rise in the future. The motive is twofold; cores that are bought prior to the actual demand are cheap, hence displaying large profit margins if later sold. In addition, buying pre-demand cores prevents or diminishes competition within core acquisition. However, the speculating comes with a price tag. Apart from the obvious cost associated with storing cores until demand arises, there is an imminent risk that the anticipated demand never occurs, thus making the acquired cores less worth. This situation especially concerns CAN bus mechatronic devices that have yet to be remanufactured.

Additional challenges within the collection phase

A challenge identified by company E concerns the identification and sorting of cores at its suppliers, which, in this case, are OEM retailers. The deliveries often contain unwanted parts and mechatronic devices (e.g. turbochargers and dashboards). This is believed to be caused by a lack of routines at the retailer where the personnel, even though having specified which parts were to be sent for remanufacturing, do not put the right parts in the right core bins. The effect on company E is that warehouse space is allocated, both to store the unsorted goods as well as to store the unwanted goods before being transported away for scrapping. It is also time-consuming to perform the sorting. However, this identification issue is not considered as severe and can sometimes even provide useful information about new part numbers currently not in the remanufacturing program.

A challenge acknowledged by company F is the decreasing number of scrap yards, which are core suppliers to company F. One plausible explanation for the decrease is that insurance companies certify fewer scrap yards now than before. The consequence for company F is twofold. Firstly, fewer suppliers result in less competition among those remaining and hence higher core prices. Secondly, but not less important, the fewer larger suppliers remaining tend to focus their business on scrapping cars rather than dealing with cores.

The second challenge affecting company F is the recently increasing scrap prices, which follow the prices of raw materials, which, in this case, are metals. When scrap prices are high, scrap yards would rather sell the dismantled cars as scrap than sell the individual mechatronic devices to core brokers or remanufacturers.

A further challenge for company F is the competition from low-labour cost countries. When there are brand-new spare parts available at a cost close to, or even cheaper, than a remanufactured part, the demand for remanufactured parts decreases. This is also acknowledged by companies B and E.

In addition, company F has experienced actions from OEMs where they have tried to hinder competition from core brokers and remanufacturers, either by dumping prices of new parts or by clean-sweeping the market from cores. Similar actions have been taken against company D, where OEMs refuse to sell spare parts needed to perform remanufacturing of their products. Company B experiences that OEMs delay technical information (e.g. test parameters) about the products on purpose, which, due to the effort, put into reverse engineering results in higher remanufacturing costs, which is further elaborated on in [8].

Expected reverse logistics challenges of CAN bus mechatronic devices

A concern for companies B, D, E and F is the handling of cores containing CAN bus mechatronic devices.

Three perspectives on this matter have been brought up during the interviews. These are the following:

- *Disassembly*: it is important that OEM retailers and scrap yards have routines for how to remove parts without damaging them.
- *Storage*: cores containing electronics are sensitive to moisture; hence, it is important that CAN bus cores are stored in dry and preferably warm environments.
- *Transportation*: there is a concern that cores not being handled and stored properly during transportation will be damaged on their way to the remanufacturer or core broker.

A challenge company F will be facing, and is currently facing to a certain extent, is the large gap between the sales price of the remanufactured product and the cost of the core. For example, a remanufacturer buys a used ECU for $3 to $5 from the core broker which is then sold remanufactured for up to $500. For a core broker to sell a core for $3 to $5, it has to be bought for $1 to make profit, which is scrap price. This is claimed to be caused by the lack of competition among remanufacturers of automotive electronics.

During the interview with company F, it was mentioned that dealing with electronic mechatronic devices will further complicate the supply versus demand challenge since an entirely mechanical device will surely fail during a car's life, while the fail pattern of an electronic device is much more stochastic - it might even last the vehicle's entire lifetime.

In addition to the findings of this paper, another paper by Freiberger et al. [8] outlines challenges, possible solutions and technological progress for the reverse engineering process of CAN bus mechatronic devices. That paper includes the reverse engineering process demonstrated on an electro-hydraulic power steering, which is a CAN bus mechatronic device used in a Volkswagen Polo [8].

Discussions

The business model that the remanufacturing company uses might have an effect on the remanufacturing system and especially the core acquisition. For example, Sundin and Bras [22] stated that functional sale reduces the uncertainty regarding returns by giving the remanufacturing company better knowledge of the timing and quantity of the return. This is also acknowledged by Thierry et al. [10] who stated that quantity and timing of the returns are easy to predict at the end of a leasing or rental contract even though the quality still can be uncertain. It is easier to have rental programs for OEMs to perform remanufacturing than for those independent remanufacturers included in this study. However, the independent remanufacturers could move towards being contracted by the OEMs in order to benefit from a functional sales (e.g. rental) business model.

How close collaboration the remanufacturing company has with the retailers and distributors also affects the possibility for a remanufacturing company to coordinate the collection of returned products [18,19]. Different relationships and the effect they have on the situation for the remanufacturing company were deeply discussed by Östlin et al. [18] and Lind et al. [20,21]. A type of relationship that also was discussed by Guide [11] and Östlin et al. [16] is when the remanufacturing company remanufactures used products (cores) that the customer sends to them. Then the challenge with balancing supply and demand does not exist, though this might add additional challenges to the production planning.

Conclusions

This paper addresses reverse logistics challenges experienced by six SMEs in the automotive remanufacturing industry. Indeed, these companies, five remanufacturers and one core broker, face challenges traditional manufacturing companies do not have to deal with. The challenges previously identified by researchers, e.g. Lundmark et al. [13], which are faced by remanufacturers in general, were also confirmed as challenges by the interviewed companies remanufacturing automotive mechatronic devices. For instance, a reflection of the uncertainties regarding the demand and the difficulties in securing core supply was given by company F, which keeps a stock of cores despite not yet having a concrete demand for remanufacturing.

Additional challenges varying in significance and frequency were identified during the interviews. One concerns the reckless handling of the cores, both during disassembly and during transport to the remanufacturer. A possible solution that seems simple to implement, at least theoretically, is to inform mechanics as well as the responsible logistics company how to handle and package cores. This issue was also discussed during the interviews, now concerning future challenges when remanufacturing CAN bus mechatronic devices. The majority of the interviewed companies believe that careful dismantling, storage and transportation of cores containing electronics will be of high importance.

This paper, along with a previous literature review by Lundmark et al. [13] and the work on inter-organisational relationships by Lind et al. [20,21], builds a foundation for the future work to design, verify and implement reverse supply chains that diminish the impact of the challenges identified in this paper and hence better suit CAN bus mechatronic device remanufacturers. In addition, remanufacturing needs to deal with more process-oriented challenges such as testing and diagnosing as described for the CAN REMAN project in Freiberger et al. [8] in order to boost and facilitate CAN bus mechatronic device remanufacturing even more.

Methods

The empirical data have been collected through semi-structured [23], face-to-face interviews ranging between 2 and 3 h and were conducted by native-speaking interviewees from both Germany and Sweden. The interviews were held at the companies' sites in late 2009 and late 2010.

Abbreviations

CAN: Controller Area Network; DPF: diesel particulate filter; ECU: electronic control unit; OEM: original equipment manufacturer; SME: small and medium-sized enterprise.

Competing interests

The authors declare that they have no competing interests.

Authors' contributions

Both authors have collaborated in writing the manuscripts of this paper. ES planned and designed the interviews together with the Ph.D. students who conducted the interviews. ES conducted the pilot interview. OD summarized the Ph.D. students' interview data and made the first draft of the manuscript. Both authors read and approved the final manuscript.

Authors' information

ES (M.Sc., Ph.D.) is an associate professor since 2008 at the Division of Manufacturing Engineering, Department of Management and Engineering, Linköping University, Sweden. His main research interests are remanufacturing, design for remanufacturing, product/service systems and EcoDesign. His Ph.D. thesis is entitled 'Product and process design for successful remanufacturing'. OD (M.Sc.) obtained his Master of Science degree in Mechanical Engineering from 2009. He currently holds a position as a service method engineer at Scania CV AB, Södertälje, Sweden, since 2011. His M.Sc. thesis is entitled 'Verification of hybrid operation points'. He had presented a previous version of this paper at the First International Conference on Remanufacturing in Glasgow 2011.

Acknowledgements

The authors would like to thank the participating companies in this research, the Swedish Governmental Agency for Innovation Systems (VINNOVA) as the main funder of this research, and the Ph.D. students, Kirsten Bohr and Peter Lundmark, for their empirical contribution to this paper.

References

1. Steinhilper, R: Remanufacturing: The Ultimate Form of Recycling. Fraunhofer IRB, Stuttgart (1998)
2. Seitz, MA, Peattie, K: Meeting the closed-loop challenge: the case of remanufacturing. Calif. Manag. Rev. 46(2), 74–89 (2004)
3. Giutini, R, Gaudette, K: Remanufacturing: the next great opportunity for boosting US productivity. Business Horizon 46(6), 41–48 (2003)
4. Sundin, E, Lee, HM: In what way is remanufacturing good for the environment? In: Proceedings of the 7th International Symposium on Environmentally Conscious Design and Inverse Manufacturing (EcoDesign), Kyoto, 30 Nov–2 Dec 2011, pp. 551–556
5. Lund, RT, Hauser, WM: Remanufacturing - an American perspective. Int. Conf. Responsive Manufac. 5, 1–6 (2010)
6. Department of Commerce, Office of Transportation and Machinery: On the Road: U.S. Automotive Parts Industry Annual Assessment. Department of Commerce, Washington, D.C (2010)
7. Davis, R, Burns, A, Bril, R, Lukkien, J: Controller Area Network (CAN) schedulability analysis: refuted, revisited and revised. Real-Time Syst. 35(3), 239–272 (2007)

8. Freiberger, S, Albrecht, M, Käufl, J: Reverse engineering technologies for remanufacturing of automotive systems communicating via CAN bus. J. Remanufac. **1**, 6 (2011)
9. Östlin, J: On remanufacturing systems - analysing and managing material flows and process organisation. PhD thesis. Linköping University, Department of Management and Engineering, Linköping (2008)
10. Thierry, M, Salomon, M, van Nunen, J, van Wassenhove, L: Strategic issues in product recovery management. Calif. Manag. Rev. **37**(2), 114–135 (1995)
11. Guide, VDR: Production planning and control for remanufacturing: industry practice and research needs. J. Oper. Manag. **18**(4), 467–483 (2000)
12. Fleischmann, M, Bloemhof-Ruwaard, JM, Dekker, R, van der Laan, E, van Nunen, JAEE, Van Wassenhove, LN: Quantitative models for reverse logistics: a review. Eur. J. Oper. Res. **103**(1), 1–17 (1997)
13. Lundmark, P, Sundin, E, Björkman, M: Industrial challenges within the remanufacturing system. In: Proceedings of Swedish Production Symposium, Stockholm, 2–3 December 2009, pp. 132–138
14. Fleischmann, M, Krikke, HR, Dekker, R, Flapper, SDP: A characterisation of logistics networks for product recovery. Omega **28**(6), 653–666 (2000)
15. Sundin, E: How can remanufacturing processes become leaner? In: CIRP Intl Conference on Life Cycle Engineering, Leuven, 31 May–2 June 2006
16. Östlin, J, Sundin, E, Björkman, M: Product life-cycle implications for remanufacturing strategies. J. Clean. Prod. **17**(11), 999–1009 (2009)
17. Guide Jr, VDR, Van Wassenhove, LN: The evolution of closed-loop supply chain research. Oper. Res. **57**(1), 10–18 (2009)
18. Östlin, J, Sundin, E, Björkman, M: Importance of closed-loop supply chain relationships for product remanufacturing. Int. J. Prod. Econ. **115**(2), 336–348 (2008)
19. Guide Jr, VDR, Van Wassenhove, LN: The reverse supply chain. Harvard Business Rev. **80**(2), 25–26 (2002)
20. Lind, S, Olsson, D, Sundin, E: Exploring inter-organizational relationships within the remanufacturing of automotive components. In: International Conference on Remanufacturing, Glasgow, 26–29 July 2011
21. Lind, S, Olsson, D, Sundin, E: Exploring inter-organisational relationships within the remanufacturing of automotive parts. J. Remanufac., (2013, in press)
22. Sundin, E, Bras, B: Making functional sales environmentally and economically beneficial through product remanufacturing. J. Clean. Prod. **13**(9), 913–925 (2005)
23. Jacobsen, JK: Intervju - konsten att lyssna och fråga. Studentlitteratur, Lund (1993)

11

A cost model for optimizing the take back phase of used product recovery

Niloufar Ghoreishi[1]*, Mark J Jakiela[1] and Ali Nekouzadeh[2]

Abstract

Taking back the end-of-life products from customers can be made profitable by optimizing the combination of advertising, financial benefits for the customer, and ease of delivery (product transport). In this paper we present a detailed modeling framework developed for the cost benefit analysis of the take back process. This model includes many aspects that have not been modeled before, including financial incentives in the form of discounts, as well as transportation and advertisement costs. In this model customers are motivated to return their used products with financial incentives in the forms of cash and discounts for the purchase of new products. Cost and revenue allocation between take back and new product sale is discussed and modeled. The frequency, method and cost of advertisement are also addressed. The convenience of transportation method and the transportation costs are included in the model as well. The effects of the type and amount of financial incentives, frequency and method of advertisement, and method of transportation on the product return rate and the net profit of take back were formulated and studied. The application of the model for determining the optimum strategies (operational levels) and predicting the maximum net profit of the take back process was demonstrated through a practical, but hypothetical, example.

Keywords: Take Back, Product Acquisition, Remanufacturing, Modeling, Cost Benefit Analysis

Introduction

Taking back used products is the first step in most of the end of life (E.O.L) recovery options which include remanufacturing, refurbishment, reuse, and recycling. "Take back" includes all the activities involved in transferring the used product from the customers' possession to the recovery site. In general optimizing of the take back (also called product acquisition) has received limited attention in research and operations. Guide and Van Wassenhove categorized take back processes into two groups: waste stream and market driven [1]. In a waste stream process, the collecting firm cannot control the quality and quantity of the used products: all the E. O.L. products will be collected and transferred. In a market driven process, customers are motivated to return the end of life product by some type of financial incentive. This way, the (re)manufacturer can control the quantity and quality of the returned products

through the amount and type of incentives and increase its profit [2-4].

In general the taking-back firm can control the process by setting strategies regarding financial incentives, advertisement, and collection/transportation methods [2,3,5-8]. Usually, offering higher incentives (in the form of cash or discounts toward purchasing new products) will increase the return rate and lead to acquisition of higher quality used products. Higher incentives sometimes can encourage the customers to replace their old products with a new one earlier [9]. Another way to control the quality of the used product is to have a system for grading the returned products based on their condition and age and paying the financial incentives accordingly [4]. Proper advertisement and providing a convenient method for the customers to return the E.O.L product can increase the return rate as well [9].

In the existing models of the take back process all the involved costs are bundled together as the take back cost and the return rate is modeled as a linear function of the take back cost [9] or as a linear function (with a threshold) of the financial incentive [4]. We developed a

---
* Correspondence: ng1@seas.wustl.edu
[1]Mechanical Engineering and Materials Science Department, Washington University in St. Louis, 1 Brooking Dr., St. Louis Missouri 63130, USA
Full list of author information is available at the end of the article

market driven model of a take back process by considering different aspects of take back including financial incentives, transportation methods, and advertisement separately to provide more theoretical insights about the process. Three different types of financial incentives (cash, fixed value, and percentage discount) were modeled. This includes considering the effect of discount incentives on the sale of new (or remanufactured) products and allocating the relevant costs and revenues among the take back process and the sale process of the new products. The relation between the incentives and return rate is considered as a market property reflecting consumers' willingness to return products. This should be measured or estimated. The model enables operational level decisions over a broader choice of variables and options compared to existing approaches. A practical example is used to show how this modeling framework can determine the optimum options and values of the take back process and provide significant insights for analyzing and also managing the take back process.

Model

We consider three important aspects of take back in our model: the financial incentives, the transportation and the advertisement. Each of these aspects incurs a cost to the process, and in return, can increase the revenue by increasing the number and average quality of returned products. Some of the take back costs are associated with each individual product and so are scaled with the number of returned products and some are fixed costs associated with the whole take back process. The value of a returned product at the recovery site is termed a. a is the price that the recovery firm is willing to pay for the used product at the site. If the take back is performed by the recovery firm then a would be a transfer price [10,11] which separates the cost benefit analysis of the take back from the rest of the recovery process. We modeled the net profit of take back during a certain period of time. If the take back process is intended for a period of time, this period could be the entire time of the take back process, and if it is intended to be a long lasting process, this period is a time window large enough to average out the stochastic fluctuations in the return rate.

Financial incentives

Three strategies were considered for motivating the customers to return their used products:

1- Paying a cash value $c.

2- Offering a discount of value d, for purchasing new products (usually of similar type).

3- Offering a percentage discount of %p, for purchasing new products.

These incentives affect the total cost, the number of return, and the average quality of the returned products. Increasing these incentives may increases the net profit by increasing the number of returned products and their average quality, or may decrease the net profit by increasing the cost of take back. Therefore, it is an optimization problem to find the type and amount of incentive to maximize the net profit. It is reasonable to expect the number of returns, N_R, varies by the amount of incentives and also varies differently for different types of incentives:

$$N_R = N_{Rc}(c) = N_{Rd}(d) = N_{Rp}(p) \qquad (1)$$

However, we may assume that N_R is a function of a more general variable called motivation effectiveness, which is considered as the amount of motivation induced in the customers by a motivation strategy. The magnitude of motivation effectiveness, mte, is defined as the equivalent amount of cash that generates the same level of motivation in the customers to return the used product. Therefore, we may simply write:

$$N_R = N_R(mte) \qquad (2)$$

Different customers respond differently to the same amount of mte. A customer returns the used product if the motivation effectiveness of the incentive (mte) is higher than his or her threshold motivation effectiveness for returning the used product. Therefore, $N_R(mte)$ represents the number of customers that their threshold motivation effectiveness is less than mte (the cumulative density function for the threshold motivation effectiveness among the customers).

The attractiveness of the discount is less than or equal to the same amount of cash, because the discount can be used only to buy specific products [12-16]. We define c_d as the cash equivalent of discount d; the number of customers that return the used product with discount incentive d is equal to the number of customers that return the used product with cash incentive c_d. Then we define α, the ratio of cash to discount incentive, via:

$$c_d = d\alpha(d) \qquad (3)$$

The value of α depends on the new products that the discount is applicable to and varies between 0 and 1. Generally, if customer X has a higher cash incentive threshold than customer Y to return the used product, he has most likely a higher discount incentive threshold as well. Therefore, it is reasonable to assume a linear regression between the d and c_d and replace α (d) by its average value simply termed α. Therefore mte for three different motivation strategies is modeled by:

$$mte = c, \quad mte = \alpha d, \quad \text{or} \quad mte = \alpha Ap \qquad (4)$$

where A is the average price of the new products to which the discount can be applied.

Transportation

Once a customer is motivated to return the used product, the product must be transported to the recovery site. Gathering the used product from the customers can be very costly. In many situations, it may be possible to reduce the transportation cost by asking the customer to contribute partially or fully to the transportation of their products. This usually comes at the cost of reducing the motivation effectiveness of the financial incentives because it requires the customers to spend time and energy to return the used product. Therefore, the motivation effectiveness depends on the convenience of the transportation in addition to the financial incentives. To quantify the convenience of the transportation, we introduce the parameter f, termed the convenience factor of transportation method. In general mte is assumed as a function of f in our modeling framework:

$$mte = mte(f, c), \quad mte = mte(f, \alpha d), \quad \text{or } mte = mte(f, \alpha A p) \quad (5)$$

Transportation imposes a cost termed TC to the take back process. Transportation cost is a function of the number of returns. A linear relation [17] between the transportation cost and the number of returns is the simplest method for modeling this cost [18]:

$$TC = N_R t + tg \quad (6)$$

where t is the transportation cost per returned item (slope of the variable cost) and tg is the fixed cost of transportation (does not scale with the number of returns).

Advertisement

Advertisement includes any action for informing the customers about the take back policy. Optimum advertisement strategy depends on many social and psychological factors which are beyond the scope of this paper. Here, we only determine the aspects of advertisement that are important for cost benefit analysis of the take back procedure. Advertisement cost is categorized into two groups: W_1, the one-time cost of advertisement associated with preparing and designing the ad., including its content and its presentation (e.g. posters, audio clips or video clips), and W_2, cost of running the ad. (e.g. posting, publishing, distributing or broadcasting). We may refer to W_2 as the advertisement expenditure.

Among all the customers that possess the used product, only the ones that are aware of the take back procedure may return the used product (if they are

motivated enough). Therefore, we may rewrite the number of returns as:

$$N_R(mte, W_2) = N\Omega(W_2)\Gamma(mte) \quad (7)$$

Where N is the total number of customers holding the used product, Ω is the fraction of total customers that are informed by the advertisement and Γ is the fraction of informed customers that return the used product in response to motivation effectiveness of the take back procedure. Ω depends on the frequency of running the advertisement and therefore, is a function of W_2. Equation (7) implicitly assumes that the demography of the informed customers and consequently how they respond to the motivation effectiveness is independent of the number of informed customers. The following expression was derived as an estimate for the Ω function (see Appendix):

$$\Omega(W_2) = \Omega_{ss}(1 - e^{\frac{W_2}{W_{sc}}}) \quad (8)$$

W_{sc} and Ω_{ss} are characteristic parameters of advertisement method; they are different for different advertisement options. The Ω function presented in equation (8) is derived analytically for a general advertisement method. More accurate functions may be derived by fitting the empirical data (if available) for each specific advertisement method. Other advertisement models like Vidale-Wolfe model [19], Lanchester model [20], or empirical models [21] may be used as well.

Advertisement, if designed accordingly, can have a motivating effect by informing the customers about the environmental and global benefits of their product return effort including reducing waste and reducing the consumption of energy and natural recourses. To quantify the motivation effect of advertisement, we introduce the parameter g. Therefore, mte can be written in general as a function of financial incentive, the convenience factor of transportation and the motivation effect of advertisement.

$$mte = mte(f, c, g), \quad mte = mte(f, \alpha d, g), \quad \text{or } mte = mte(f, \alpha A p, g) \quad (9)$$

A suggested model for motivation effectiveness

mte should be determined for all the possible combinations of the financial incentive, the convenience factor of transportation and the motivation effect of advertisement, for the three financial incentive strategy. However, this requires extensive amount of data points and makes the calibration procedure very expensive and even impractical. In this section we rationalize a simple model for mte without further empirical validation. Alternative models may be used based on empirical data.

In equation (4) we modeled the motivation effect of the three financial incentives by estimating the cash equivalent of a discount incentive. In order to quantify the convenience of the transportation, we should first determine its effect on the motivation effectiveness. If a customer participates partially in transporting the used product, he or she has to spend some time and energy which reduces the effective value of the financial incentive. Defining mte_t as the reduction in motivation effectiveness associated with the transportation method we may write:

$$mte = c - mte_t, \; mte = \alpha d - mte_t, \; \text{or } mte = \alpha Ap - mte_t \quad (10)$$

The energy and time that a customer has to spend on transportation is almost the same for different customers, but different customers value their time and energy differently. Usually the customers that return their used product at higher financial incentives are busier or less interested in returning their product and so are more sensitive to the convenience of transportation. Therefore a correlation between mte_t and mte is expected. Assuming a linear relation between mte_t and mte:

$$mte_t = \beta c, \; mte_t = \beta \alpha d, \; \text{or } mte_t = \beta \alpha Ap \quad (11)$$

we may rewrite equation (10) as:

$$mte = (1 - \beta)c, \; mte = (1 - \beta)\alpha d, \; \text{or } mte = (1 - \beta)\alpha Ap \quad (12)$$

where β represents the inconvenience of transportation and varies between 0 and 1; it is zero if the take back firm undergoes all the transportation activities. The convenience factor of transportation, f, may be quatified as:

$$f = (1 - \beta) \quad (13)$$

And consequently the equation (12) can be rewritten as:

$$mte = fc, \; mte = f\alpha d, \; \text{or } mte = f\alpha Ap \quad (14)$$

In contrast, there is no reason to believe a significant correlation between the motivation effect of the advertisement and the motivation effect or the type of the financial incentive. Therefore, we may assume that g represents the average increase in the motivation effectiveness associated with the advertisement. Therefore, equation (14) can be rewritten as:

$$mte = fc + g, \; mte = f\alpha d + g, \; \text{or } mte = f\alpha Ap + g \quad (15)$$

In general g depends on the quality of the ad and providing a more effective ad usually costs more. Therefore, the motivation effect of advertisement may be considered as a function of W_1:

$$g = g(W_1) \quad (16)$$

Cost model

In the discount incentive strategies the cost benefit analysis of take back and the sale of new products are coupled together. Therefore, the cost model of the cash incentive strategy differs substantially from the cost model of discount incentive strategies. In the following, different cost models were derived for different incentive strategies.

Cash incentive strategy

The cost that is scaled with the number of returns (cost per returned item) consists of the amount of cash incentive, c, and the transportation cost, t. The revenue which is generated by the value of returned product, a, also scales with the number of returns. Advertisement costs, W_1 and W_2 and the fixed cost of transportation, tg, do not scale with the number of returns. Therefore, the net profit of take back, Ψ_c, can be modeled as:

$$\psi_c = N_R.[a - c - t] - W_1 - W_2 - tg - tb \quad (17)$$

Where tb is the implementation cost of take back, modeled as a fixed cost. A variable term may be considered for the implementation cost as well; for example larger number of returns usually corresponds to larger capacity of the take back process and consequently higher implementation cost. In this model a is the average value of taken back products. Taken back products are expected to have better quality (in average) at higher incentives [4]. To include this effect, we considered a as a function of mte in the model. Note that the decision of customers for returning their used product depends on the all the incentives which are included in the motivation effectiveness, mte. Substituting for number of returns from equation (7) and for mte from equation (15) the net profit in a cash incentive strategy is:

$$\psi_c = N.\Gamma(fc + g).\Omega(W_2).[a(fc + g) - t - c] - W_1 - W_2 - tg - tb \quad (18)$$

Discount incentive strategies

If the take back is performed by the OEM (Original Equipment Manufacturer) firm, the financial incentives may be offered in the form of discount (fixed value of percentage) toward buying a new product. The discount incentive reduces the net profit of the new products by selling a fraction of them at the discounted price. On the other hand, the discounted price makes the product affordable for some additional customers and may increase the net profit by increasing the number of sales or redistributing the sale profile toward more profitable products. As both changes in the net profit of new products are caused by the take back procedure, the reduction of profit, associated with reduced price, is considered as a take back cost and the extra revenue associated with the increased amount of sales is considered as take back revenue. To model the effect of discount coupons on the sale profile of new products we first categorize the customers who would return their used product into the following groups:

1- Current customers who planned to buy a certain product (with or without the discount). These customers simply use the coupon to pay less for the new product they would have bought anyway.

2- New customers who have been motivated by the discount incentive to return their used product and buy a new product at discounted price. Their choice of new product may or may not depend on the amount of discount incentive.

3- Customers who returned their used product but for any reason do not buy any new product to redeem their coupon.

Customers of group 1 are the less favorable customers for the take back procedure and do not bring any extra revenue to the company as a consequence of the take back strategy. Customers of group 2 are new customers that are motivated by the discount and so any generated revenue associated with their purchase can be attributed to the take back procedure. Finally customers of group 3 do not impose any motivation cost on the take back procedure.

The motivation cost, MC, in this method can be assumed as the total value of redeemed coupons minus the extra generated revenue in the sale of new products caused by discount motivation:

$$MC = \sum_{j=1}^{M} m_j d - \sum_{j=1}^{M} n_j s_j \qquad (19)$$

where M is the total number of discountable products referred by index j; s_j is the sale profit of new product j; n_j is the change in number of sale of the new product j, caused by discount incentive; m_j is the number of discount coupons used for the new product j. Including the motivation cost, the net profit of discount incentive strategy is:

$$\psi_d = N.\Gamma(mte).\Omega(W_2).[a(mte) - t]$$
$$-d\sum_{j=1}^{M} m_j + \sum_{j=1}^{M} n_j s_j - W_1 - W_2 - tg - tb \qquad (20)$$

The customers' decision regarding returning the used product depends on the motivation effectiveness, but, once the customers returned the product their decisions for choosing the new product depend only on the

amount of discount. We define η_i as the proportion of the discount coupons that are used for the new product j. Therefore:

$$m_j = N_R \eta_j = N\Gamma(mte)\Omega(W_2)\eta_j(d) \qquad (21)$$

Assuming that η_o and m_o show the proportion and the number of coupons that are not used (customers of group 3), respectively:

$$\eta_0 + \sum_{j=1}^{M} \eta_j = 1$$
$$m_0 + \sum_{j=1}^{M} m_j = N_R \qquad (22)$$

Note that the number of issued coupons is the same as the number of returned products, NR. We also define ξ_j as the proportion of the sale of each new product without the take back procedure. Usually, the discount incentives of the take back procedure increase the sale of new product and we define Λ as the ratio of the new customers (estimated by the increased in the number of sale) to the total customers who buy a new product with coupon. Therefore, number of new customers (who buy a new product because of discount) is $(N_R-m_o)\Lambda$ and the number of customers that would have bought a new product without the discount is $(N_R-m_o)(1-\Lambda)$.

n_j and m_j are related to each other for each new product j. For each new product j, n_j is m_j minus the number of customers that would have bought a new product without discount. These customers were distributed proportional to ξ_j before discount incentive, so:

$$n_j = m_j - \xi_j(N_R - m_o)(1 - \Lambda) = N_R[\eta_j - \xi_j(1 - \eta_0)(1 - \Lambda)] \qquad (23)$$

Substituting equations (15), (21), (22) and (23) in equation (20), the net profit in discount incentive strategy can be rewritten as:

$$\psi_d = N.\Gamma(\alpha fd + g).\Omega(W_2).$$
$$\left(a(\alpha fd + g) - t - d(1 - \eta_o(d)) + \sum_{j=1}^{M} [\eta_j.(d) - \xi_j(1 - \eta_o(d))(1 - \Lambda)]s_j \right) \qquad (24)$$
$$-W_1 - W_2 - tg - tb$$

Therefore, to include the effect of discount in the net profit, we need to estimate Λ, the proportion of new customers and η_i, the distribution of discount coupons among the new products. These parameters are measurable once the take back procedure is implemented. However, in order to use the model for feasibility analysis of the take back procedure, accurate estimates of Λ and η_i is required. In equation (24) it is implicitly assumed that the number of new customers increases proportionally by the number of returns, and consequently the fraction of new customers is modeled with a constant number. For a more accurate model, Λ may be considered as a function of *mte*. However, this accuracy comes at the cost of more complex model calibration.

Comparing equation (24) with equation (17) helps to understand how changing the financial incentive from cash to discount affects the net profit of the take back. First the cash incentive cost, c, is replaced by the discount incentive cost. The discount incentive, d, is reduced by a constant factor to account for the unused coupons. As discussed before, changing the incentive from cash to discount decreases the profit by reducing the motivation of customers to return the used product and increases the net profit by increasing the sale of new products. Scaling down the discount incentive by parameter α is how the first effect appeared in the cost model. It reduces the number of returns and consequently the net profit of take back. The second effect appeared as a summation term in the right side of equation (24). The term inside the square brackets is difference between the sale (for each new product) of new products with and without the coupon. The number of sale without the coupon is the number of customers that would have purchased the product without the coupon, $(1-\Lambda)$, distributed among the new products.

The net profit of take back for the percentage discount strategy, ψ_p, can be derived using a similar approach as for the fixed discount strategy. With a percentage discount, the amount of discount is not fixed and depends on the sale price of new products. The motivation cost, *MC*, is:

$$MC = \sum_{j=1}^{M} m_j v_j p - \sum_{j=1}^{M} n_j s_j \qquad (25)$$

where v_j is the sale price of new product j and p is the percentage of discount. Therefore, the net profit of take back with a percentage discount is:

$$\psi_p = N.\Gamma(mte).\Omega(W_2).[a(mte) - t]$$
$$-p \sum_{j=1}^{M} m_j v_j + \sum_{j=1}^{M} n_j s_j - W_1 - W_2 - tg - tb \qquad (26)$$

Similar to a fixed value discount, m_j can be modeled as:

$$m_j = N_R \eta_j = N\Gamma(mte)\Omega(W_2)\eta_j(p) \qquad (27)$$

The average price of discountable products, A, can be determined as:

$$A = \frac{\sum_{j=1}^{M} m_j v_j}{\sum_{j=1}^{M} m_j} = \frac{\sum_{j=1}^{M} \eta_j(p) v_j}{\sum_{j=1}^{M} \eta_j(p)} \qquad (28)$$

We used A previously to estimate the motivation effectiveness of a percentage discount. In the percentage discount strategy, buying more expensive products is more motivated compared to the fixed value discount strategy as the amount of discount increases by the price of product. Therefore, the η_j functions and Λ are different from the fixed value discount and need to be estimated or measured separately. The relationship between m_j and n_j is the same as in the fixed value discount strategy. The net profit of a percentage discount strategy can be rewritten using equations (23) and (28) as:

$$\psi_p = N.\Gamma(\alpha fAp + g).\Omega(W_2).$$
$$\left(a(\alpha fAp + g) - t - Ap(1 - \eta_o(p)) + \sum_{j=1}^{M} [\eta_j(p) - \xi_j(1 - \eta_o(p))(1 - \Lambda)]s_j \right) \qquad (29)$$
$$-W_1 - W_2 - tg - tb$$

Note that in general A is a function of p. A list of all model variables is provided in Table 1. This list also includes intermediate variables that do not appear in the final equations of the net profit.

Results

The model developed in previous sections provides a general framework to optimize the take back procedure by determining the type and amount of financial incentives, optimum options of transportation and advertisement, and the optimum spending on advertisement. In this section we present a hypothetical real world take back problem that is characterized in this general framework. The model will be used to estimate the net profit of the take back and determine optimum values and choices of parameters.

Take back problem and its characteristic parameters

Cellular phones are among the products considered suitable for multiple life cycles [22]. Our goal is to outline a take back procedure for collecting a particular type of used hand set from the market for a recovery firm. The optimum recovery option and marketing the recovered product (or material) is out of the scope of this problem. In the following we explain the parameters and options we considered. Although, the parameter values are hypothetical and are not measured for a specific case, they represent a set of possible options and values.

It is assumed that the recovery firm is willing to pay from \$30 to \$50 for each used handset at the recovery site based on the average condition. The average value of returned product, a, is modeled as:

$$a = \begin{cases} 30 + 1.5mte & mte < 20 \\ 50 & mte > 20 \end{cases} \qquad (30)$$

Three transportation options have been considered:
1- Pick up from the customers convenient location (residential or business location).

Table 1 Parameters of the model

a	Average value of returned product at the recovery site
c	Amount of cash incentive
d	Amount of discount incentive (fixed value discount)
p	Percentage of discount incentive
N_R	Number of returned products
mte	Motivation effectiveness
c_d	Cash equivalent of discount
α	Ratio of cash to discount incentive
A	average price of the new products to which the discount can be applied
f	Convenience factor of transportation
t	Transportation cost per returned product
tg	Fixed cost of transportation
W_1	Onetime cost of advertisement (Preparing the ad.)
W_2	Advertisement expenditure (e.g. posting, publishing, distributing, broadcasting)
N	Total number of customers holding the used product
Ω	Fraction of (total) customers that are informed about take back
Γ	Fraction of (informed) customers that return the used product
Ω_{ss}	Parameter of advertisement method
W_{sc}	Parameter of advertisement method
m_j	Number of coupons used for new product j.
m_o	Number of coupons that have never been used
N_{ad}	Number that are reached by advertisement
N_{ss}	Maximum that can be reached by advertisement
g	Motivation effectiveness of advertisement
mte_t	Reduction in motivation effectiveness caused by transportation method
β	Inconvenience of transportation
tb	Fixed cost of take back
M	Total number of discountable products
m_j	Number of discount coupons used for the new product j
n_j	Change in number of sale of the new product j
s_j	Sale profit of new product j
ξ_j	Proportion of the sale of new products without the take back procedure
η_j	Proportion of discounts used for new product j
Λ	Proportion of new customers due to discount
m_o	Number of the coupons that are not used
η_o	Proportion of the coupons that are not used
ψ_c	Profit of take back with cash incentive
ψ_d	Profit of take back with fixed value discount incentive
ψ_p	Profit of take back with percentage discount incentive
v_j	Sale price of new product j

2- Providing the customers with the postage paid envelopes.

3- Asking the customers to hand deliver their handsets at particular locations.

Table 2 Parameters of transportation options

Transportation Options	t	tg	f
Option 1: Pick Up	15	5000	1
Option 2: Postages Paid Mail	4	2000	0.85
Option 3: Collecting at Branches	2	500	0.6

The transportation costs, t and tg and the convenience factor, f, of each method is summarized in Table 2.

Five options have been considered for advertisement:

1- Broadcasting a video clip on a T.V. channel

2- Broadcasting a vocal clip on a radio channel

3- Internet advertisement

4- Advertising in local newspapers

5- Announcing (by LCD panels or posters) in related retail stores

Characteristic parameters of each method of advertisement are given in Table 3. The values of the advertisement parameters are roughly estimated based on the available data on costs (e.g. air time rates) and estimates of the number of people that will be impacted by the ad.

N, the total number of customers that posses the used handset is assumed to be 70,000 and the Γ function is modeled as:

$$\Gamma(mte) = \frac{mte^3 + 20}{1.2mte^3 + 10mte^2 + 1000} \tag{31}$$

This function is drawn in Figure 1. This estimate of the Γ function is based on the following assumptions: 1-with no financial incentive still a small fraction of customers (~2%) who are motivated by the overall environmental aspects of take back would return their hand sets. 2-incentives up to $4 would have no significant motivation effect and the return rate would start to increase for incentives of $5 or more. 3-return rate increases almost linearly in the beginning and then yields toward a saturation value. 4-$25 motivation effectiveness is a fair exchange value and about half of the customers would return their handsets at this price.

For discount strategies it is assumed that the customer can buy 3 new handsets (Table 4) with their discount. The η_j proportions are assumed to vary linearly (after an initial threshold, x_{ts}) with the amount of discount:

Table 3 Parameters of different advertisement options

	W_1	g	Ω_{ss}	W_{sc}
Option 1: TV ad.	8000	7	0.9	400000
Option 2: Radio ad.	1000	5	0.5	40000
Option 3. Internet ad.	400	5	0.35	30000
Option 4. Local Newspaper	500	3	0.3	8000
Option 5. Retail Store ad.	700	4	0.4	25000

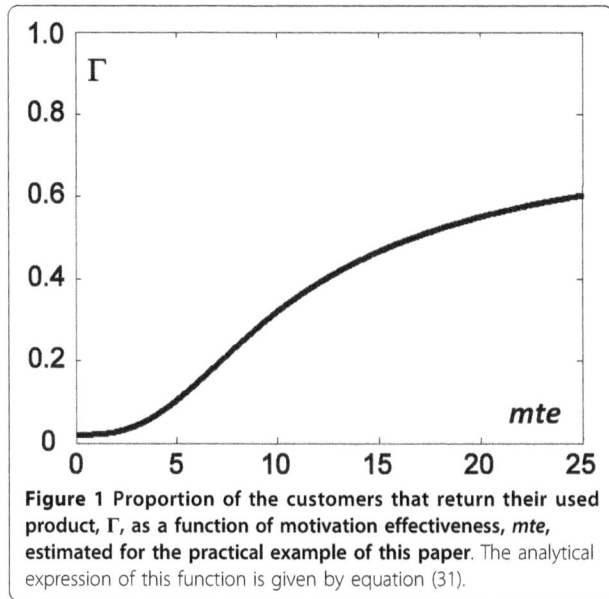

Figure 1 Proportion of the customers that return their used product, Γ, as a function of motivation effectiveness, mte, estimated for the practical example of this paper. The analytical expression of this function is given by equation (31).

Table 5 Parameters of η_j functions

	x_{ts}	λ_1	λ_2	λ_3	ρ_1	ρ_2	x_{sc}
d	5	-0.005	0.003	0.002	0.03	0.17	10
p	0.05	-0.4	0.1	0.3	0.02	0.18	0.2

$$\eta_j(x) = \begin{cases} \xi_j(1 - \eta_0(x)) & x < x_{ts} \\ \xi_j(1 - \eta_0(x)) + \lambda_j(x - x_{ts}) & x > x_{ts} \end{cases} \quad j = 1, 2, 3 \quad (32)$$

where x is the amount of discount (d or p). When the discount is small it does not affect the customers' decision for selecting the new product and the discounts are distributed among the new products proportional to their global sale distribution, ξ_j. The proportion of customers who have returned the used product without using their discount coupon is assumed to decline exponentially:

$$\eta_0 = \rho_1 + \rho_2 \exp(-x/x_{sc}) \quad (33)$$

Parameters of the η_j functions are provided in Table 5. Finally the fraction of new customers, Λ, is assumed to be 0.5 and the ratio of cash to discount incentive, α, is assumed to be 0.8.

Model prediction for the optimum strategy and net profit

Finding the optimum strategy in this problem involves determining the type of financial incentive (cash, fixed value or percentage discount), the amount of financial incentive, the optimum transportation method, the optimum advertisement method and the optimum volume of advertisement (W_2) to maximize the profit. The

Table 4 Specifications of new discountable products

New Handsets	v_j	s_j	ξ_j
HS1	90	30	0.3
HS2	110	35	0.45
HS3	150	55	0.25

advertisement cost, W_2, and the amount of incentives, x (c, d, or p), are continuous parameters. Therefore, for each combination of incentive strategy, transportation method, and advertisement method, we calculated the profit of take back, ψ, as a 2D function of x and W_2 and determined the maximum amount of net profit, ψ, and its associated W_2 and x. These maximum profits were compared to find the maximum net profit of the take back and its associated incentive strategy, transportation and advertisement methods.

Figure 2 shows the net profit of take back, ψ, and the number of returns, N_R, as a function of advertisement cost, W_2 and percentage of discount, p, for a percentage discount incentive, method 2 of advertisement (radio advertisement) and method 2 of transportation (postage paid mailing). Increasing the amount of advertisement (W_2) and percentage of discount incentive, initially increases the profit because of increasing the amount of returns, and after a maximum point, decreases the profit because of increased costs of motivation or advertisement. It has a maximum shown by the black circle over the 2D domain of its two variables. The number of returns increases monotonically (as expected) by increasing the amount of advertisement and incentive and approaches a maximum value. The net profit of take back of all 15 combinations of advertisement method and transportation method is shown in Figure 3 for cash, fixed value discount, and percentage discount incentives in panels A, B and C respectively. Quantitative comparison of these net profits concludes that a percentage discount incentive, method 2 of advertisement, and method 2 of transportation generates the maximum net profit of about $685,000 in a year (time duration of modeling) based on the estimated values we chose for the parameters of this problem. The maximum net profit of fixed value discount and percentage discount strategies are close to each other (panels B and C) which means that the type of discount does not have a significant effect on the net profit. The maximum net profit of cash incentive strategy is significantly lower than the discount strategies. This means that a significant portion of the profit in discount strategies is resulted from the sale of new products, particularly to the new customers. The maximum net profit in cash incentives is about $404,000 associated with method 2 of advertisement and method 2 of transportation. For each combination of incentive strategy, advertisement method, and transportation method, the maximum net

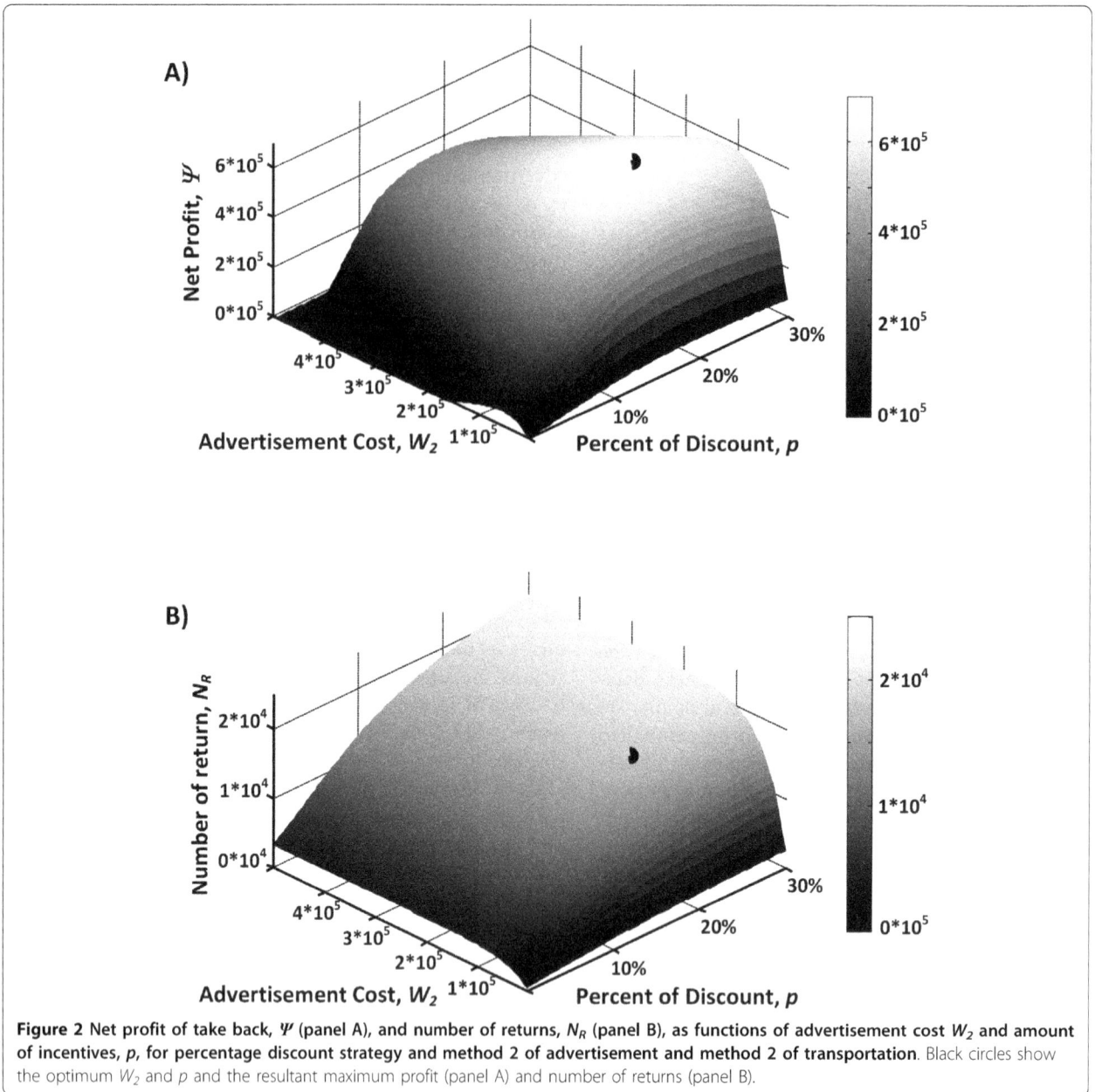

Figure 2 Net profit of take back, Ψ (panel A), and number of returns, N_R (panel B), as functions of advertisement cost W_2 and amount of incentives, p, for percentage discount strategy and method 2 of advertisement and method 2 of transportation. Black circles show the optimum W_2 and p and the resultant maximum profit (panel A) and number of returns (panel B).

profits resulted from an optimum advertisement cost and an optimum amount of incentives. Figure 4 shows the optimum W_2 and d, and the resultant number of returns N_R, for the fixed value discount strategy. Comparing these optimum values provides more insight on how different transportation and advertisement methods can maximize the profit. For example the optimum cost of TV advertisement (method 1) is much larger than other plans clearly because TV advertisement is more expensive. This method of advertisement, however, can generate a net profit more than many other advertisement plans. This extra cost is compensated partly by better motivation effect of an ad, which enables lowering

the financial incentives (Figure 4 panel A), and partly by increasing the number of returns (Figure 4 panel C), as it covers a broader number of customers. Also it is noticeable that the resultant optimum number of returns does not vary significantly in different transportation methods but varies significantly by advertisement methods. This means that if a transportation method is less convenient for customers the firm has to compensate for that by increasing the financial incentives (Figure 4 panel A) to increase the motivation effectiveness in order to reach a certain number of returns.

As would be the case in a practical example, many of the characteristic parameters of the procedure are

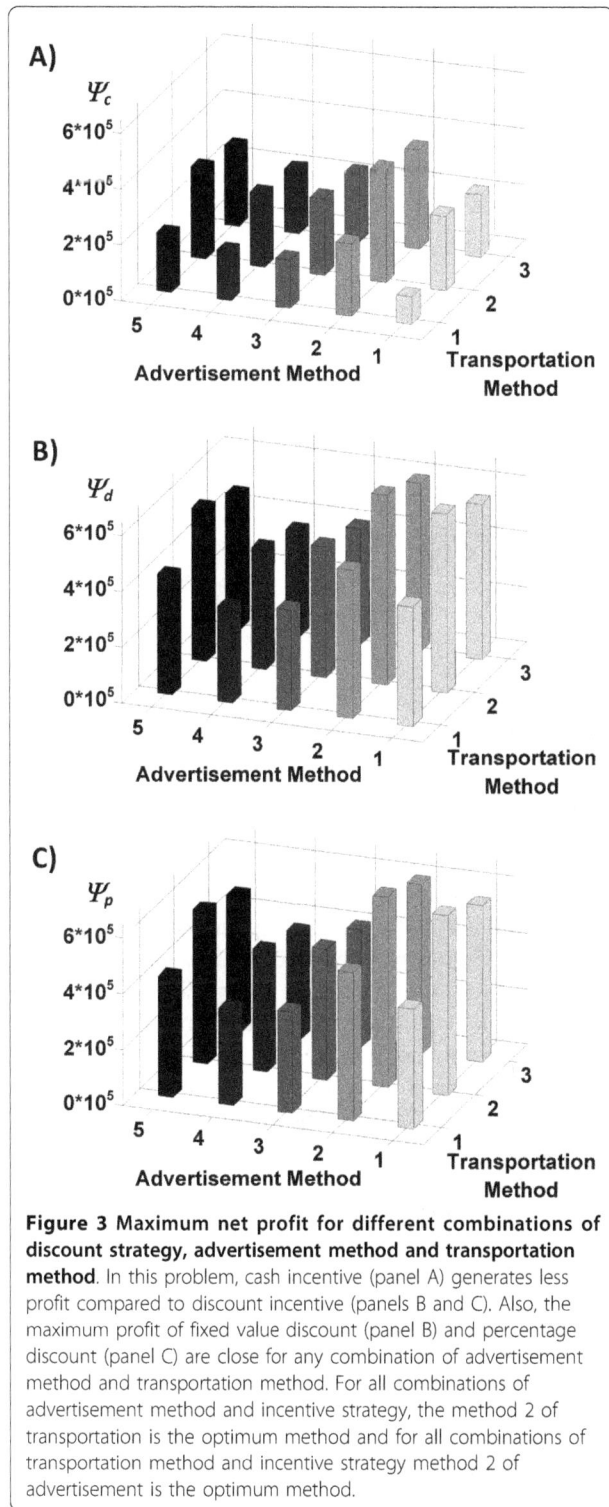

Figure 3 Maximum net profit for different combinations of discount strategy, advertisement method and transportation method. In this problem, cash incentive (panel A) generates less profit compared to discount incentive (panels B and C). Also, the maximum profit of fixed value discount (panel B) and percentage discount (panel C) are close for any combination of advertisement method and transportation method. For all combinations of advertisement method and incentive strategy, the method 2 of transportation is the optimum method and for all combinations of transportation method and incentive strategy method 2 of advertisement is the optimum method.

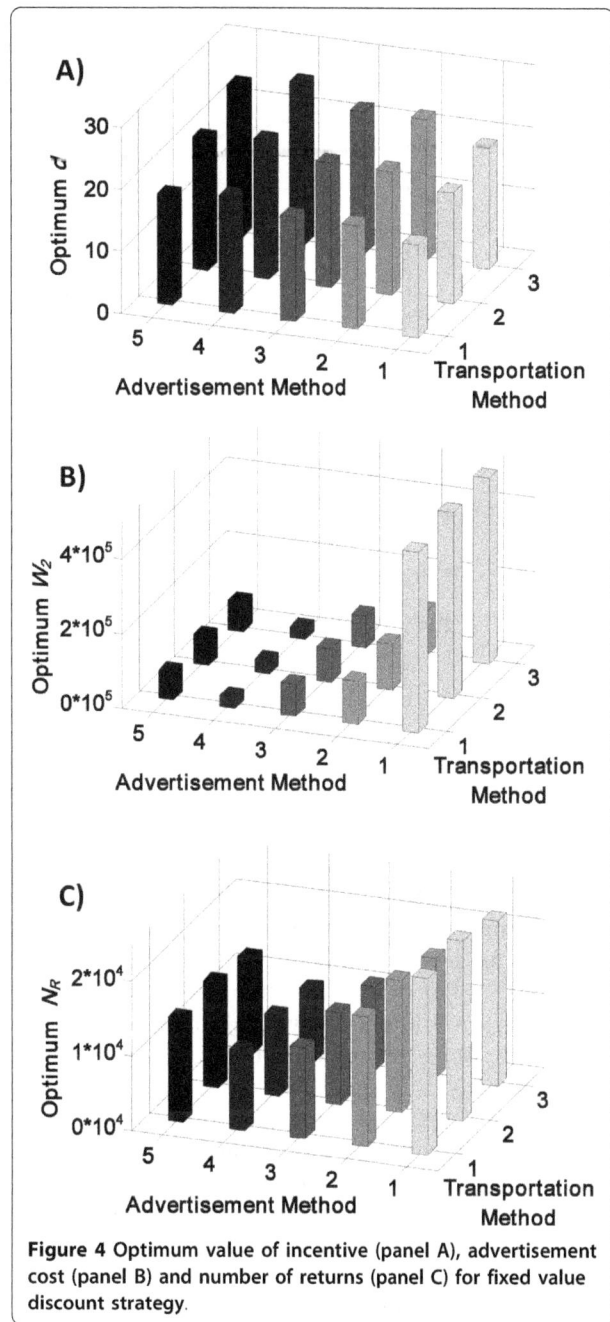

Figure 4 Optimum value of incentive (panel A), advertisement cost (panel B) and number of returns (panel C) for fixed value discount strategy.

estimated. The model predictions for the maximum net profit and optimum values of parameters are estimates as well. Using this model we can predict sensitivity of the maximum profit to any characteristic parameter of the take back procedure for analyzing the associated

risk. In this example we simulated the sensitivity of maximum profit with respect to three characteristic parameters: W_{sc}, α and Λ. Figure 5 shows how the maximum net profit and the optimum financial incentive vary by varying W_{sc} and α over a large range. Panel A shows net profit as a function of W_{sc} when method 2 of advertisement is considered. A 10 times increase of W_{sc} from ($10,000 to $100,000) reduces the net profit by less than 40%. Note that the estimated value of W_{sc} is $40,000 in Table 3. Interestingly, this large variation of W_{sc} does not affect the optimum type and amount of

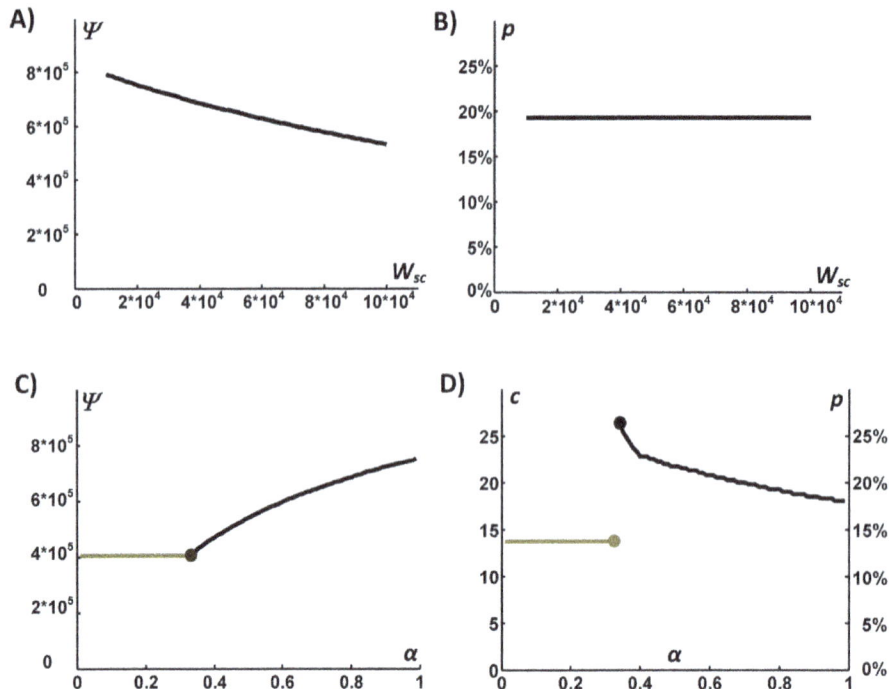

Figure 5 Sensitivity of the maximum profit and the optimum amount of incentive with respect to the cost scale of advertisement, W_{sc} (panels A and B) and the ratio of cash to discount incentive, α (panels C and D). Optimum type and amount of incentive is not sensitive to W_{sc} (panel B), but is sensitive to α (panel D). Change in total net profit is minor with respect to both parameters (panels A and C). Note both W_{sc} and α vary over a very large range.

financial incentive (panel B). It means that if the number of customers that are informed by each run of advertisement are less than what has been estimated (i.e. the actual W_{sc} is larger than its estimated value), the optimum compensation strategy would be to inform more customers by increasing the amount of advertisement, W_2, rather than to increase the financial incentives and motivate more (of the informed) customers to return their used product. Panels C and D (Figure 5) show the maximum net profit and the optimum financial incentive for different values of α (the ratio of cash to discount incentive). If α is less than about 0.35 the cash incentive is the optimum strategy and therefore, net profit and amount of incentive do not vary with α (gray segments). If α is larger than 0.35, percentage discount is the optimum strategy. By increasing α the net profit increases (up to about 70% in this example) and the optimum amount of discount decreases. Increase of the net profit is caused partially by reduction in the discount incentives and partially by the increase of the number of returns and sale of new products. Note that although an optimum value of p reduces (by increasing α) the motivation effectiveness, and consequently the number of returns increases.

The sensitivity of the model respect to Λ is shown in Figure 6. If there is no new customer ($\Lambda < 0.03$) the cash incentive is the optimum strategy and the maximum profit is about $404,000 which corresponds to about $14 cash incentive (Figure 6B) and $105,000 advertisement (Figure 6C). However, even if there is a small fraction of new customers ($\Lambda > 0.03$) the discount incentives strategies are more profitable. For $0.03 < \Lambda < 0.65$ the percentage discount and for $\Lambda > 0.65$ the fixed value discount is the optimum type of financial incentive. The net profit increases almost linearly by increasing Λ and is more sensitive to Λ than to α and W_{sc}. At $\Lambda = 0.65$, where the optimum incentive strategy switches from percentage discount to fixed value discount, there is a jump in the optimum advertisement cost (Figure 6C) which causes the jump in the number of returns (Figure 6D). At $\Lambda = 0.65$ the global minimum switches from one local minimum to another local minimum, where the same profit (Figure 6A) can be achieved through larger number of returns (Figure 6D) that justifies the significant increase in the advertisement cost (Figure 6C). Therefore, if the estimated value of Λ is around 0.65 then the optimum amount of advertisement would be sensitive highly to Λ; it should be either $125,000 to set the take back process for the smaller number of returns (18,000) or $500,000 to set the process at the larger number of returns (24,500). Note that the financial incentive does not change significantly across this jump (Figure 6B).

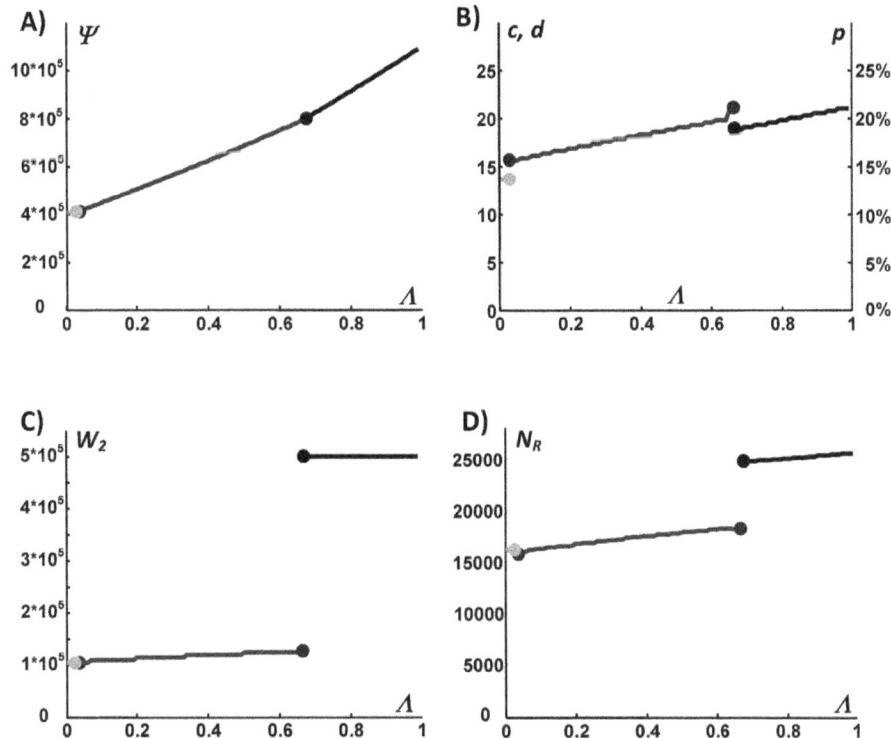

Figure 6 Sensitivity of the maximum profit (panel A) and the optimum amounts of incentive (panel B), advertisement cost (panel C) and number of returns (panel D) with respect to the fraction of new customers, Λ. The optimum incentive strategy changes from cash incentive (light gray) to percentage discount incentive (dark gray) at $\Lambda = 0.03$, and from percentage discount to fixed value discount incentive (black) at $\Lambda = 0.65$.

Discussion

Determining the number of returns and its variation with respect to different parameters of the take back procedure is required in a cost benefit analysis of a take back problem. Number of returns depends on many parameters and in general should be measured or estimated for all combinations of these parameters (i.e. in a multidimensional domain of variables), which is not practical. In a simple model, the number of returns may be considered simply as a function of one variable [4,9] usually termed the financial incentive or more generally the take back cost per returned product. Such a simple model, although provides overall theoretical insights about he take back process, but is not sufficient for many practical applications. It is not clear how the number of returns, which is a function of several variables, can be calibrated in terms of one variable. For example, increasing either the transportation cost or the financial incentive by $5, increases the take back cost by $5, but the resultant change in the number of returns can be significantly different. To overcome this limitation of the simple models, we first determined a set of factors that can significantly affect the number of returns like the transportation method, advertisement expenditure, and type and amount of financial incentives. Based on a solely theoretical analysis of the take back process, we derived a more detailed model for take back process that present several aspects of take back process. We tried to keep the model as simple as possible by imposing some reasonable assumptions. This model provided a general framework for different aspects of take back process and determined what empirical data is required for model calibration/validation.

Number of returns is modeled in terms of two functions; it is equal to the number of customers that are informed about the take back policy times the proportion of informed customers that return their used product. Number of informed customers depends on the method and volume of advertisement and is modeled as the Ω function. Proportion of informed customers that would return their used product depends on financial incentives and transportation method in addition to the method of advertisement; it is modeled as the Γ function. Γ function is a market characteristic of the take back process and should be determined using function approximation methods and the data obtained through surveys or pilot implementations. A general form of the Ω function was derived based on a basic analysis of advertisement. It should be mentioned that a detailed analysis of the advertisement is out of the scope of this paper; we only identified a set of parameters that are associated with advertisement and affect the number of returns through Ω or Γ functions. To determine Γ function, we first introduced the concept of motivation effectiveness, *mte*, and modeled Γ as a function of *mte* and then quantified and modeled the effect of different parameters of the take back (e.g. convenience of transportation and type of financial incentives) in terms of how they change the motivation effect of financial incentive. For example we assumed that offering financial incentive in the form of discount scales down the motivation effect of financial incentive (compared to equal amount of cash) by an average factor termed α. This enabled estimating Γ as a simplified single variable function while effects of other significant factors are included. Depending on the nature of the take back problem this model can be modified for the specific conditions of the problem. For example assume that the recovery firm requires the number of used products to be between N_{min} and N_{max}. This means that the number of taken back products should be larger than N_{min} and the taken back products beyond N_{max} does not generate any revenue. Therefore, in equations (10), (16) and (21) the value of used product, *a* should be multiplied by the minimum of N_R and N_{max} and in determining the maximum profit at each combination of reward strategy, advertisement method, and transportation method the domain of advertisement cost (W_2) and financial incentive (*c*, *d* or *p*) should be limited to the regions where N_R is greater than N_{min}.

Although as pointed out by Guide et al. [4], offering multiple incentives based on the condition of product can potentially increase the profit, it may not be the optimum strategy in all take back problems. In many practical cases customers may not be able to determine the condition of their used product and make their own decision about the return without knowing what they get in exchange. This usually affects the return rate adversely and may reduce the profit. However, most likely, the average quality of the returned products increases by increasing the incentive. This effect is included in the model by assuming the average value of returned products is a function of motivation effectiveness.

In this modeling framework the mutual effect between take back procedure and new product sale in discount strategies has been dissected and included in determining the net profit of take back. We allocated the total amount of discount as a cost to the take back procedure. We also allocated the increase in the profit of new product sale (because of discount) as revenue to the take back procedure. In doing this it is implicitly assumed that the take back and recovery procedures are performed by different segments of the same firm.

However, even if the take back is offered by a different firm, the discount strategy can be considered as a financial incentive. Generally the take back firm should be able to purchase the new products from the new product manufacturer below their retail value at a wholesale price and resell them to the take back customers at a discounted price. The cost model is applicable to this case as well; the value of Λ should be set to one and the sale profits are the difference between the retail price of new product and the wholesale price minus any handling fee associated with the resell.

Conclusion

The amounts and types of advertisement and transportation can significantly affect the net profit of take back. The type and amount of financial incentive is similarly influential. The developed modeling framework enables the determination of the optimum strategies for advertisement and transportation. It also compares cash and discount incentives, and determines if the extra sale of new product associated with the discounts can generate sufficient revenue to compensate for the reduced motivation of discount incentives (compared to cash). For the take back process studied in this paper, the model predicts that the maximum profit of the discount incentive strategy is about 70% higher than the cash incentive strategy, even though it requires a higher amount of financial incentives. The model also provides insights about the take back process and can be used for sensitivity analysis and feasibility study. For example, for the take back problem presented, the model predicts that the return rate and consequently the net profit are initially more sensitive to the frequency of advertisement (or advertisement cost W_2) than the amount of financial incentive (Figure 2). Therefore, if the system parameters and consequently the optimum advertisement cost are unspecified, it would be a wise operational decision to implement the take back process initially with a higher advertisement frequency, until more accurate data is acquired.

Appendix

An estimate can be found for the number of customers that are exposed to the advertisement (Ω function) based on available information about the statistics of advertisement method. Assume N_{ad} is the number of customers (or in general people) that are exposed to the advertisement at least one time. Not all customers can be reached by a specific advertisement method. For example, the customers who do not read the newspaper containing the ad, or do not watch or hear the TV or radio program that broadcasts the ad, will not be exposed to the ad independent of the number of the times the ad posts or broadcasts. The maximum number of customers that are potentially exposed to the ad over frequent postings or broadcasts is defined as N_{ss}. Also the average fraction of customers that are exposed to the ad in one run is defined by λ^*. Both N_{ss} and λ^* are statistical parameters of the advertisement method and are assumed to be known.

As N_{ad} is the number of customers that have seen the ad (after a known number of iterations) at least once, the number of customers that have not seen the ad, and may be exposed to the ad in the next iteration is $N_{ss} - N_{ad}$. Therefore, ΔN_{ad}, the change in N_{ad} after each iteration of the ad is:

$$\Delta N_{ad} = \lambda^*(N_{ss} - N_{ad})   \tag{A1}$$

The advertisement cost W_2 is proportional to the number of times the ad is broadcast or published. Let's assume that the cost of running the ad is ΔW_2 per each run. We may rewrite equation (A1) as:

$$\frac{\Delta N_{ad}}{\Delta W_2} = \frac{\lambda^*}{\Delta W_2}(N_{ss} - N_{ad}) = \lambda(N_{ss} - N_{ad})   \tag{A2}$$

where λ is defined as:

$$\lambda = \frac{\lambda^*}{\Delta W_2}   \tag{A3}$$

Although N_{ad} is a discrete function, when $\lambda \ll 1$ we may approximate it by a continuous function of W_2 and write:

$$\frac{d N_{ad}}{dW_2} = \lambda(N_{ss} - N_{ad})   \tag{A4}$$

and therefore:

$$N_{ad}(W_2) = N_{ss}\left(1 - e^{-\lambda W_2}\right) = N_{ss}\left(1 - e^{-\frac{W_2}{W_{sc}}}\right)   \tag{A5}$$

where W_{sc} is defined as the reciprocal of λ and from a physical point of view is the cost of the advertisement that is required to inform about 63% $(1-e^{-1})$ of the potential audience of the advertisement method. Dividing both sides by N we can find an estimate for Ω:

$$\Omega(W_2) = \Omega_{ss}\left(1 - e^{-\frac{W_2}{W_{sc}}}\right)   \tag{A6}$$

where Ω_{ss} is the maximum fraction of customers that can be informed by this method of advertisement. Ω_{ss} and W_{sc} are the two parameters that are different for different advertisement methods.

Acknowledgements
Authors are thankful to Dr. Garry Brandenburger and Dr. Guy Genin for their insightful comments.

Author details
[1]Mechanical Engineering and Materials Science Department, Washington University in St. Louis, 1 Brooking Dr., St. Louis Missouri 63130, USA
[2]Biomedical Engineering Department, Washington University in St. Louis, 1 Brooking Dr., St. Louis Missouri 63130, USA

Authors' contributions
N.G. reviewed the literature of product acquisition and had the leading role in developing the model. She designed the practical example and wrote the code for the computer simulations. MJ. defined the research subject and directed the research from the start to the end. He provided important advices throughout the study and helped in editing the manuscript. A.N. served as a consultant in developing the theoretical model and helped in writing and revising the manuscript and preparing the figures. All authors read and approved the final manuscript.

Competing interests
The authors declare that they have no competing interests.

References
1. Guide VDR, Van Wassenhove LN: **Managing Product Returns for Remanufacturing.** *Production and Operations Management* 2001, 10:142-155.
2. Guide VDR, Srivastava R: **Inventory Buffers in Recoverable Manufacturing.** *Journal of Operations Management* 1988, 16:551-568.
3. Guide VDR, Srivastava R, Kraus M: **Product Structure Complexity and Scheduling of Operations in Recoverable Manufacturing.** *International Journal of Production Research* 1997, 35:3179-3199.
4. Guide VDR, Teunter RH, Van Wassenhove LN: **Matching Demand and Supply to Maximize Profits from Remanufacturing.** *Manufacturing & Service Operations Managements* 2003, 5:303-316.
5. Galbreth MR, Blackburn JD: **Optimal Acquisition and Sorting Policies for Remanufacturing.** *Production and Operations Management* 2006, 15:384-392.
6. Guide VDR: **Production Planning and Control for Remanufacturing: Industry Practice and Research Needs.** *Journal of Operations Management* 2000, 18:467-483.
7. Gupta SM, Nakashima K: **Optimal ordering policy for product acquisition in a remanufacturing system.** *Book Optimal ordering policy for product acquisition in a remanufacturing system (Editor ed.^eds.). City* 2008, Paper 85.
8. Nakashima K, Gupta SM: **Analysis of remanufacturing policy with consideration for returned products quality.** *Book Analysis of remanufacturing policy with consideration for returned products quality (Editor ed.^eds.), vol. Paper 11. City* 2010, 486-491.
9. Klausner M, Hendrickson C: **Reverse-Logistic Strategy for Product Take-Back.** *INTERFACES* 2000, 30:156-165.
10. Edlin AS, Reichelstein S: **Specific Investment Under Negotiated Transfer Pricing: An Efficiency Result.** *Accounting Review* 1995, 70:275-292.
11. Vaysman I: **A Model of Negotiated Transfer Pricing.** *Journal of Accounting & Economics* 1988, 25:349-385.
12. Jeffrey SA, Shaffer V: **The Motivational Properties of Tangible Incentives.** *COMPENSATION & BENEFITS REVIEW* 2007, 39:44-50.
13. List JA, Shogren JF: **The Deadweight Loss of Christmas: Comment.** *The American Economic Review* 1998, 88:1350-1355.
14. Ruffle BJ, Tykocinski O: **The Deadweight Loss of Christmas: Comment.** *The American Economic Review* 2000, 90:319-324.
15. Waldfogel J: **The Deadweight Loss of Christmas.** *The American Economic Review* 1993, 83:1328-1336.
16. Wei KC: **Modeling the impact of incentives on vehicle sales volume.** *Control Applications; Sep 05-07, 2001; Mexico City, Mexico* 2001, 1135-1140.
17. O'Sullivan A, Sheffrin SM: *Economics: Principles in Action* Pearson Prentice Hall; 2007.
18. Vidal CJ, Goetschalckx M: **A global supply chain model with transfer pricing and transportation cost allocation.** *European Journal of Operational Research* 2001, 129:134-158.
19. Vidale ML, Wolfe HB: **An Operations Research Study of Sales Response to Advertising.** *Operations Research* 1957, 5:370-381.

20. Erickson GM: *Dynamics Models of Advertising Competition: open- and closed-loop extensions* Norwell, Massachusetts: Kluwer Academic Publishers; 1991.

21. Cowling K, Cable J, Kelly M, McGuinness T: *Advertising and Economic Behaviour* London, UK: The McMillan Press LTD; 1975.

22. Kerr W: **Remanufacturing and eco-efficiency: A case study of photocopier remanufacturing at Fuji Xerox Australia.** *Book Remanufacturing and eco-efficiency: A case study of photocopier remanufacturing at Fuji Xerox Australia (Editor ed.^eds.)* City: IIIEE Communications; 2000, 2005.

12

On the alignment of lot sizing decisions in a remanufacturing system in the presence of random yield

Tobias Schulz[1*] and Ivan Ferretti[2]

Abstract

In the area of reverse logistics, remanufacturing has been proven to be a valuable option for product recovery. In many industries, each step of the products' recovery is carried out in lot sizes which leads to the assumption that for each of the different recovery steps some kind of fixed costs prevail. Furthermore, holding costs can be observed for all recovery states of the returned product. Although several authors study how the different lot sizes in a remanufacturing system shall be determined, they do not consider the specificity of the remanufacturing process itself. Thus, the disassembly operations which are always neglected in former analyses are included in this contribution as a specific recovery step. In addition, the assumption of deterministic yields (number of reworkable components obtained by disassembly) is extended in this work to study the system behavior in a stochastic environment. Three different heuristic approaches are presented for this environment that differ in their degree of sophistication. The least sophisticated method ignores yield randomness and uses the expected yield fraction as certainty equivalent. As a numerical experiment shows, this method already yields fairly good results in most of the investigated problem instances in comparison to the other heuristics which incorporate yield uncertainties. However, there exist instances for which the performance loss between the least and the most sophisticated heuristic approach amounts to more than 6%.

Keywords: reverse logistics, remanufacturing, lot sizing, disassembly, random yield

Introduction

The reuse field has grown significantly in the past decades due to its economical benefits and the environmental requirements. Remanufacturing which represents a sophisticated form of reuse (see, for instance, Atasu et al. [1]) focusses on value-added recovery and has been introduced for many different products ranging from car engines (as has been reported in [2]) over photocopiers (as in [3]) to water pumps for diesel engines (as in [4]). Within the process of remanufacturing, products that are returned by the customers to the producer are disassembled to obtain functional components. The obtained components are afterwards cleaned and reworked until a "good-as-new" quality is assured. Having met the required quality standards, these components can be used for the assembly of a remanufactured product that is delivered to the customers with the same warranty as a newly produced one. In addition to the economic profitability, as a part of the embedded economic value can be saved by remanufacturing, there is an increasingly legislative restriction that assigns the producers the responsibility for their used products, for instance the Directive 2002/96/EC related to Waste Electrical and Electronic Equipment and the Directive 2002/525/EC related to End of Life Vehicles. Because of that, remanufacturing has become an important industry sector to achieve the goal of sustainable development (see, for instance, [5]). Therefore, the management and control of inventory systems that incorporate joint manufacturing and remanufacturing options has received considerable attention in recent literature contributions.

One of the main topics in these contributions is the assessment of joint lot sizing decisions for remanufacturing and manufacturing which has been thoroughly

* Correspondence: Tobias.Schulz@ovgu.de
[1]Faculty of Economics and Management, Otto-von-Guericke University of Magdeburg, Germany
Full list of author information is available at the end of the article

investigated in recent years. One of the first authors who established a basic modeling approach was Schrady [6] who developed a simple heuristic procedure for determining the lot sizes of repair and manufacturing lots. He assumes in his work that a constant and continuous demand for a single product has to be satisfied over an infinite planning horizon. Furthermore, a constant return fraction is established that describes the percentage of used products that return to the producer. By using that assumption a constant and continuous return rate is ensured. Presuming fixed costs for remanufacturing and manufacturing as well as different holding costs for repairable and newly manufactured products, a simple EOQ-type formula (with EOQ being the economic order quantity) is proposed that minimizes the sum of fixed and holding costs per time unit. As a result, an efficient cyclic pattern is established which is characterized by the fact that within each repair cycle a number of repair lots of equal size succeed exactly one manufacturing lot. By solving the proposed EOQ-formula which can be applied because an infinite production and repair rate is presumed as well, the number of repair lots and the length of a repair cycle can be determined. Teunter [7] generalized the results of Schrady in a way that he examined different structures of a repair cycle. His analysis concludes that it is not efficient if more than one repair lot and more than one manufacturing lot are established in the same repair cycle. This result extends the efficient cycle patterns by a cycle in which several manufacturing lots of equal size are followed by exactly one repair lot. The assumption of equal lot sizes is among other aspects critically studied in the contribution of Minner and Lindner [8]. They show that a policy with non-identical lot sizes can outperform a policy with identical lot sizes. However, the structure of an efficient repair cycle prevails also when the assumption of equal lot sizes is lifted.

Next to the analysis of the basic model context several extensions have been proposed that relax some of the assumptions made so far. Teunter [9], for instance, relaxes the assumption of an instantaneous manufacturing and repair process in order to derive more general expressions for the number of manufacturing and repair lots and their corresponding lot sizes. Since only a heuristic procedure was introduced on how to determine these values, Konstantaras and Papachristos [10] extended Teunter's work by developing an algorithm that leads to the optimal policy for certain parameter classes. By incorporating stochastic lead times and thereby including the possibility of back-orders, Tang and Grubbstrom [11] extend the basic model. Two general options are recommended on how such a system can be dealt with, a cycle ordering model and a dual

sourcing ordering policy. Both approaches are compared in a numerical study that indicates certain parameter specifications under which one approach outperforms the other. Furthermore, several papers have been published by Richter and Dobos (e.g. [12] and [13]) that relax the assumption of a constant rate of return. In their work they derived for several situations that a so called pure strategy is always optimal. In this context, a pure strategy means that either every returned product is repaired or everything is disposed of immediately. Therefore, a mixed strategy in which a part of the returned products is repaired and the rest is disposed of is always dominated by one of the pure strategies. Finally, the assumption of continuous demand and return rates has been relaxed by several authors. Consequently, the former EOQ-type model becomes a dynamic lot sizing problem. The contribution of Teunter et al. [14] extended well-known dynamic lot sizing heuristics such as the Silver Meal or the Part Period algorithm in order to test their performance in a remanufacturing environment. In their work, the adapted Silver Meal approach revealed an average percentage deviation of around 8% compared to the optimal solution. Schulz [15] extended among other things their approach by incorporating ideas known from the static environment and could reduce the average error to around 2%.

Common to all contributions is that they do not consider the remanufacturing process explicitly. Although some authors speak of remanufacturing, they analyze a remanufacturing system in the same way as a repair system. This may lead to wrong conclusions as it is not regarded that the remanufacturing process itself consists of two different subprocesses, a disassembly process in which the returned products are disassembled and a rework process in which the obtained components are brought to an as-good-as-new quality (for a definition see Thierry et al. [3] as well as Atasu et al. [1] for a more recent one). By explicitly incorporating both subprocesses in this contribution, the decisions that need to be made regarding disassembly and rework are decoupled which generalizes the basic models used so far.

Next to this generalization, this contribution will further relax the assumption of a deterministic yield, i.e. the number of components obtained by disassembly is not known with certainty beforehand. To present a practical application of this problem, the remanufacturing process for a car engine can be analyzed. When a batch of returned engines is disassembled, the remanufacturer does not know in advance how many remanufacturable components can be obtained. This is because the quality of the returns cannot always be assessed before disassembly. Hence, such a process can only be analyzed thoroughly when both processes disassembly and rework are evaluated separately. Considering stochastic yields has attained significant interest in the scientific literature as the basic work of Yano and Lee [16] as well as the overview of Grosfeld-Nir and Gerchak [17] present. However, most of the contributions mentioned in [17] describe purely manufacturing environments which cannot be entirely translated to a remanufacturing system as this inherits greater risks to be dealt with [18]. Nevertheless, stochastic yields have also been studied in a remanufacturing environment. Inderfurth and Langella, for instance, have concentrated their analysis specifically on the yield risk within the disassembly process [19]. Yet, they focussed on a multi-product multi-component problem setting in which a given discrete demand for components needs to be satisfied by either disassembling used products or manufacturing new components. The authors develop in their contribution heuristic methods on how to deal with such a problem in which they neglected the presence of fixed costs for the disassembly and the remanufacturing process. In another work, Ferrer [20] evaluates four different scenarios in a single period remanufacturing environment that differ in their process capabilities. For each scenario the optimal policy has been derived. In a numerical study, all four scenarios have been tested and compared.

After this short introduction, the problem assumptions and the nomenclature used in the remainder of the paper are illustrated in the subsequent section. Thereafter, two solution procedures are presented to find the optimal solution in the deterministic yield scenario before the scope is widened to a stochastic yield problem. For this problem setting, three heuristic approaches are introduced that facilitate the decision making process in such an environment. Next, a numerical experiment is conducted in order to test the heuristics' performance. Finally, a conclusion and an outlook are given in the last section.

Problem setting and model formulation

A company engaged in the area of remanufacturing that remanufactures several used products (e.g. engines) coming back from their customers shall be the background for the problem setting. To keep the analysis simple, the focus shall be restricted to only one specific remanufactured product named A. Figure 1 presents the general structure of this simplified system which is modeled as a multi-level inventory system containing three stages. Further simplifications are made regarding the fact that there are neither lead nor processing times. Moreover, no disposal option is included in the problem setting.

The customers' demand for the final product A is assumed to be constant and depletes the finished goods inventory continuously by a constant rate of λ units per

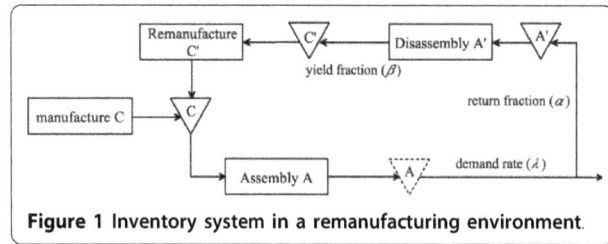

Figure 1 Inventory system in a remanufacturing environment.

time unit. In order to satisfy that demand, the company manufactures the final product by using component C which represents the most important component of the product. For the sake of simplicity, only the most important component C is included in the analysis. However, the proposed model could be easily extended to a multi-component setting. The assembly process is supposed to be a flow line process at which the final product is assembled continuously and immediately delivered to the customers. When the customers have no further use for their product A (e.g. it is broken or its leasing contract ends) they have the opportunity to return the product to the company. However, only a fraction (named α) of those products in the market returns to the producer. For the subsequent analysis, the return flow of used products (which are denoted A') fills the used product inventory by the constant and continuous rate of $\lambda\alpha$. By disassembling A' the worn component C' is obtained. Although the process of disassembly typically consists of manual work, fixed costs prevail for setting up required disassembly tools and/or measuring devices that allow an improved assessment of the reusability of components before disassembly. Within this model K_d represents the fixed costs for a disassembly batch while h_d is the holding cost incurred for storing one unit of A' for one time unit. Due to different stages of wear, not all returned products contain a reworkable component C'. The ratio of the number of reworkable items obtained from the disassembly of A' to the rate of product returns $\lambda\alpha$ is denoted by β. Assuming that at most one reworkable component C' can be obtained by disassembling one unit of A' the ratio β must not exceed one while being non-negative. As the released components C' cannot be used directly for the assembly of the final product A since they usually do not meet the designated quality standards, these components have to be remanufactured. Since the remanufacturing process incurs fixed costs of K_r for setting up the cleaning and mechanical rework tools, a batching of reworkable components takes place as well. Hence, some reworkable components need to be stored before the next remanufacturing batch is started resulting in costs of h_r per unit and time unit. It is furthermore assumed that each component that is remanufactured is brought to an as-good-as-new condition. All successfully reworked components are held in a serviceables inventory at a cost of h_s per unit and time unit. In order to secure the final product assembly of A, some components of C have to be manufactured in addition (as α and β are usually smaller than one). The decision relevant fixed costs are denoted by K_m representing the cost for setting up a manufacturing lot for component C. Newly manufactured components are held in the same serviceables inventory as remanufactured ones and it is supposed that the holding costs do not differ between both sourcing options. A detailed discussion on the topic on how to set the holding cost parameters can be found in [21]. In general, the holding costs (when interpreted as costs for capital lockup) of all levels are connected by the following inequality since more value is added to the component on each level, i.e. $h_d < h_r < h_s$.

Balancing fixed and holding costs shall be achieved by applying an average cost approach to this model. This is commonly done for one-level inventory systems as for the well known EOQ-model formulation but can be easily extended to a multi-level environment by respecting the stipulated assumptions of the EOQ-model (e.g. infinite planning horizon with constant costs over time). As a result, an optimal cyclic pattern is obtained by minimizing the average cost per time unit. In order to control the entire system, three decision variables are required. Firstly, the length of the disassembly cycle T determines the lot size of each disassembly batch ($\lambda \alpha T$) under the assumption that there is only one disassembly lot per cycle. This assumption is made for the sake of simplicity as an additional decision variable (number of disassembly lots per cycle) would complicate the analysis significantly. However, if we consider high fixed costs of disassembly, we conjecture that this assumption of one disassembly lot per cycle assures the optimality of the introduced deterministic policy. Furthermore, by fixing the number of remanufacturing lots R per disassembly cycle, their equal lot size can be computed by $\lambda \alpha \beta T / R$. Finally, the number of manufacturing lots M per disassembly cycle determines the lot sizes of the manufacturing lots to be $\lambda (1 - \alpha\beta) T / M$. The subsequent chapter presents the optimal solution of a completely deterministic setting in which all parameters are known with certainty.

Deterministic yields

In this section, a model is introduced that permits the evaluation of the optimal number of manufacturing and remanufacturing lots in a disassembly cycle. Before expanding the scope to stochastic yields from disassembly which represents the core issue of this contribution, the deterministic setting is studied in order to gain insight into the interrelations of the whole system. Figure 2 illustrates the behavior of the relevant inventory levels

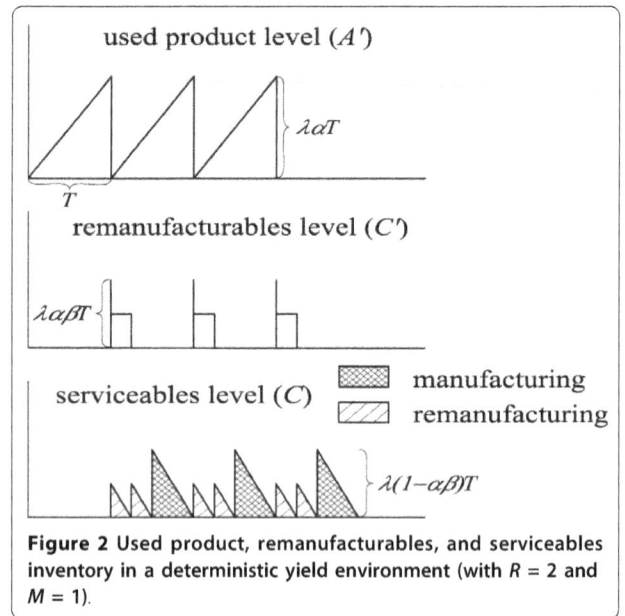

Figure 2 Used product, remanufacturables, and serviceables inventory in a deterministic yield environment (with $R = 2$ and $M = 1$).

for three consecutive disassembly cycles. As a matter of fact, the optimal decision variables (T, R, and M) remain constant over time in a deterministic environment. As shown in the figure below the manufacturing lots are positioned always after the remanufacturing lots in the serviceables inventory. This is obvious as this strategy strictly dominates the strategy of starting a cycle on the serviceables level with a manufacturing lot due to the increased holding costs on the remanufacturables level.

By minimizing the total average cost per time unit, this specific example shows the optimal cycle length T for two remanufacturing lots ($R = 2$) which split the remanufacturables inventory inflow equally and one manufacturing lot ($M = 1$) which satisfies the remaining demand of the assembly process for component C. To analyze the total cost function (TC^D) only two main types of costs have to be considered, the fixed costs F^D and the holding cost H^D in which the index D indicates the deterministic setting. A detailed discussion on how this formula can be obtained is presented in [22]. In addition, this contribution proves that equal lot sizes are optimal in this setting. The total cost function in the deterministic setting can be formulated as follows:

$$TC^D = \frac{F^D}{T} + \frac{\lambda T H^D}{2} \tag{1}$$

with $F^D = K_d + R K_r + M K_m$ and

$$H^D = \alpha h_d + \frac{R-1}{R} \alpha^2 \beta^2 h_r + \left(\frac{\alpha^2 \beta^2}{R} + \frac{(1-\alpha\beta)^2}{M} \right) h_s.$$

In order to minimize the total cost function which is a mixed-integer non-linear optimization problem two procedures can be applied. The first procedure is a simple enumerative procedure. Since R and M need to be integer valued only a finite number of calculations (in which R and M are set to integer values) have to be compared if R and M are restricted to certain intervals. The original objective function simplifies for given values of R and M to a non-linear convex function that only depends on T. Such a problem can be solved easily by using the subsequent equations:

$$T^{D*} = \sqrt{\frac{2F^D}{\lambda H^D}} \tag{2}$$

$$TC^{D*} = \sqrt{2\lambda F^D H^D}. \tag{3}$$

The formulas presented above are comparable to the determination of the economic order interval. However, the optimality of this solution approach can only be guaranteed if the optimal total cost TC^{D*} is determined for every combination of realization of R and M which leads to a quite large number of calculations. Nevertheless, a good solution can be obtained in a fast manner by restricting the number of possible realizations.

After introducing an enumerative procedure another promising approach will be presented next. By relaxing the original objective function (1) such that R and M need not to take on integer values, one can prove that the total cost function has only a single local minimum in the relevant area (for T, R, $M > 0$). Appendix C of [22] focusses on this specific aspect. Yet, by evaluating the Hessian matrix in this area, it can be shown that the total cost function is not entirely convex in all variables. This leads to the significant problem that a simple rounding procedure cannot be used to obtain the optimal solution for the integer valued number of remanufacturing and manufacturing lots. Therefore, a solution algorithm could be implemented that can globally determine the minimum cost of this mixed-integer non-linear optimization problem. The BARON algorithm, as implemented in the GAMS software package, proved to be a valuable tool for this problem setting. In general, BARON implements deterministic global optimization algorithms of the branch-and-reduce type in order to determine the optimal solution for a mixed-integer non-linear optimization problem. For a detailed description of the algorithm please refer to [23].

The subsequent chapter extends the deterministic model of this section to incorporate stochastic yields.

Stochastic yields

One of the main problems for many practical applications in the area of remanufacturing is that they have to deal with stochastic yields which means that the amount of remanufacturable components obtained from disassembling used returned products is not known with certainty (see also [19]). Due to the significance of that problem in a remanufacturing planning environment, we will now put forth the extension of the deterministic model that was introduced in the last section to incorporate stochastic yield fractions resulting from the disassembly process. Although being uncertain, it can be assured that the lowest possible yield fraction β_l cannot be smaller than zero as negative yields would not be reasonable. The largest possible yield fraction β_u, however, cannot exceed the value of one since this describes the situation that from every disassembled used product more than one remanufacturable component is obtained which is ruled out by the assumptions made. Within the range from β_l to β_u a specific distribution function can be defined which will be denoted in the following analysis by $\phi(\beta)$. As the number of returned products disassembled per cycle corresponds to $\lambda \alpha T$, the independence of $\phi(\beta)$ with respect to T reflects the fact that the subsequent analysis assumes stochastic proportional yields (for a definition see [16]). Therefore, the formerly used total cost function for a deterministic yield scenario (formula (1)) has to be extended in order to incorporate any possible yield outcome. Hence, the total cost of a given stochastic yield scenario can only be formulated as an expected total cost (which will be further on denoted as TC^S) that is presented in the following equation:

$$TC^S = \frac{F^S}{T} + \frac{\lambda T H^S}{2} \tag{4}$$

with $F^S = K_d + \int_{\beta_l}^{\beta_u} (R(\beta)K_r + M(\beta)K_m) \cdot \varphi(\beta)d\beta$ and

$$H^S = \alpha h_d + h_s \int_{\beta_l}^{\beta_u} \frac{1}{M(\beta)} \cdot \varphi(\beta)d\beta - 2\alpha h_s \int_{\beta_l}^{\beta_u} \frac{1}{M(\beta)} \cdot \beta \cdot \varphi(\beta)d\beta +$$
$$\alpha^2 \int_{\beta_l}^{\beta_u} \left(h_r - \frac{h_r}{R(\beta)} + \frac{h_s}{R(\beta)} + \frac{h_s}{M(\beta)} \right) \cdot \beta^2 \cdot \varphi(\beta)d\beta.$$

The fact that for any possible yield realization β an integer number of R and M has to be defined complicates the analysis of the total cost function TC^S significantly. In this setting, $R(\beta)$ describes the optimal number of remanufacturing lots for a given yield fraction β. Likewise, $M(\beta)$ represents the optimal number of manufacturing lots if the yield fraction β is fixed. Due to the fact that β is not known with certainty the total cost per time unit can only be formulated as an expectation over all different yield realizations. In contrast to the total cost function of the deterministic case (1), F^S and H^S can be regarded as an expectation of their corresponding deterministic equivalents F^D and H^D. As finding the optimal solution for any problem setting cannot be guaranteed, which will be shown later in this chapter, three different heuristic policies will be presented in the succeeding paragraphs that differ in their degree of sophistication. The first and least complex policy is introduced in the following:

Policy I

The easiest option on how to handle a stochastic problem is to neglect the underlying stochastics in order to derive a deterministic equivalent of the stochastic problem. The first policy introduced proceeds exactly in this manner as it neglects the fact that R and M depend on the yield realization β. Thus, only one value for R and M needs to be derived that is valid for every yield realization between β_l and β_u. To obtain these values, one can insert a specific yield fraction into the deterministic total cost function of the last section (1) and apply the recommended solution procedures to obtain R and M. As any yield fraction can be inserted that lies in the range of possible yield realizations and therefore many different combinations of R and M may prevail, we limit the focus of policy I on inserting only the mean yield fraction into the deterministic model since the mean yield is one of the most important characteristics of the underlying yield distribution. As a result we obtain the values of R^D and M^D that replace $R(\beta)$ and $M(\beta)$ for every possible yield realization β in formula (4). The expected total cost of the first policy (TC_I) can therefore be easily calculated by the subsequent equation:

$$TC_I = TC^S(T, R^D, M^D). \qquad (5)$$

Since policy I is a very simple approach, the decision maker can improve the expected total cost by incorporating the underlying stochastics in the decision making process which is introduced in policy II.

Policy II

Contrary to the first policy, the second policy does not neglect the dependence of R and M on the realization of the random yield fraction β. Nevertheless, in order to keep this policy simple, the disassembly cycle length T is kept constant which reduces the policies' complexity significantly. For the sake of simplicity the length of the disassembly cycle T will be set to the optimal deterministic cycle length T^{D*} obtained by formula (2) assuming that the mean yield fraction has been inserted as the deterministic equivalent for the underlying yield distribution. The assumption of fixing the cycle length to a specific value can be further used to draw some basic conclusions that can only be drawn for a given cycle length. The stochastic yield realization β determines for

each disassembly cycle the number of remanufacturable items. As the number of remanufacturable and manufactured components per cycle always adds up to the value of λT, the number of manufactured items depends as well on the yield realization. However, for both options of demand fulfillment it can be observed that if more items are processed (either by manufacturing or remanufacturing) the number of respective lots in a cycle does not decrease. Therefore, when comparing two different yield realizations with all other parameters set equally it can be said: For the larger yield realization the number of remanufacturable components increases which means that the number of remanufacturing lots per cycle does not decrease. On the other hand, the number of newly manufactured components decreases with larger yield realizations which means that the number of manufacturing lots per cycle does not increase. Figures 3 and 4 compare both heuristic policies introduced so far for three consecutive disassembly cycles. On the left hand side (Figure 3), it can be observed for policy I that regardless of the realized yield fraction the same number of R and M is applied in every cycle ($R = 2$ and $M = 1$). Figure 4 on the right hand side shows policy II that reacts for the same cycle length T on the different realizations of β which is supported by the fact that for a small yield realization the number of remanufacturing lots is smaller than for a large yield realization ($R = 1$ in the first cycle compared to $R = 3$ in the third cycle). An opposing behavior can be observed for the number of manufacturing lots per cycle that does not increase the smaller the yield realization is.

These general conclusions cannot only be formulated verbally but also in a mathematical form by introducing

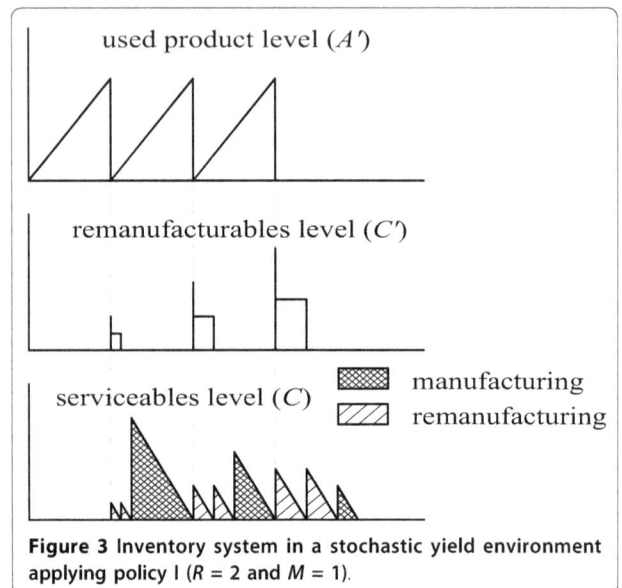

Figure 3 Inventory system in a stochastic yield environment applying policy I ($R = 2$ and $M = 1$).

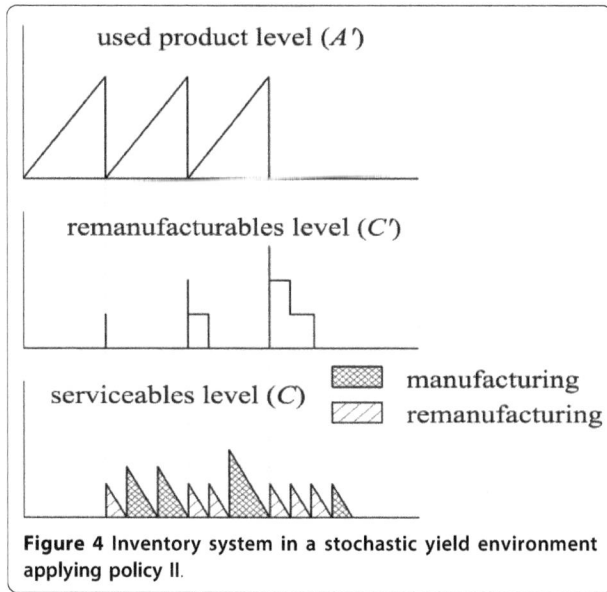

Figure 4 Inventory system in a stochastic yield environment applying policy II.

$$\beta(M) = \frac{1}{\alpha} - \frac{1}{\alpha T} \cdot \sqrt{\frac{2K_m M(M-1)}{\lambda h_s}}. \tag{7}$$

Because this function is monotonously decreasing in M, the insight that a larger yield fraction does not lead to less manufacturing lots in a cycle is approved. Consequently, the lowest and highest values for R and M can be determined by exploiting the two formulas given above. Thus, for the lowest possible yield fraction β_l R_{min} and M_{max} can be computed by the following procedure (analogously R_{max} and M_{min} can be computed for the highest possible yield fraction β_u):

$$R_{min} = \min_R \{\beta(R) \geq \beta_l\} \qquad M_{max} = \max_M \{\beta(M) \geq \beta_l\} \tag{8}$$

As the disassembly cycle length is fixed to a given value, the distribution function of the stochastic yield fraction can be subdivided into several intervals. Each interval j contains all yield realizations between its lower bound l_j and its upper bound u_j. The main characteristic of such an interval is the fact that within this interval only one number of remanufacturing and manufacturing lots induces the optimal solution for any possible yield fraction within this interval. The optimal number of remanufacturing and manufacturing lots in a certain interval j are furthermore denoted by R_j and M_j, respectively. For the identification of the respective interval bounds the following pseudocode can be used:

Start $j = 1$, $l_j = \beta_l$, $R_j = R_{min}$, $M_j = M_{max}$, $\beta(0) = \infty$
While $min\{\beta(R_j + 1), \beta(M_j - 1)\} < \beta_u$ **do**
    if $\beta(R_j + 1) < \beta(M_j - 1)$ **then**
        $u_j = \beta(R_j + 1)$
        $j = j + 1$, $l_j = u_{j-1}$, $R_j = R_{j-1} + 1$, $M_j = M_{j-1}$
    else
        $u_j = \beta(M_j - 1)$
        $j = j + 1$, $l_j = u_{j-1}$, $R_j = R_{j-1}$, $M_j = M_{j-1} - 1$
    end if
end do
$u_j = \beta_u$, $J = j$
end

After the initialization in which the first interval $j = 1$ is opened ($l_1 = \beta_l$) and given the values R_{min} and M_{max} the procedure evaluates if the transition to $R_{min}+1$ or $M_{max}-1$ is closer to β_l. For the lower of these two values, the upper bound of the first interval u_1 is fixed to the transition rate and the next interval is opened ($l_2 = u_1$). This procedure stops when both next transitions to $R+1$ and $M-1$ are larger than the highest possible yield fraction β_u. At this point the total number of intervals into which the yield distribution can be separated is determined by the index j which is set to the number of intervals J. As a result, the total yield distribution is separated into several intervals which is depicted for an example in Figure 5. In this example (with $\beta_l = 0$ and

so-called transition yield fractions which have the property that either the number of remanufacturing lots or the number of manufacturing lots changes when optimizing the deterministic equivalent problem. For the calculation of the specific yield fraction that is characterized by a switch of the optimal policy from R to $R+1$ remanufacturing lots, one needs to equate the deterministic total cost functions for R and $R+1$ as presented in the following equation:

$$\frac{F^D(R)}{T} + \frac{\lambda T H^D(R)}{2} = \frac{F^D(R+1)}{T} + \frac{\lambda T H^D(R+1)}{2}$$

with $F^D(R) = K_d + RK_r + M\,K_m$ and

$$H^D(R) = \alpha h_d + \frac{R-1}{R}\alpha^2\beta^2 h_r + \left(\frac{\alpha^2\beta^2}{R} + \frac{(1-\alpha\beta)^2}{M}\right)h_s.$$

This equation can be solved with respect to β in order to obtain the transition yield fraction $\beta(R)$ at which the optimal decision in the deterministic case switches from R to $R+1$ for a given cycle length T:

$$\beta(R) = \frac{1}{\alpha T} \cdot \sqrt{\frac{2K_r R(R+1)}{\lambda(h_s - h_r)}}. \tag{6}$$

Not only is this function monotonously increasing in R which corresponds to the findings above that the number of remanufacturing lots does not decrease for larger values of β but it also does not depend on the number of manufacturing lots per cycle M. Thus, the same analysis can be carried out independently for the transition from M to $M-1$ manufacturing lots per cycle by equating both total cost functions in order to obtain the transition yield fraction $\beta(M)$:

$\beta_u = 1$) it can be observed that the solution of policy I would have been $R = 3$ and $M = 4$ as this would solve the deterministic equivalent to optimality for $\beta = 0.5$.

As the interval bounds vary with a changing disassembly cycle length T, the expected total cost function for policy II can be formulated as follows using the optimal disassembly cycle length T^{D*} obtained by inserting the mean yield fraction into equation (3):

$$TC_{II} = \frac{F^S}{T^{D*}} + \frac{\lambda T^{D*} H^S}{2} \tag{9}$$

with $F^S = K_d + \sum_{j \in J} \left(R_j K_r + M_j K_m \right) \cdot \int_{l_j}^{u_j} \varphi(\beta) d\beta$ and

$H^S = \alpha h_d + h_s \sum_{j \in J} \frac{1}{M_j} \cdot \int_{l_j}^{u_j} \varphi(\beta) d\beta - 2\alpha h_s \sum_{j \in J} \frac{1}{M_j} \cdot \int_{l_j}^{u_j} \beta \cdot \varphi(\beta) d\beta +$

$\alpha^2 \sum_{j \in J} \left(h_r - \frac{h_r}{R_j} + \frac{h_s}{R_j} + \frac{h_s}{M_j} \right) \cdot \int_{l_j}^{u_j} \beta^2 \cdot \varphi(\beta) d\beta.$

In comparison to formula (4) only a finite number of R and M have to be considered in order to determine the solution using policy II. The formerly required $R(\beta)$ which represents the optimal number of remanufacturing lots for any given yield fraction β can be replaced with R_j after separating the yield distribution into intervals in which only one R is optimal for each yield realization. Consequently, the same simplification holds for the number of manufacturing lots M. However, this solution can be further improved by varying the disassembly cycle length T which shall be done in the most sophisticated heuristic approach of this contribution.

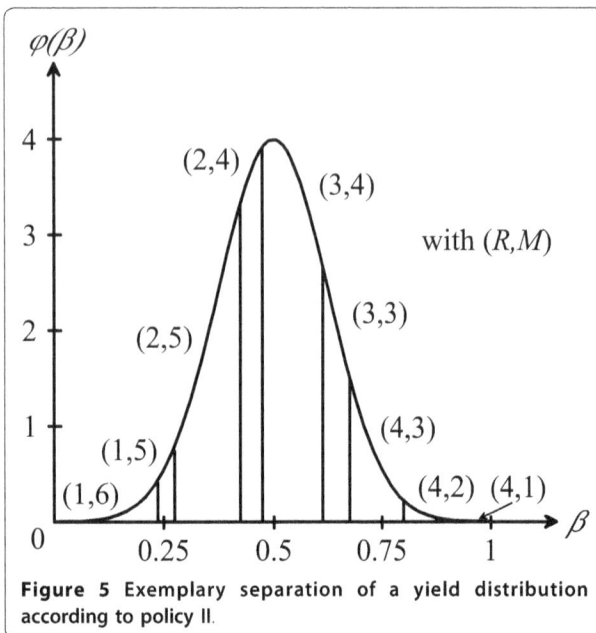

Figure 5 Exemplary separation of a yield distribution according to policy II.

Policy III

As the convexity of the expected total cost function of policy II (9) regarding the only remaining variable T cannot be proven for any possible yield distribution we face the fact that obtaining the optimal solution for this system cannot be guaranteed. Nevertheless, a simple local search heuristic can be implemented that alters the disassembly cycle length T from its initial value of policy II in order to check whether the expected total cost increases or decreases. The expected total cost function is evaluated by applying the procedure of policy II for any chosen parameter T. The local search procedure stops when both an increase or a decrease of T results in an increasing expected total cost meaning that at least a local minimum has been found that improves the solution of policy II at the expense of an increased complexity.

The following chapter elaborates a numerical experiment in which all three introduced heuristic approaches are tested in order to evaluate their performance in a stochastic yield environment.

Numerical experiment

The main objective of the numerical experiment conducted in this section is to evaluate the error that can be made when the simplest approach (policy I) is used compared to the more complex ones (policies II and III). In order to estimate the error, several numerical tests have been conducted using randomly generated instances. To our knowledge, no scientific contribution contains reliable and complete real life data for this specific problem setting. As the number of adequate test instances cannot be guaranteed in this case, Rardin and Uzsoy [24] recommend to create an experimental design based on random test instances. Although they discuss the pitfalls of random test instances in detail, we have applied this procedure to provide a first insight into each policy's performance. All parameters required for the test instances were drawn from a discrete uniform distribution $DU(a, b)$ with a as the lower bound and b representing the upper bound of the distribution. Some parameters were multiplied after the random draw with a constant term in order to obtain reasonable values. Table 1 lists all parameters that were randomly drawn in this experiment:

The return fraction α, for instance, can take on values between 30% and 90%, only limited by the fact that the percentage must be an integer multiple of 5%. Regarding the fixed costs, we restricted the possible region on integer values between 0 and 50 for the disassembly process and 1 to 100 for setting up a remanufacturing or a manufacturing lot. For the disassembly lot, we established smaller values as these processes are done manually in some industrial applications and do not necessarily

Table 1 Parameters generated randomly in numerical experiment

Parameter	Generation method
Demand rate	$\lambda \sim DU(1, 10) \cdot 100$
Return fraction	$\alpha \sim DU(6, 18) \cdot 0.05$
Fixed cost for disassembly	$K_d \sim DU(0, 50)$
Fixed cost for remanufacturing	$K_r \sim DU(1, 100)$
Fixed cost for manufacturing	$K_m \sim DU(1, 100)$
Holding cost for used product	$h_d \sim DU(1, 10) \cdot 0.01$
Holding cost for remanufacturable component	$h_r \sim DU(5, 15) \cdot 0.01$
Holding cost for serviceable component	$h_s \sim DU(10, 20) \cdot 0.01$

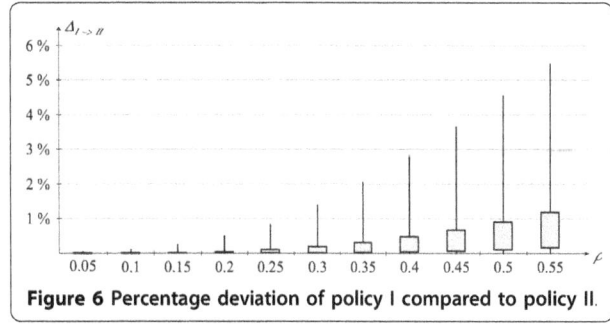

Figure 6 Percentage deviation of policy I compared to policy II.

require a specific setup. With respect to the holding costs we implicitly assumed that the holding cost increase from level to level as more effort has been put into the components. This means that each randomly generated instance has to fulfill the presumed inequality $h_d < h_r < h_s$. From these probability distributions, 1,000 instances were drawn and tested for different yield distributions. Generally, the yield distribution followed a symmetric beta-distribution within the limits $\beta_l = 0$ and $\beta_u = 1$. The parameter that altered the yield distribution was the coefficient of variation ρ that was changed in the limits between 0.05 and 0.55 which is motivated by our experience with an automotive remanufacturer regarding its yield fractions. While a ρ of 0.05 indicates that almost the entire probability mass is centered around the distribution's mean, a coefficient of variation of 0.55 indicates for a beta-distribution within the interval 0 to 1 an approximately uniform yield distribution.

All three introduced heuristic approaches were tested for all instances. Figure 6 illustrates, for instance, the percentage deviation of the total costs of policies I and II. $\Delta_{I->II}$ denotes this percentage deviation and is calculated by $\Delta_{I->II} = TC_I / TC_{II} - 1$. In detail, this deviation shows the percentage loss in performance if policy I (at which only the mean yield fraction is considered to represent the entire yield distribution) is applied instead of policy II. The deviation with respect to the coefficient of variation of the underlying yield distribution is presented with the aid of box plots that do not only show the maximum and minimum deviation but also where half of the deviations are located inside the shaded area around the median.

For very small coefficients of variation that are characterized by the fact that almost the entire probability mass is centered around the mean yield, the deviation between policy I and policy II is almost negligible. The reason for that is easy to be found. Although the yield distribution is defined in the interval between 0 and 1, the range of realizations that have a significant probability is very small. If the optimal number of remanufacturing and manufacturing lots per cycle that is determined

by policy I is also optimal for a wide range of yield fractions around the distribution's mean both policies arrive at the same result. However, if the coefficient of variation grows larger the deviations increase as well. For an approximately uniformly distributed yield, for instance, the maximum deviation between policy I and II is around 5.4%. On the other hand, the minimum deviation is 0% which means that the optimal cycle pattern of policy I is still optimal for every yield realization between 0 and 1 even for such a widespread distribution. Although many instances have been tested, the effect of every parameter on the deviation cannot be observed without doubt. Yet, some general trends can be derived from the experiments. For instance, it seems to be the case that the percentage gap in the total cost between policy I and II increases in most scenarios for instances with an increasing return rate α. Additionally, the different fixed costs seem to influence this gap as well. For high fixed costs for disassembly and remanufacturing (K_d and K_r) as well as for small fixed costs for manufacturing (K_m) the observed percentage gap increases for a large coefficient of variation of the yield distribution. The same analysis can be conducted for the different holding cost parameters, too. The percentage gap between policy I and II increases if the holding costs h_d, h_r, and h_s deviate significantly. Furthermore, it can be said as larger the difference between R_{min} and R_{max} as well as the difference between M_{min} and M_{max} is as larger is the percentage gap. Finally, no considerable influence on the percentage gap can be observed for the demand per time unit λ.

Figure 7 presents the deviation of policy II from policy III which means that the cycle length T is varied in order to decrease the total cost function even further. By $\Delta_{II->III}$ this deviation is represented. Regarding the coefficients of variation the same can be observed as for the first examined deviation. For small coefficients of variation there is almost no improvement possible by changing the cycle length. On the other hand, for larger coefficients the percentage gap grows larger which means that an adaption of T can improve the total cost function. However, these improvements are relatively

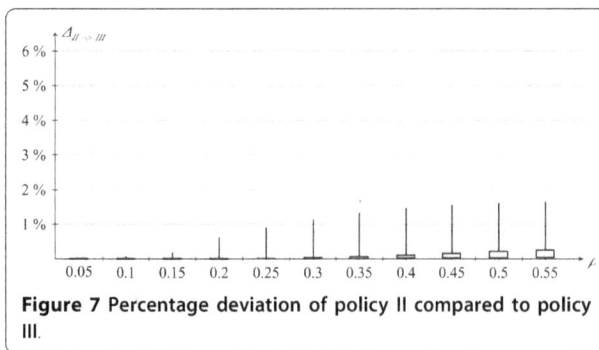

Figure 7 Percentage deviation of policy II compared to policy III.

small (in 97.4% of all cases smaller than 1% for $\rho = 0.55$). Regarding the cost deviation between policy II and III, it is even more difficult (in comparison to the deviation between policy I and II) to define parameter areas at which the deviation is typically high or low. Yet, two general trends can be noticed. The largest deviations can be observed for instances with a large α and a wide spread of the holding cost levels. However, this observation cannot be generalized for all instances with this parameter constellation.

Another interesting question that can be analyzed with this numerical experiment is whether the optimal cycle length increases or decreases in comparison to the cycle length of policy I and II that remains constant for all coefficients of variation. In 69.1% of all instances the cycle length decreased while it increased in the remaining 31.9%.

Therefore, no general conclusion can be drawn regarding this aspect as no specific parameter constellation can be identified that increases or decreases the cycle length in general.

Conclusion and outlook

This contribution outlined an approach on how to handle deterministic and stochastic yield fractions within a multi-level remanufacturing system that considers the disassembly process explicitly. While being restricted to a single disassembly lot per cycle, simple derivations are made with respect to the three necessary parameters, the optimal disassembly cycle length as well as the optimal number of remanufacturing and manufacturing lots per disassembly cycle. By examining both the stochastic and the deterministic case, the error that can be made by neglecting the underlying stochastics is evaluated. The numerical experiment has confirmed a quite straight-forward assumption. The less variability of the random yield fraction is faced, the smaller is the error that is made by using the mean yield policy I instead of the more sophisticated ones. However, there exist situations in which using the simple policy I results in performance losses of more than 5%. Nevertheless, in most

cases the decision maker will obtain fairly good results if he neglects the underlying yield distribution and follows the deterministic mean yield fraction approach of policy I. In this sense, a problem setting has been identified in which the influence of stochastic yields does not complicate the decisions to be made significantly.

Next, an outlook regarding future research efforts shall be given. The proposed model can be extended in several ways. For both the stochastic and the deterministic one, the option of allowing more than one disassembly lot per disassembly cycle is a promising extension of the model presented in this contribution. Especially for instances showing a small fixed cost of disassembly this might provide a valuable option to decrease the average costs per time unit. Furthermore, it can be studied how a multi-product multi-component setting affects the decision making process in both environments since aspects like multiplicity (one component can be obtained by the disassembly of different product types) have to be incorporated. Another interesting topic that can be included in the analysis is a disposal option. This might be a worthwhile option if the fixed costs of remanufacturing are quite high and the yield realization is very small. In the proposed model context, at least one remanufacturing process has to be set up in such a disassembly cycle. However, if there is a disposal option, the obtained components can be disposed of and the total customer demand will be satisfied by newly manufactured components, i.e. the optimal number of remanufacturing lots R can be 0. As a last extension, all heuristic approaches can be tested not only for proportional stochastic yields but also for non-proportional yields. In order to achieve this objective, the yield fraction distribution cannot be modeled as a beta-distribution any more but needs to be modeled for instance with a binomial distribution.

As discussed above, a number of different uncertainties can be found in a real life remanufacturing system. This contribution has revealed that a possible yield uncertainty can be neglected in a multitude of problem instances when considering joint remanufacturing and manufacturing lot sizing decisions. This is a rather untypical result when real life industrial applications such as the remanufacturing of car engines face stochastic yields in their process. To obtain good solutions for the lot sizing problem presented above, the only required information regarding the yield distribution is its mean value. By planning the lot sizes with this mean value, the error of neglecting stochastic yields can be reduced in most cases to less than 2%.

Acknowledgements

The authors wish to express their thanks to the Ministry of Education, University and Research of Italy (MIUR) for the support given for the development of this work under the Interlink program (Supply Chain Sustainability project). Furthermore, the authors wish to express their thanks to the helpful comments of the referees.

Author details

[1]Faculty of Economics and Management, Otto-von-Guericke University of Magdeburg, Germany [2]Mechanical and Industrial Engineering Department, University of Brescia, Italy

Authors' contributions

TS elaborated the deterministic and stochastic solutions and carried out the numerical study. IF defined the problem setting, participated in the elaboration of the general method and validated the results by implementing a simulation model. All authors read and approved the final manuscript.

Competing interests

The authors declare that they have no competing interests.

References

1. Atasu A, Guide VDR Jr, van Wassenhove LN: **So what if remanufacturing cannibalizes my new product sales?** *California Management Review* 2010, **52**(2):56-76.
2. Seitz M, Wells PE: **Challenging the implementation of corporate sustainability: The case of automotive engine remanufacturing.** *Business Process Management Journal* 2006, **12**(6):822-836.
3. Thierry M, Salomon M, van Nunen J, van Wassenhove LN: **Strategic issues in product recovery management.** *California Management Review* 1995, **37**(2):114-135.
4. Tang O, Teunter R: **Economic lot scheduling problem with returns.** *Production and Operations Management* 2006, **15**(4):488-497.
5. Webster S, Mitra S: **Competitive strategy in remanufacturing and the impact of take-back laws.** *Journal of Operations Management* 2007, **25**(6):1123-1140.
6. Schrady DA: **A deterministic inventory model for reparable items.** *Naval Research Logistics Quarterly* 1967, **14**:391-398.
7. Teunter RH: **Economic ordering quantities for recoverable item inventory systems.** *Naval Research Logistics* 2001, **48**(6):484-495.
8. Minner S, Lindner G: **Lot sizing decisions in product recovery management.** In *Reverse Logistics - Quantitative models for closed-loop supply chains.* Edited by: Dekker R, Fleischmann M, Inderfurth K, van Wassenhove LN. Springer; 2004:157-179.
9. Teunter RH: **Lot-sizing for inventory systems with product recovery.** *Computers & Industrial Engineering* 2004, **46**(3):431-441.
10. Konstantaras I, Papachristos S: **A note on: Developing an exact solution for an inventory system with product recovery.** *International Journal of Production Economics* 2008, **111**(2):707-712.
11. Tang O, Grubbstrom RW: **Considering stochastic lead times in a manufacturing/remanufacturing system with deterministic demands and returns.** *International Journal of Production Economics* 2005, **93-94**:285-300.
12. Richter K: **The extended EOQ repair and waste disposal model.** *International Journal of Production Economics* 1996, **45**(1-3):443-448.
13. Richter K, Dobos I: **Analysis of the EOQ repair and waste disposal problem with integer setup numbers.** *International Journal of Production Economics* 1999, **59**(1):463-467.
14. Teunter RH, Bayindir ZP, van den Heuvel W: **Dynamic lot sizing with product returns and remanufacturing.** *International Journal of Production Research* 2006, **44**(20):4377-4400.
15. Schulz T: **A new silver-meal based heuristic for the single-item dynamic lot sizing problem with returns and remanufacturing.** *International Journal of Production Research* 2011, **49**(9):2519-2533.
16. Yano CA, Lee HL: **Lot sizing with random yields.** *Operations Research* 1995, **43**(2):311-334.
17. Grosfeld-Nir A, Gerchak Y: **Multiple lotsizing in production to order with random yields: Review of recent advances.** *Annals of Operations Research* 2004, **126**(1-4):43-69.
18. Toffel MW: **Strategic management of product recovery.** *California Management Review* 2004, **46**(2):120-141.
19. Inderfurth K, Langella IM: **Heuristics for solving disassemble-to-order problems with stochastic yields.** *OR Spectrum* 2006, **28**(1):73-99.
20. Ferrer G: **Yield information and supplier responsiveness in remanufacturing operations.** *European Journal of Operational Research* 2003, **149**(3):540-556.
21. Teunter RH, van der Laan E, Inderfurth K: **How to set the holding cost rates in average cost inventory models with reverse logistics.** *Omega* 2000, **28**(4):409-415.
22. Schulz T, Ferretti I: **On the alignment of lot sizing decisions in a remanufacturing system in the presence of random yield.** Technical report, Faculty of Economics and Management, Otto-von-Guericke University Magdeburg; 2008.
23. Sahinidis NV, Tawarmalani M: **Accelerating branch-and-bound through a modeling language construct for relaxation-specific constraints.** *Journal of Global Optimization* 2005, **32**(2):259-280.
24. Rardin RL, Uzsoy R: **Experimental evaluation of heuristic optimization algorithms: A tutorial.** *Journal of Heuristics* 2001, **7**:261-304.

13

An analysis of remanufacturing practices in Japan

Mitsutaka Matsumoto[1*] and Yasushi Umeda[2]

Abstract

Purpose: This study presents case studies of selected remanufacturing operations in Japan. It investigates Japanese companies' motives and incentives for remanufacturing, clarifies the requirements and obstacles facing remanufacturers, itemizes what measures companies take to address them, and discusses the influence of Japanese laws related to remanufacturing.

Methods: This study involves case studies of four product areas: photocopiers, single-use cameras, auto parts, and ink and toner cartridges for printers. Results and conclusions are based on the authors' discussions and interviews with 11 remanufacturers–four original equipment manufacturers (OEMs) and seven independent remanufacturers (IRs). In the discussions and the interviews, we asked the companies their motives for remanufacturing and asked the measures they take to overcome the obstacles of remanufacturing. This study highlighted three requirements for remanufacturing: (1) collection of used products, (2) efficient remanufacturing processes, and (3) demand for remanufactured products.

Results: Where OEMs are the main remanufacturers of products covered by this study, their motives are long-term economic and environmental incentives. Where IRs are the main remanufacturers, it is often because OEMs shun remanufacturing, fearing to cannibalize new product sales. Companies' efforts to meet the above mentioned three requirements were observed and documented: (1) establishing a new collection channel; (2) developing reverse logistics to collect used products; (3) designing products for remanufacturing (DfReman); (4) accumulating know-how to establish remanufacturing processes; and (5) controlling product quality to stimulate demand for remanufactured products. This study also notes that (6) OEMs who engage in remanufacturing build consumer demand by incorporating remanufactured components into new products. This point has not been particularly noted in previous studies, but it has an important implication for OEMs' remanufacturing. The authors found that Japan's Home Appliances Recycling Law and End-of-Life Vehicle Law have promoted material recycling but have been insufficient to stimulate remanufacturing within the country.

Conclusions: This study clarified the differences between OEMs' and IRs' remanufacturing. Both IRs and OEMs are important for remanufacturing. Institutional measures to encourage appropriate competition between OEMs and IRs and to enhance consumers' acceptance of remanufactured products is important to promote remanufacturing.

Keywords: Remanufacturing, Practices in Japan, Business obstacles, Service

Introduction

This study presents and analyzes case studies of selected remanufacturing operations in Japan. Remanufacturing can limit environmental impacts, and is a key strategy to for sustainable manufacturing and in turn for addressing the needs of sustainable development [1]. A multinational comparison of remanufacturing practices and relevant legislations is indispensable in assessing the measures to promote remanufacturing worldwide. Few existing international literature have analyzed remanufacturing practices in Japan. This study aims to examine this issue.

Remanufacturing is the process of restoring broken assemblies to a "like-new" functional state by rebuilding and replacing their component parts [2]. Remanufacturing has spread worldwide to sectors as disparate as auto parts, electric home appliances, personal computers, cellular phones, photocopiers, single-use cameras, cathode ray tubes, automatic teller machines, vending machines, construction machineries, industrial robots, medical

* Correspondence: matsumoto-mi@aist.go.jp
[1]Center for Service Research, National Institute of Advanced Industrial Science and Technology (AIST), Umezono, Tsukuba, Japan
Full list of author information is available at the end of the article

equipment, heavy-duty engines, aircraft parts, and military vehicles. Japanese remanufacturing practices are advanced in some product areas, notably photocopiers and single-use cameras, and lagging in fields like auto parts, where remanufacturing is commonplace elsewhere [3]. Factors that determine whether remanufacturing prevails include the engagement of products' original equipment manufacturers (OEMs) and independent remanufacturers (IRs), consumers' awareness and preferences for remanufactured products, related legislations, and relevant social institutions. It is significant to verify which factors promote and hinder remanufacturing through case studies.

This study analyzes the following aspects through case studies in Japan. First, it investigates Japanese companies' motives and incentives for remanufacturing and explores the conditions to prompt OEMs and IRs to remanufacture. Previous studies have paid disproportionate attention to the advantages and incentives OEMs have in remanufacturing. However, in many industrial segments where OEMs lack incentive or have a negative attitude toward remanufacturing, IRs lead in remanufacturing. Therefore, understanding companies' motives for remanufacturing is essential to promote remanufacturing.

Second, this study clarifies the requirements and obstacles of remanufacturers and discusses what measures companies take to address them. Justifiably called a "Hidden Giant" [4], the remanufacturing industry has good market potential. However, companies have to overcome certain obstacles to achieve it. This study highlights three requirements for remanufacturing: (1) collection of used products, (2) efficient remanufacturing processes, and (3) demand for remanufactured products, as discussed in Section 2. Case studies presented here investigate measures taken by companies to meet these requirements.

Third, this study discusses Japanese legislation related to remanufacturing and its influence. Such arguments are crucial to design legislation and institutions that support remanufacturing.

The paper is organized as follows. Section 2 reviews the existing literature. Section 3 describes the case studies, their method, and instructive conclusion. Section 4 discusses relevant legislation and its influence on remanufacturing. Section 5 describes discussing issues in the paper. The final section summarizes key findings and their contribution to remanufacturing.

Literature review

Over the past few decades, increasing interest in remanufacturing has prompted several studies. These studies have emphasized OEMs, which have numerous advantages over IRs and perhaps greater incentive to remanufacture. Lund and Skeels [5] and Lund [6] pointed out the advantages unique to OEMs: feedback on product reliability and durability, competition in lower-priced markets, a manufacturer's reputation for quality, and gaining advantages over IRs in data, tooling, and access to suppliers. Similarly, Haynsworth and Lyons [7] envisioned how OEMs could realize the potential for remanufacturing through appropriate marketing and product design and by developing a product distribution and return system. Many studies have confirmed that remanufacturing is profitable for OEMs [8,9]. Some studies even consider profitability as given, since resources used in manufacturing products are reused and production costs of remanufactured products are less than new production [10,11].

On the other hand, OEMs face unique obstacles. Although remanufacturing may reduce sales of new products, profits on sales of new products often exceed profits on those of remanufactured products [12-14]. There are several counter-arguments to support this claim. First, new and remanufactured products are targeted toward different market segments, minimizing their potential conflict [15]. Second, economic incentives are not OEMs' primary motive for remanufacturing. Studies have cited considerations such as ethical responsibility [16], corporate brand protection [17], intellectual property protection [18], and other considerations (see also [15,19,20]).

Previous studies have described other requirements and obstacles faced by companies in developing a new remanufacturing business. Lund and Skeels [5] pointed the following issues: (1) product selection, (2) marketing strategy, (3) remanufacturing technology, (4) financial aspects, (5) organizational factors, and (6) legal considerations. Steinhilper [21] proposed eight criteria to be evaluated in establishing the suitability of products for remanufacturing: (1) technical criteria (type or variety of materials and parts, suitability for disassembly, cleaning, testing, reconditioning), (2) quantitative criteria (amount of returned products, timely and regional availability), (3) value criteria (value added from material/production/assembly), (4) time criteria (maximum product life time, single-use cycle time), (5) innovation criteria (technical progress regarding new products and remanufactured products), (6) disposal criteria (efforts and cost of alternative processes to recycle the products and possible hazardous components), (7) criteria regarding interference with new manufacturing (competition or cooperation with OEMs), and (8) other criteria (market behavior, liabilities, patents, intellectual property rights). Other relevant arguments were provided in e.g., Hammond et al. [7], Guide and Van Wassenhove [22], Ijomah et al. [19], Subramoniam et al. [23,24], and Matsumoto [25].

This study highlights three factors raised by Geyer and Jackson [26] and Lundmark et al. [20]. According to these authors, the remanufacturing system consists of three parts–collection, the remanufacturing process itself, and redistribution–each having its distinct challenges. For a company to undertake remanufacturing, it must (1) develop a collection system for used products, (2) develop efficient remanufacturing processes, and (3) cultivate demand for remanufactured products. This study investigates companies' efforts to meet these requirements.

In some cases, legislation is indispensable in enabling companies to operate as remanufacturers, and in other cases it creates barriers to remanufacturing. Hammond et al. [8] found that in auto parts remanufacturing, increased part proliferation and new governmental regulations in the United States caused major changes within the industry. Webster and Mitra [27] analyzed the effects of governmental subsidies on sustainable operations and found that they encourage remanufacturing activities. Zuidwijk and Krikke [28] analyzed the strategic response of the industry to the Waste Electric and Electronic Equipment Directive (WEEE Directive) in the European Union (EU). Gerrard and Kandlikar [29] assessed the impact of the End-of-Live Vehicles Directive (ELV Directive) in the EU and found that while it led car OEMs to take steps toward recycling and disassembly, progress in designing the process for reuse and remanufacturing was limited. This study introduces relevant legislation in Japan–the Home Appliance Recycling Law and the End-of-Life Vehicles Recycling Law–and discusses their influences on remanufacturing.

Methods and results: Case studies of remanufacturing businesses in Japan
Methods
Following discussions and interviews with Japanese companies engaged in remanufacturing, the authors focused on case studies related to four types of products: photocopiers, single-use cameras, auto parts, and ink and toner cartridges for printers.

These product areas were selected for the following reasons. Photocopiers and single-use cameras were selected because they are the two most successful cases of OEM remanufacturing in Japan. Studying these cases provides insights into OEMs' remanufacturing practices.

Remanufactured auto parts are prevalent worldwide [3]. In Japan, as in other countries, the main remanufacturers are IRs rather than OEMs. This case study is helpful to learn IRs' remanufacturing practices, and OEMs incentives and disincentives to remanufacture. In addition, since auto parts remanufacturing in Japan is less prevalent than in the United States and EU, the

reasons, obstacles, and companies' efforts to overcome the obstacles are investigated.

The printer cartridge case exemplifies the conflict between profits on OEMs' sales of new products and IRs' sales of remanufactured products. The share of remanufactured products in Japan's printer ink cartridge market has increased rapidly since early 2000s. The efforts of the successful IRs are studied.

The case studies are based on the authors' discussions and interviews with major Japanese OEMs and IRs in the targeted product areas. The authors have occasions for discussions and interviews with OEMs in the country[a]. To interview IRs, the authors visited their companies and interviewed the managers on site. Case studies involving IRs derive from the authors' interviews with the presidents of five companies and executive directors of two companies. Interviews were semi-structured and lasted from one to several hours. Table 1 lists the companies on which case studies are based.

In the discussions and interviews, the questions asked were as follows. First, we asked about the basic features of the companies' businesses. The topics included the companies' profiles, remanufacturing practices, market size, market shares, businesses strategies, and areas of competence. Then, we inquired about the companies' motives and incentives for remanufacturing. Moreover, we asked about their major motives for remanufacturing. To IRs, in addition to these questions, we asked about the attitudes of OEMs (whose products they remanufacture) toward remanufacturing. Next, we asked what measures the companies take to meet the following three requirements for remanufacturing [20,26]:

1) Collection of used products
2) Development of efficient remanufacturing processes
3) Cultivation of demand for remanufactured products.

Interview data was supplemented with observations and secondary data. The IR participants' views of OEMs' attitudes toward remanufacturing were supplemented by the authors' interpretations because some interviewees talked implicitly. The results of the case studies follow.

Case studies
Photocopier machines
The remanufacturing of photocopy machines is a well-known example of remanufacturing. Three major OEMs of photocopy machines in Japan–Fuji Xerox, Ricoh, and Canon–have been undertaking remanufacturing activities [30-32]. Practices in other countries have also been studied [33-35].

Table 1 Companies discussed with or interviewed

Products	Company	Company size	Note
Photocopiers	Fuji Xerox Co., Ltd.	OEM, large	Discussions with managers
	Ricoh Co., Ltd.	OEM, large	Discussions with managers
	Canon Co., Ltd.	OEM, large	Discussions with managers
Single-use cameras	Fuji Film Co., Ltd.	OEM, large	Discussions with managers
Auto parts	Shin-Etsu Denso Co., Ltd.	IR, small	Three interviews with the president
	U-PARTS Inc.	IR, small	Twice interviews with the president
	Kaiho Sangyo Co., Ltd.	IR, small	An interview with the president
	Tsuneishi C values	IR, small	An interview with the executive director
	Asahi-parts	IR, small	An interview with the president
	NGP group	IR, small	A discussion with a manager
Printer ink cartridges	Ecorica Inc.	IR, small	An interview with the executive director
Toner cartridges	Association of Japan Cartridge Remanufacturers	Association of IRs	A discussion with a manager of a member company

Fuji Xerox, Ricoh, and Canon, account for about 90% of Japan's photocopier market. Until the 1970s, photocopiers were so expensive that they generally were rented by their users; however, after prices fell, sellouts and leasing became commonplace. Fuji Xerox started remanufacturing in 1990s, and Ricoh and Canon began selling remanufactured machines in the 2000s. Ricoh's and Canon's remanufactured products are made of reused components. According to Ricoh, 93% by weight of a typical remanufactured photocopy machine is composed of reused parts, its price is 50% to 70% less than prices of new products, and profits from remanufactured machines are larger than those from newly produced machines.

In Fuji Xerox's remanufacturing process, reused components are incorporated in new products. Thus, all products may include reused components and there is no distinction between new and remanufactured products. As far as the authors know, Xerox in the United States and Europe is not remanufacturing in this fashion. In Xerox's remanufacturing, as with Ricoh and Canon, remanufactured products are distinguished from new products, which are made exclusively from new components. The merit of the Fuji Xerox approach is that demand for reused components is not restricted by customers' product selection. On the contrary, when reused components are used solely in remanufacturing, if many customers prefer new products and avoid remanufactured products, components are not reused. Moreover, Fuji Xerox is said to have the highest ratio of reused components among the three companies.

The companies' motives for remanufacturing came from concerns about the environment and from corporate social responsibility. In addition, these companies are convinced that component reuse brings economic benefits. However, the benefits they expect are long-term. Fuji Xerox, for example, made a large investment

to renovate and adapt to remanufacturing that it took more than 10 years to recoup.

Photocopiers need frequent maintenances, which makes it easier for OEMs to manage product life cycles and thus to collect used products. Many products are leased to customers. OEMs take return delivery of leased products from the leasing companies in abundance. In case of sellout products, since in general, customers buy a new product in replacement of the old one, OEMs can reclaim their discarded product. The three OEMs form partnerships to collect and return each other's used products. However, third party intermediaries also buy used products, and as a result OEMs cannot re-acquire all used products. The ratios of takeback among OEMs vary.

The companies implement design for remanufacturing (DfReman) of products to facilitate remanufacturing, which substantially enhances the efficiencies of their remanufacturing processes. The companies have also been renovating remanufacturing processes and accumulating know-how. For example, Fuji Xerox developed a subparts cleaning method using chilled carbon dioxide gas. It is used to clean the frames of the photocopier machines and it substantially shortened cleaning time in remanufacturing.

Photocopiers are business equipments, and customers' aversion to remanufactured products in business equipment is generally lower than that in consumer products. In addition, product leasing could lower customers' aversion to remanufactured products. Ricoh and Canon offer remanufactured products with prices lower than those of new products. Fuji Xerox installs reused components in all products and there is no distinction between new and remanufactured products. Thus, demand for reused components is not restricted by customers' product selection. The companies' thorough quality controls have earned trust from customers, who show little dissatisfaction regarding products with reused components.

In sum, the following points were observed.

- Motive: long-term economic and environmental incentives
- Collection of used products: Companies accept returns from leasing companies in abundance and companies collaborate in collecting returns.
- Efficiency of remanufacturing processes: Efficiency has been achieved through DfReman, process renovation, and know-how accumulation.
- Cultivation of demand: Companies provide products to business and leasing customers, and they more readily accept remanufactured products than individual and sellout customers. Companies provide remanufactured products with lower prices (Ricoh and Canon). Reused components are installed in all products, and thus demand for used components is not restricted by customers' product selection (Fuji Xerox). Thorough quality controls have been carried out.

Single-use cameras

Remanufacturing of single-use cameras is another often-studied example of remanufacturing [11,36-38]. Single-use cameras began to appear in 1986, and three OEMs–Fuji Film, Kodak, and Konica–have dominated the Japanese market. In 1987, OEMs began to collect used products and recycle them. Fuji Film developed an automated production line for single-use cameras in 1992 and launched research and development into product designs that facilitated recycling and remanufacturing (Figure 1). In 1998, it developed a remanufacturing line that fully automated all processes–product disassembly, parts cleaning, parts inspection, parts replenishment, reassembly, and final testing. The parts–flash,

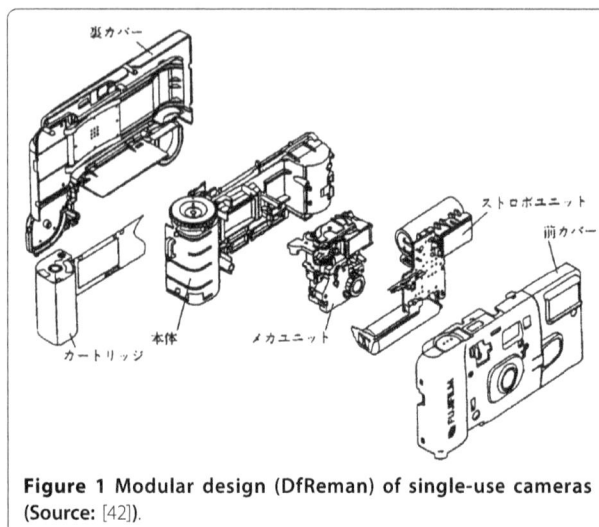

Figure 1 Modular design (DfReman) of single-use cameras (Source: [42]).

battery, plastic, mechanical parts–are reused, and if parts wear out, new parts are replenished. The company reports that more than 82% by weight of all camera components are reused or recycled.

Economic and environmental incentives are Fuji Film's motive to remanufacture. Before the company began remanufacturing, the waste disposal costs at film developing centers had been expensive. The economic incentive is long-term, as with photocopiers, and it took approximately 10 years to recoup its investment. Before OEMs undertook product remanufacturing, Japanese consumers had criticized them for wasting materials. Concerns about the environment and its customer image motivated Fuji Film to remanufacture. A conflict between sales of new and remanufactured products for the company never occurred because used parts are incorporated in all products and there is no distinction between new and remanufactured products.

Fuji Film's customers bring about 90% of single-use cameras to its centers to have their film developed. To transport the cameras back to its remanufacturing factory in Ashigara, Kanagawa Prefecture, Japan, the company simply reversed its pre-existing logistics for distributing supplies and chemicals from Ashigara to its development centers. The reverse logistics system was key to remanufacturing. The OEMs formed partnerships to collect and return each other's used products.

Fully automated remanufacturing is ideal for quality assurance and high efficiencies. DfReman of products was a prerequisite for the automation. Some parts are used only once, whereas others are used up to five times.

Consumers accepted remanufactured products well. The company carries out thorough quality control, and there have been few complaints regarding reused components. In addition, consumers' aversion to remanufactured products did not occur because there is not distinction between new products and remanufactured products, and the company could avoid demand cultivation problems.

The case is summarized as follows.

- Motive: long-term economic and environmental incentives
- Collection of used products: Fuji Film reversed the flow of its pre-existing forward logistics system
- Efficiency of remanufacturing process: The company developed a fully-automated remanufacturing process. DfReman was essential in developing the line.
- Cultivation of demand: Thorough quality controls were carried out. Used components are incorporated in all products, consumers' aversion to remanufactured products did not occur.

Auto parts

Auto parts are the most prevalent target of remanufacturing in the world. Up to two-thirds of remanufacturing businesses globally is estimated to involve auto parts [3]. In Japan, however, remanufacturing of auto parts is less common than in other developed countries. One reason is that the prevalence of automobiles in Japan is more recent compared with the United States and many European countries and thus auto parts remanufacturing has a briefer history. Auto parts remanufacturing saves material and energy. Manufacturing a new starter, for example, requires more than nine times the quantity of new material and about seven times more energy than remanufacturing a starter [3].

Japan's auto parts remanufacturers are primarily IRs. OEMs are generally reluctant remanufacturers because remanufacturing conflicts with sales of new parts. Profit margins on new auto parts are high–in some cases over 90%–whereas margins on remanufactured parts are lower.

In Japan, as in other countries, the remanufactured auto parts primarily include engines, turbo chargers, alternators, starters, compressors, transmissions, and steering units. The case study of Shin-Etsu Denso, one of the largest auto parts remanufacturers in Japan, shows the importance of assuring collection of used products, efficient remanufacturing processes, and demand for remanufactured products. This company remanufactures alternators and starters and ships about 100,000 of each annually.

To collect used products, the company supplies car maintenance shops with remanufactured products in exchange with used products. In addition, the company continually purchases and stocks used products from car dismantling companies. It stocks about 300,000 used products, which are essential for its business. Figure 2 shows the flow of the company's remanufacturing processes. Although the company has developed and accumulated know-how involving each of its processes, its

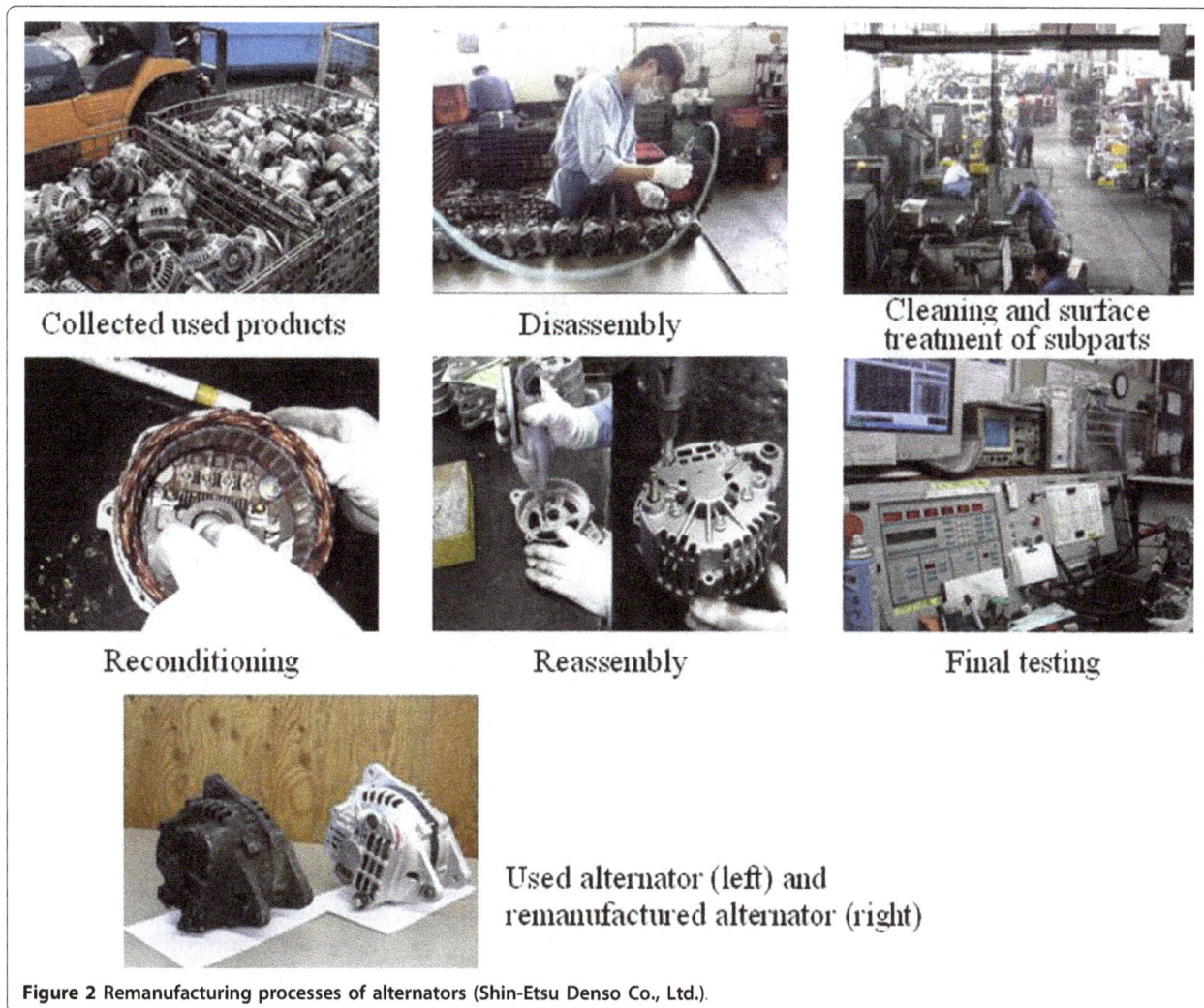

Collected used products

Disassembly

Cleaning and surface treatment of subparts

Reconditioning

Reassembly

Final testing

Used alternator (left) and remanufactured alternator (right)

Figure 2 Remanufacturing processes of alternators (Shin-Etsu Denso Co., Ltd.).

president indicated that know-how in cleaning and surface treatment of subparts is primarily important. His estimate reinforces previous studies showing that cleaning process is the most costly and knowledge-intensive process in auto parts remanufacturing [8]. Products can be remanufactured two to four times.

Until the early 1990s, there had been little demand for remanufactured auto parts in Japan, and Shin-Etsu Denso shipped most of its remanufactured products to the United States and Europe. However, since the late 1990s, the Japanese demand has increased, and today about 45% of its shipments (measured in yen) are for the domestic market. The company's thorough quality control (Figure 2) has enhanced users' confidence in remanufactured products and has helped to stimulate demand. Auto parts remanufacturers are cooperating with suppliers of reused auto parts, i.e., the car dismantling companies. The companies are forming networks to share information about inventories [14]. Car maintenance shops—the main buyers of reused and remanufactured parts—pass orders to member companies in the network. So partnerships with reused auto parts suppliers help remanufacturers to stimulate demand. The Japanese end-users' low recognition of remanufactured products is problematic for increasing demand further; an author's previous study found that nearly 60% of Japanese drivers know little about reused (including remanufactured) auto parts [39].

The auto parts case study presents the following observations:

- Motive: IRs' motives primarily come from economic incentives. Regarding OEMs, they face profit conflicts between remanufacturing and selling new auto parts and are reluctant to remanufacture.
- Collection of used products: Shin-Etsu Denso collects used products from car maintenance shops (by shipping remanufactured products in exchange with used products) and from car dismantling companies.
- Efficiency of remanufacturing processes: Companies have been developing and accumulating knowhow about processes, especially the cleaning and surface treatment of subparts.
- Cultivation of demand: Companies have emphasized quality control in order to build users' trust in and demand for remanufactured products. Remanufacturers cooperate with reused parts suppliers to fetch orders from car maintenance companies. Publicizing remanufactured auto parts is significant to further increase demand in Japan.

Printer ink cartridges and toner cartridges
In Japan, 200 million *ink cartridges* are sold annually, primarily for use in personal printers. Remanufactured cartridges account for 15 million in sales. Ecorica, an IR founded in 2003, is Japan's largest ink cartridge remanufacturer, shipping approximately 10 million remanufactured products annually (other 5 million products are provided by other IRs). Ecorica collects used cartridges from end-users, and remanufactures and sells them. Of the 30 million *toner cartridges* sold annually in Japan, mainly for office printers, 5 to 6 million are remanufactured. There are a number of independent toner cartridge remanufactures; 33 IRs formed the Association of Japan Cartridge Remanufacturers, and member companies account for 90% of remanufactured toner cartridges sold in Japan.

Some OEMs do remanufacturing of printer cartridges. But most of remanufacturing in the product areas is done by IRs, and OEMs generally respond negatively to such activities by IRs. In their normal business model, companies sell printers at low prices and earn most of their profit from selling ink and toner cartridges. In 2004, soon after Ecorica began to remanufacture ink cartridges, Epson, Japan's second-highest-share OEM of printers, sued Ecorica for intellectual property infringement, but Epson lost the case in 2008. Recently, OEMs began to recover used ink cartridges. Six OEMs—Epson, Canon, Hewlett-Packard, Brother, Dell, and Lexmark—collaborated to collect used ink cartridges. However, since remanufacturing ink cartridges costs more than manufacturing new cartridges, OMEs are unenthusiastic about remanufacturing. These companies are not active remanufacturers and merely recycle the collected cartridges. According to an IR interviewee, even an OEM which remanufactures cartridges is unenthusiastic about remanufacturing because it sells only on internet sites, not in shops, and at prices similar to that of new cartridges. It is possible that OEMs collect used ink cartridges to discourage IRs from remanufacturing. The executive director of Ecorica maintains that in recent years OEMs have designed products to make the remanufacturing process more difficult.

To collect used ink cartridges, Ecorica had placed 6,000 collection boxes in electronics retail stores nationwide in 2008. That year the company recovered 20 million used ink cartridges (5% of the ink cartridges sold in Japan). It remanufactured 15 million cartridges and sold 10 million of them (5 million was backlogged). Although OEMs efforts to collect used *ink cartridges* have had limited influence on IRs' business, their efforts to collect used *toner cartridges* have been more significant. It is no longer easy for IRs to collect used cartridges, which restricts growth of the remaining market.

The five processes of remanufacturing—inspection, disassembly, reconditioning, reassembly, and final testing—are labor-intensive. Ecorica has invested in the development of ink, the quality of which is crucial to its business, and in developing techniques to decode IC chips in cartridges.

Ecorica sells remanufactured ink cartridges at 20% to 30% below new product prices. It is attempting to increase consumers' recognition and demand by enhancing quality control and after-sales services. According to the executive director, to achieve further growth in the market requires increased collection of used cartridges alongside increasing demand for remanufactured ink cartridges.

The case is summarized as follows:

• Motive: IRs' motives for remanufacturing primarily come from economic incentives. OEMs derive profits on their printer products from the sale of ink and toner cartridges. Since profits on remanufactured cartridges are less than those on new cartridges, OEMs are indifferent remanufacturers and are hostile toward IRs' remanufacturing.
• Collection of used products: Ecorica (IR) opened a new collection channel, placing boxes in retail stores to collect used ink cartridges. Collection of toner cartridges is an effort to increase sales has been difficult.
• Efficiency of processes: IRs have invested in remanufacturing and have accumulated know-how such as developing ink and techniques to decode IC chips in cartridges.
• Cultivation of demand: Ecorica is attempting to increase consumers' recognition and demand by enhancing quality control and after-sales services.

Summary of case study results

OEMs' incentives to remanufacture and remanufacturers' efforts to meet the requirements of remanufacturing are summarized in Table 2 for each of the four types of products.

Review and results: Relevant Japanese legislation and its influence on remanufacturing

In Japan, legislations relevant to recycling of products are the Home Appliance Recycling Law and the End-of-Life Vehicle Recycling Law. Enacted in 2001, the former provides rules for collection and recycling of air conditioners, television sets, refrigerators, freezers, and washing machines. In effect since 2005, Japan's End-of-Life Vehicle Recycling Law requires OEMs to be responsible for collecting and recycling chlorofluorocarbons, airbags, and shredder dust for EOL vehicles. OEMs have a contract with car dismantling, shredding, and collecting companies, which handle take-back and recycling. Car owners pay the recycling fees when they buy the car.

These two laws have promoted material recycling and have helped mitigate Japan's landfill shortage. In discussions with OEMs, the authors found that the laws have motivated OEMs to implement environmentally conscious product designs that facilitate material recycling. For example, OEMs have designed products to facilitate product disassemblies and have attempted to decrease the variety of materials used in products. However, it was expected that the laws also would encourage OEMs

Table 2 Motives and companies' efforts to overcome the obstacles of remanufacturing businesses

Products	Main business segment	Motives for remanufacturing	Efforts to overcome the obstacles of remanufacturing businesses		
			Collection of used products	Development of efficient remanufacturing processes	Cultivation of demand
Photocopier machines	OEMs	Long-term economic and environmental incentives	Accepting returns from leasing companies in abundance Collaborating in collecting returns	DfReman Process renovation Development and accumulation of know-how	Strong quality control Incorporation of used components in new products (Fuji Xerox)
Single-use cameras	OEMs	Long-term economic and environmental incentives	Development of reverse logistics	DfReman Process renovation (automation)	Strong quality control In corporation of used components in new products
Auto parts	IRs	IRs: Economic incentives OEMs: Low (negative) economic incentives	Collecting used products in exchange with product shipment Purchasing from car dismantling companies	Development and accumulation of know-how	Strong quality control Cooperating with inventory networks of reuse auto parts to fetch orders from car maintenance shops
Printer cartridges	IRs	IRs: Economic incentives OEMs: Low (negative) economic incentives	Opening a new collection channel, placing boxes to collect used products (Ecorica)	Investing in remanufacturing Accumulation of know-how	Increasing consumers' recognition of the products Strong quality control Enhancing after services

to undertake remanufacturing as well as product servicing, and they seem not to have had that effect.

The Home Appliance Recycling Law requires consumers to pay the fees[b] when they dispose off products, not at the time of purchase. Although an expected increase in illegal dumping never materialized following the law's enactment, exports of end-of-life (EOL) products to foreign countries, primarily developing countries, have increased because consumers avoid recycling fees by handing EOL products to exporters rather than to retailers and OEMs. Japan generates about 20 million units of EOL home appliances. About one-third of these units are exported to foreign countries[c], and about half are returned to OEMs [40].

The End-of-Life Vehicle Recycling Law requires car owners to pay EOL recycling fees at the time of purchase, but the fees are refunded if owners sell cars to secondhand dealers (including exporters) rather than deliver them to car dismantling companies. Again, the law increased exports of EOL autos. In Japan, about 5 million cars are discarded annually. About 3.5 million are disposed of by domestic car dismantling companies, and 1.5 million are exported. Car dismantling companies are increasingly active in dealing with reused auto parts. However, during interviews with the authors, these companies indicated that increased exports of EOL autos impedes their collecting EOL products and is a significant obstacle in their reuse businesses.

Regarding reuse and remanufacturing operations in the worldwide scope, although both laws have increased exports of EOL products from Japan and have impeded remanufacturing within the country, most of the products exported are reused at the destined countries after being repaired there [41]. In other words, the laws could have stimulated reuse and remanufacturing in other countries. Further arguments are needed regarding EOL product exports and product reuse and remanufacturing in developing countries.

Discussion

In the case studies, we first examined Japan's major remanufacturers, particularly OEMs, and their motives for remanufacturing. OEMs remanufacturing photocopiers and single-use cameras, whereas IRs focus on remanufacturing auto parts and printer cartridges. Previous studies have indicated that OEMs have advantages over IRs in remanufacturing. However, OEMs face unique obstacles. For example, sales of remanufactured products may reduce their sales of new products, which customarily yield higher profit margins than remanufactured products. In such instances, OEMs have little incentive or have a negative attitude toward remanufacturing, as shown in the auto parts and printer cartridge case studies. Moreover, even though photocopiers and

single-use cameras are successful examples of OEM remanufacturing, establishing remanufacturing systems required OEMs to make large initial investments; it took over 10 years for Fuji Xerox and Fuji Film to recoup their initial investments. IRs might not need to make initial investments as large as OEMs. In general, OEMs pursue higher quality control levels than IRs for products from the initial stage of the business. This makes OEMs' initial investment expensive.

OEMs lack of incentive to remanufacture presents IRs with an opportunity, and IRs are expected to lead Japanese remanufacturing business. If IRs successfully create a market for remanufactured goods and stimulate consumers' demand, OEMs could be forced to become remanufacturers despite their reservations. Auto parts remanufacturing, for example, is more prevalent in the United States and Europe than in Japan, and some OEMs in these countries are active remanufacturers. The same could occur in Japan if end-users demand more remanufactured products, and demand could be cultivated through IRs' remanufacturing practices. This is expected to happen for many products worldwide.

Regarding the effects of relevant Japanese legislation on remanufacturing, Japan's Home Appliance Recycling Law and End-of-Life Vehicle Recycling Law have promoted material recycling, but have failed to stimulate remanufacturing. Even worse, both laws have increased exports of EOL products and have impeded IRs' remanufacturing operations in the country. Thus, there is a pressing need for institutional measures that stimulate remanufacturing. An important point in designing institutional measures is that, because IRs could lead remanufacturing even if OEMs are reluctant to remanufacture, and counteracting IRs' remanufacturing drives OEMs to begin remanufacturing [16], policy-making to encourage appropriate competition between OEMs and IRs could effectively stimulate remanufacturing. It is expected that remanufacturing will be stimulated through OEMs' and IRs' competition and through consumers' acceptance of remanufactured products.

Regarding the perspectives of remanufacturing in Japan, the markets for remanufactured products and reused products (i.e., secondhand products) have grown steadily in the last 10 to 20 years. This growth indicates that Japanese consumers have increasingly accepted remanufactured products. This Japanese market trend of remanufacturing growth seems destined to continue, at least in the product areas where remanufacturing already occurs. One possible obstructive factor for continued growth is the decreasing price of new products, particularly those imported from newly developing countries such as China. Remanufactured products often have to face competition from such products, and if consumers prefer cheaper new products to remanufactured products, the remanufactured market will shrink. To date, in auto parts and printer ink cartridge products, remanufactured products have been accepted by consumers more than the cheap, new, imported products. However, we need to monitor the direction of the market. To extend the scope of products remanufactured, it would be effective to refer to and consider adopting other countries' remanufacturing practices.

Conclusion

This study has analyzed cases of selected remanufacturing operations in Japan. We focused on remanufacturing in four product areas: photocopiers, single-use cameras, auto parts, and ink and toner cartridges for printers.

The study investigated companies' motives and incentives for remanufacturing. OEMs' motives are long-term economic and environmental incentives. However, OEMs often shun remanufacturing, fearing to cannibalize new product sales.

We highlighted three requirements for successful remanufacturing: (1) develop collection systems for used products; (2) develop efficient remanufacturing processes; and (3) cultivate demand for remanufactured products. Companies' efforts to meet these requirements were observed: (1) establishing a new collection channel, (2) developing reverse logistics to collect used products, (3) designing products for remanufacturing (DfReman), (4) accumulating know-how to establish remanufacturing processes, and (5) controlling product quality to stimulate demand for remanufactured products. Another important implication of this study is that (6) incorporating used components into new products increases the demand for remanufactured products. In Fuji Xerox's photocopier and Fuji Film's single-use camera businesses, used components are incorporated in all new products, with no distinction made between remanufactured and new products. The advantage of this mode of remanufacturing is that (1) the supply of remanufactured products is not restricted by the timing of returns of used products, (2) reuse ratios for components are not dictated by customer demand, and (3) OEMs avoid conflict between sales of new and remanufactured products.

Endnotes

[a]One of the occasions for the discussions and interviews with OEMs was the Inverse Manufacturing Forum, a Japanese industry-government-academia forum of which the authors are committee members and many OEMs are, or once were, the member companies.

[b]The law requires consumers to pay for collection and recycling; retailers collect the used appliances, and OEMs are responsible for recycling them. Under the law, OEMs determine the recycling fees, which currently are ¥2,500 for air conditioners, ¥2,700 for televisions, ¥4,600 for refrigerators, and ¥2,400 for washing machines (¥110 = €1).

[c]The main destination of the exports was once mainland China via Hong Kong, and today many are exported to Vietnam and the Philippines [41]. The exported EOL products are used in the destinations, but after use, many are processed in informal sectors and it partially causes the e-waste problem.

Abbreviations
DfReman: design for remanufacturing; EOL: end-of-life; IR: independent remanufacturer; OEM: original equipment manufacturer

Acknowledgements
This research is partially financially supported by Grant-in-Aid for Scientific Research (No. 20246130), JSPS, Japan.

Author details
[1]Center for Service Research, National Institute of Advanced Industrial Science and Technology (AIST), Umezono, Tsukuba, Japan [2]Department of Mechanical Engineering, Graduate School of Engineering, Osaka University, Suita, Osaka, Japan

Authors' contributions
MM and YU carried out discussions and interviews with photocopier OEMs, single-use camera OEM, and auto parts remanufacturer. MM independently carried out interviews with reused auto parts suppliers, and printer and toner cartridge remanufacturers for printers. Case analyses and the discussion section are based on the authors' discussion. MM drafted the manuscript.

Competing interests
The authors declare that they have no competing interests.

References
1. Ijomah W: **Addressing decision making for remanufacturing operations and design-for-remanufacture.** *International Journal of Sustainable Engineering* 2009, **2**(2):91-102.
2. Ijomah W, Bennett J, Pearce J: **Remanufacturing evidence of environmentally conscious business practices in the UK.** *Proceedings of the 1st International Symposium on Environmentally Conscious Design and Inverse Manufacturing (EcoDesign 99), Tokyo* 1999.
3. Steinhilper R: *Remanufacturing: The Ultimate Form of Recycling* Stuttgart: Fraunhofer IRB. Verlag; 1998.
4. Lund R: *The Remanufacturing Industry: Hidden Giant. Boston* Boston University. final report of Argonne National Laboratory study; 1996.
5. Lund R, Skeels F: *Guidelines for an original equipment manufacturer starting a remanufacturing operation. Government Report, DOE/CS/40192, CPA-83.8* Cambridge, MA: Massachusetts Institute of Technology, Center for Policy Alternatives; 1983.
6. Lund R: **Remanufacturing.** *Technology Review* 1984, **87**(2):19-23.
7. Haynsworth H, Lyons R: **Remanufacturing by design, the missing link.** *Production and Inventory Management* 1987, **28**(2):24-29.
8. Hammond R, Amezquita T, Bras B: **Issues in the automotive parts remanufacturing industry e a discussion of results from surveys performed among remanufacturers.** *International Journal of Engineering Design and Automation* 1998, **4**(1):27-46.
9. Guide VDR, Harrson T, Van Wassenhove LN: **The challenge of closed loop supply chains.** *Interfaces* 2003, **33**(6):3-6.
10. Bras B, McIntosh M: **Product, process, and organizational design for remanufacture - an overview of research.** *Robotics and Computer Integrated Manufacturing* 1999, **15**:167-178.

11. Toffel MW: **Strategic management of product recovery.** *California Management Review* 2004, **46**(2):120-141.

12. Ferguson M, Toktay L: **The effect of competition on recovery strategies.** *Production & Operations Management* 2006, **15**(3):351-368.

13. Linton J: **Assessing the economic rationality of remanufacturing products.** *Journal of Product Innovation Management* 2008, **25**(3):287-302.

14. Matsumoto M: **Business frameworks for sustainable society: A case study on reuse industries in Japan.** *Journal of Cleaner Production* 2009, **17**(17):1547-1555.

15. Ostlin J, Sundin E, Bjorkman M: **Business drivers for remanufacturing.** *Proceedings of 15th CIRP International Conference on Life Cycle Engineering. Sydney* 2008.

16. de Brito M, Dekker R: In *A framework for reverse logistics. Reverse Logistics Quantitative Models for Closed-loop Supply Chains.* Edited by: Dekker R, Fleischmann M, Inderfurth K, Van Wassenhove NL. Springer, Berlin; 2004:.

17. Seitz MA: **A critical assessment of motives for product recovery: the case of engine remanufacturing.** *Journal of Cleaner Production* 2007, **15**(11&12):1147-1157.

18. Pagell M, Wu Z, Murthy NN: **The supply chain implications of recycling.** *Business Horizons* 2007, **50**:133-143.

19. Ijomah W, McMahon C, Hammond G, Newman S: **Development of robust design-for-remanufacturing guidelines to further the aims of sustainable development.** *International Journal of Production Research* 2007, **45**(18&19):4513-4536.

20. Lundmark P, Sundin E, Bjorrkman M: **Industrial challenges within the remanufacturing system.** *Proceedings of Swedish Production Symposium. Stockholm* 2009, 132-139.

21. Steinhilper R: **Recent trends and benefits of remanufacturing: from closed loop businesses to synergetic networks.** *Proceedings of 2nd International Symposium on Environmentally Conscious Design and Inverse Manufacturing (EcoDesign 2001). Tokyo* 2001.

22. Guide VDR, Van Wassenhove LN: In *Business aspects of closed-loop supply chains. Business Aspects of Closed-loop Supply Chains: Exploring the Issues.* Edited by: Guide VDR, Van Wassenhove LN. Carnegie Bosch Institute, Pittsburgh, Pennsylvania; 2003:17-42.

23. Subramoniam R, Huisingh D, Chinnam RB: **Remanufacturing for the automotive aftermarket-strategic factors: literature review and future research needs.** *Journal of Cleaner Production* 2009, **17**(13):1163-1174.

24. Subramoniam R, Huisingh D, Chinnam RB: **Aftermarket remanufacturing strategic planning decision-making framework: theory & practice.** *Journal of Cleaner Production* 2010, **18**:1575-1586.

25. Matsumoto M: **Development of a simulation model for reuse businesses and case studies in Japan.** *Journal of Cleaner Production* 2010, **18**(13):1284-1299.

26. Geyer R, Jackson T: **Supply loops and their constraints: the industrial ecology of recycling and reuse.** *California Management Review* 2004, **46**(2):55-73.

27. Webster S, Mitra S: **Competitive strategy in remanufacturing and the impact of take-back laws.** *Journal of Operations Management* 2007, **25**:1123-1140.

28. Zuidwijk R, Krikke H: **Strategic response to EEE returns - product ecodesign or new recovery processes.** *European Journal of Operations Research* 2007.

29. Gerrard J, Kandlikar M: **Is European end-of-life vehicle legislation living up to expectations? Assessing the impact of the ELV Directive on 'green' innovation and vehicle recovery.** *Journal of Cleaner Production* 2007, **15**:17-27.

30. Tani T: **Product development and recycle system for closed substance cycle society.** *Proceedings of International Symposium on Environmentally Conscious Design and Inverse Manufacturing (EcoDesign 1999). Tokyo* 1999.

31. Tanaka H: **Research and development of environmentally conscious components: Photocopiers.** *Proceedings of International Symposium on Environmentally Conscious Design and Inverse Manufacturing (EcoDesign 1999). Tokyo* 1999.

32. Suzuki M, Subramanian R, Watanabe T, Hasegawa H: **The application of the international resource recycling system to encouragement of electronic waste recycling - The case of Fuji Xerox.** *Proceedings of IEEE International Symposium on Electronics and the Environment. San Francisco* 2008.

33. Berko-Boateng V, Azar J, de Jong E, Yander G: **Asset recycle management: a total approach to product design for the environment.** *Proceedings of IEEE International Symposium on Electronics and the Environment. Arlington* 1993, 19-31.

34. Azar J, Berko-Boateng V, Calkins P, de Jong E, George J, Hilbert H: **Agent of change: xerox design for the environment program.** *Proceedings of IEEE International Symposium on Electronics and the Environment. Orlando* 1995, 89-94.

35. Kerr W, Ryan C: **Eco-efficiency gains from remanufacturing: a case study of photocopier remanufacturing at Fuji Xerox Australia.** *Journal of Cleaner Production* 2005, **13**(9):913-925.

36. Guide VDR, Van Wassenhove LN: **The reverse supply chain: smart manufacturers are designing efficient processes for reusing their products.** *Harvard Business Review* 2002, **80**(2):25-26.

37. Sundin E, Lindahl M: **Rethinking design for remanufacturing to facilitate integrated product service offerings.** *Proceedings of IEEE International Symposium on Electronics and the Environment. San Francisco* 2008.

38. Duflou JR, Seliger G, Kara S, Umeda Y, Ometto A, Willems B: **Efficiency and feasibility of product disassembly: A case-based study.** *CIRP Annals - Manufacturing Technology* 2008, **57**:583-600.

39. Matsumoto M, Nakamura N, Takenaka T: **Business constraints in reuse services.** *IEEE Technology and Society Magazine* 2010, **29**(3):55-63.

40. Joint meeting of Central Environment Council and Industrial Structure Council WGs: **Results of actual condition survey of the flow of particular home appliances on their waste generation, take-back and disposals.** 2006 [http://www.env.go.jp/council/03haiki/y0311-05/mat02_1-1.pdf], (in Japanese).

41. Yoshida A, Terazono A: **Reuse of secondhand TVs exported from Japan to the Philippines.** *Waste Management* 2010, **50**:1063-1072.

42. Fukano :Edited by: Umeda Y. Inverse Manufacturing. Kogyochosakai, Tokyo; 1998:, (in Japanese).

14

An integrated approach to remanufacturing: model of a remanufacturing system

Ana Paula Barquet[1*], Henrique Rozenfeld[1] and Fernando A Forcellini[2]

Abstract

Remanufacturing is the process of rebuilding used products that ensures that the quality of remanufactured products is equivalent to that of new ones. Although the theme is gaining ground, it is still little explored due to lack of knowledge, the difficulty of visualizing it systemically, and implementing it effectively. Few models treat remanufacturing as a system. Most of the studies still treated remanufacturing as an isolated process, preventing it from being seen in an integrated manner. Therefore, the aim of this work is to organize the knowledge about remanufacturing, offering a vision of remanufacturing system and contributing to an integrated view about the theme. The methodology employed was a literature review, adopting the General Theory of Systems to characterize the remanufacturing system. This work consolidates and organizes the elements of this system, enabling a better understanding of remanufacturing and assisting companies in adopting the concept.

Keywords: Remanufacturing, System, Elements

Background

Increasing market competition, environmental concerns, changing customer requirements, and the emergence of new laws for end-of-life product management have led companies to seek new ways to maintain and expand their market share [1]. In this context, the adoption of product end-of-life strategies, which include recycling, reusing, and remanufacturing, have gained increasing importance in day-to-day business.

In reuse, according to Rose [2], the product and/or components are used immediately after their first cycle, i.e., they are second-hand goods. On the other hand, according to Thierry et al. [3], the purpose of recycling is to enable the reuse of the materials of used products and their components. In this case, the built-in energy, identity, and functionality of products and components are lost. Remanufacturing, however, preserves the shape and added value of products since the remanufactured product should be used for the same purpose it had during its original life cycle [4]. In remanufacturing, the used product returns to the production line, where it is disassembled, cleaned, reconditioned, inspected, and reassembled to ensure that the remanufactured product has the same quality as a new one. Additionally, less effort and resources are required for the recovery of products and their components compared with other product end-of-life strategies [5].

Among these strategies, remanufacturing is one of the preferable alternatives [6] since the remanufacturing process preserves part of the raw materials and value added to the product during its fabrication, allowing companies to increase their productivity and profitability [7]. Hence, in view of its environmental and economic benefits, remanufacturing is gaining significant ground in the global scenario.

However, it is difficult to achieve an integrated and systematic vision of all the issues involved in remanufacturing. Remanufacturing is a complex business due to the high degree of uncertainty in the production process, mainly caused by two factors: the quantity and the quality of returned products. Clearly, the lack of an integrated perspective of remanufacturing limits the possibilities for companies to evaluate and decide about whether or not to offer remanufactured products [8].

The potential of remanufacturing is underexploited in Brazil. There are still only few Brazilian companies showing awareness of environmental issues and commitment to the fate of the used products they manufacture, although this situation is expected to change soon pursuant to the newly

---

* Correspondence: anabarquet@gmail.com
[1]São Carlos School of Engineering, University of São Paulo (USP), Av. Trabalhador São-carlense, 400, São Carlos, Sao Paulo 13566-590, Brazil
Full list of author information is available at the end of the article

enacted National Policy on Solid Waste. Acting to change this situation, these companies can increase the degree of competitiveness *vis-à-vis* foreign companies, gain new customers, and survive in the market, as well as contribute toward more sustainable production and consumption.

Companies find it difficult to implement and consolidate remanufacturing for several reasons, including a lack of knowledge about the theme, a lack of consideration about the strategic issues of remanufacturing [8], and the scantiness of studies indicating how to implement it [6]. Thus, it is clear that this theme, which is a new one especially for Brazilian companies, is in its exploratory phase, which explains the importance of a structured review of the literature about remanufacturing.

In this paper, remanufacturing is treated as a system. A system is considered a set of interdependent elements that interact to achieve an objective and perform a given function. It is the elements and their relationships to each other that determine how the system works, forming a unitary, organized, and complex whole [9].

The majority of authors discuss isolated elements of remanufacturing, making it difficult to gain an integrated view of them, which are treated separately and in different contexts. It is believed that the conceptuation of a model for the remanufacturing system can help companies understand and implement remanufacturing. In agreement with Östlin [10], characterizing the remanufacturing system contributes substantially toward understanding the problems and difficulties involved in remanufacturing.

Thus, the objective of this work is to organize the body of knowledge about remanufacturing by means of a model that offers a vision of the remanufacturing system, the elements in this system, and how they interact with each other, contributing toward an integrated view of the theme.

This paper is divided into introduction and context. The next section presents a literature review about remanufacturing, with the findings organized so as to draw up a model of the remanufacturing system. The elements of this system are characterized, and their difficulties and practices described, as well as their interdependencies and interconnections.

Method

This work was developed based on General System Theory [11] and by means of a cross-analysis of the elements and characteristics of remanufacturing found in the literature review. Since we intend to present these elements within a systemic vision, we use the General System Theory, the objective of which is to study the elements that make up a system as well as the interactions between them [11]. Studying each element separately does not lead to an exact conclusion of the system in which these elements are inserted for their interactions are fundamental to understanding the

system as a whole. As we found in the literature, this is the case of papers that deal with remanufacturing, which in the most part discuss its elements separately.

General System Theory arose from the need to understand the problems of today's complex world; however, analyzing them separately and dealing with them piecemeal to fit theoretical problems and problems resulting from modern technology does not suffice. A system, or 'organized complexity', can be defined as a set of elements governed by 'strong interactions' [11]. Uhlmann [12], based on Bertalanffy, sees a system as a set of elements interrelated to each other and to the environment. The next section presents our literature review about remanufacturing, in which concepts are organized to define and characterize the remanufacturing system and the elements that make up this system.

Remanufacturing system

An organizational system can be considered a set of dynamic and interdependent parts and functions with shared objectives. These systems are open and may belong to larger systems and contain smaller ones. They present specific objectives and complex structures. Since a system is larger than the sum of its parts, the investigation of any part of a system should involve it as a whole.

The remanufacturing system proposed here is inspired by Östlin [10]. According to this author, this system begins with the collection of the used product or parts, also named as core, followed by its remanufacturing and delivery of the remanufactured product to the client. Thus, the remanufacturing system comprises internal processes, such as the remanufacturing operation, and external processes involving the collection of cores and delivery of the remanufactured product.

Seeking to complement Östlin's proposal [10], this paper proposes the following elements and sub-elements for the remanufacturing system:

- Element 1: Design for remanufacturing
- Element 2: Reverse supply chain (RSC)
  - Sub-element 2.1: Acquisition/relationship with the core supplier
  - Sub-element 2.2: Reverse logistics (RL)
- Element 3: Information flow in the remanufacturing system
- Element 4: Employees' knowledge and skills in remanufacturing
- Element 5: Remanufacturing operation
- Element 6: Commercialization of the remanufactured product

Figure 1 illustrates the proposed remanufacturing system. The first element is the design for remanufacturing, which is part of the product development process and is responsible for the product's design, with a view to its end of

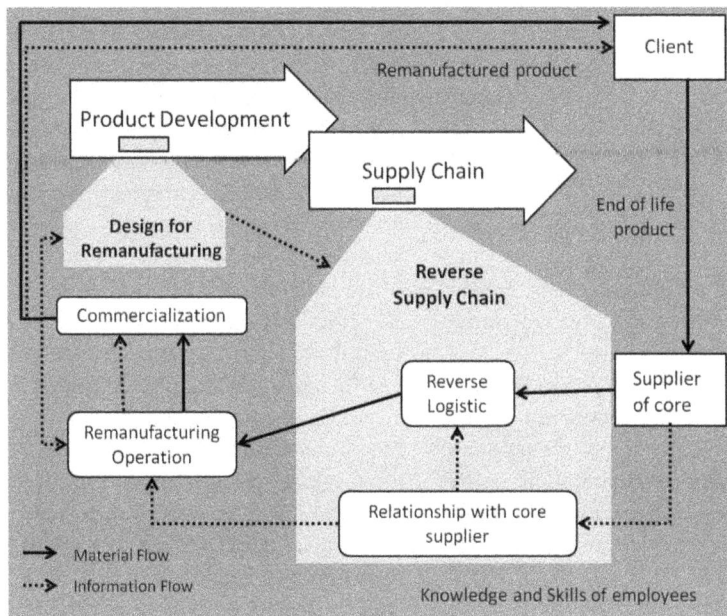

Figure 1 Remanufacturing system model (adapted from [13]).

life and how to facilitate its remanufacturing (e.g., facilitate disassembly). This element provides information for the reverse supply chain, particularly for element 5 (remanufacturing operation; e.g., disassembly sequence). The reverse supply chain, in turn, is composed of two sub-elements: acquisition/relationship with the core supplier and reverse logistics. The last element is the commercialization of the remanufactured product. Among these elements, there are both information and material flows. The main actor of this system is the end client, who becomes the supplier at the end of life of the product.

Given the importance and influence of the information flow throughout the remanufacturing system, it is also considered an element that permeates and interconnects all the others. The flow of materials was not mentioned as an element because it is mainly part of the sub-elements of the reverse supply chain. It should also be kept in mind that a crucial factor for the feasibility of the system is the employees' knowledge and skills in remanufacturing (element 4), which are required in all the processes. Each of the aforementioned elements is characterized in the following items, describing their characteristics, difficulties, and practices.

Element 1: design for remanufacturing
Characteristics
Approximately 80% of the environmental impacts of products are determined during their development, more specifically in the concept phase, which emphasizes the responsibility of product development teams to address issues related to the service life of products [14]. This

underlines the importance of designing the product considering the most suitable end-of-life strategies for their reuse after their use, such as remanufacturing [15].

A variety of pressures challenge companies to alter their product development paradigms. The sanction of legislation on manufacturer responsibility, allied to increasing global competition and the potential for the recovery of used products to make use of their residual value, encourages companies to design products with greater durability and facility to reuse them at their end of life [16].

For Andrue (in [16]), remanufacturable products have the following characteristics:

- The product contains a component or part that allows for its reuse.
- There is availability in the supply of such components or parts.
- The product and/or its parts can be disassembled and reused according to the original specifications.
- The product and/or its parts has high added value in relation to its market value and its original cost.
- The product and the process are stable.

Examples of the product's characteristics that affect remanufacturing, identified by Ijomah et al. [16], are listed below:

- Technological changes: At the time the product is to be remanufactured, its technology may have become obsolete. If the product cannot be updated during remanufacturing, its reuse is unnecessary.

- Business model of services: Development of business model that allows for a combination of products and services, e.g., a product service system.
- Environmental legislation: This type of legislation may require companies to reuse the product at its end of life and make it more expensive to discard, for instance.

During product conception, it is important to take into account strategies for updating the product due to rapid changes in technology as well as clients' needs. In this case, too, design for remanufacturing can give companies a significant market advantage [15].

Integrating the design for remanufacturing in the product development process optimizes the achievement of the benefits of remanufacturing by companies (reduction of energy, materials and wastes, among others). This is the case of companies such as Xerox Corporation (Norwalk, USA), which recognizes this factor as an opportunity to obtain competitive advantages [15].

Difficulties and practices
Many of the challenges related to remanufacturing are consequences of how the products were designed [16,17]. In order to support the company's product development process, several ecodesign methods and tools have been developed, which consider remanufacturing and product end-of-life issues [18].

Interdependence and interaction among the elements
A limiting factor for putting design for remanufacturing into practice is the low level of knowledge of product designers concerning about end-of-life strategies, such as a remanufacturing. This is due to the fact that a product's conception is usually centered on its functionality and costs, in detriment to environmental issues [16]. Thus, this indicates a relationship between the elements' design for remanufacturing and employees' knowledge and skills.

It should also be noted that to perform design for remanufacturing, product designers require specific expertise to develop the product with a view to its easy future remanufacture, in other words, to facilitate the efficient execution of the remanufacturing operation.

Element 2: reverse supply chain
Characteristics
According to Guide and Van Wassenhove [19], RSC is a set of stages needed to collect the core, followed by the application of the desired end-of-life strategy (remanufacturing, reuse, recycling or disposal) [1]. Driven by cost reductions through product reuse and by the customer's heightened perception of value, many supply chains have increased their involvement in activities that go beyond the product's service life, extending its life cycle [20].

The RSC begins with the collection of products from clients and/or companies in different links of the supply chain, and collection sources tend to be geographically dispersed. The next phase involves inspection or tests performed at the collection site, at a receiving center, or at the site where the product will be reused. At this point, a decision is made about the destination of the collected product, which presents various possibilities of reuse, such as remanufacturing [21].

An example of how a reverse supply chain works is described by Guide and Van Wassenhove [22]. To reuse used mobile phones, it is necessary first to gain access to a sufficient number of telephones of suitable quality and at the right price (acquisition of the used product). The telephones must be transported and stored (reverse logistics); after which, an end-of-life strategy is selected for the product (including remanufacturing). After they have been remanufactured (remanufacturing operation), product commercialization strategies are devised. Some recommendations to achieve efficiency in the reverse supply chain are the following [23]:

- Structure a team specialized in contacts with the used product supplier, with a view to standardizing the processes involved in the collection and increasing these products' chances for reuse;
- Make a forecast of the time of return of the product, which may be based on its sales and its service life;
- Align the reverse supply chain with the direct supply chain and achieve the effectiveness of the activities of a closed-loop supply chain.

Difficulties and practices
The phases of the RSC are treated as a series of independent stages that are dealt with separately, without considering their integrated nature. Moreover, business and academia are doing very little about the strategic issues of the RSC [1]. The majority of studies are centered on technical and operational issues since the focus on technical activities is attractive for initial investigations [22].

Because RSCs are not yet part of the core competencies of companies, it is difficult to organize and align their stages, obtain the necessary resources, and catch the attention of top management. The direct supply chain requires similar requisites, but the issues and the context of reverse supply chains are still little understood, more complex, and receive scanty attention [1].

The reverse supply chain presents some characteristics that make managing and planning its stages and activities complex. For example, when a company collects cores, its supplier is usually the end client, which makes it difficult to gain access to a sufficient number of cores at the moment of return. Another complication is the quality of these cores, which requires efficient inspection [24]. In addition, there is the need to disassemble the collected products, to set up a reverse logistics network, and to deal with the high variability in processing times [25].

Sub-elements of the RSC

The sub-elements of the RSC considered in this paper, which were adapted from the works of Guide and Van Wassenhove [22], and Blackburn et al. [26], are the following:

- Acquisition/relationship with the used product supplier: contact with the supplier for the acquisition of cores by the remanufacturer;
- Reverse logistics: the activities of transport, storage, and distribution of the products that will be reused.

Acquisition/relationship with the used product supplier

1. Characteristics. Sundin et al. [27] concluded that the remanufacturer's relationship with his supplier of cores is an extremely important aspect for the effectiveness of the business, which is consistent with the findings of Östlin et al. [28].
2. Difficulties and practices. If the collected product is destined for remanufacturing, it is important to emphasize that before the product is selected for return by the reverse flow, its remanufacturability should be 'pre-assessed' to avoid it being transported without it serving for reuse, which would incur additional costs. This trade-off between potential reuse and additional costs is difficult and requires employees with certain skills and experience [29]. Employees that perform this pre-assessment should be part of the team responsible for collecting cores because if the product's remanufacturability is evaluated at the remanufacturing company and this product is deemed unfit for reuse, its transportation to the company will have been unnecessary.

The large number of suppliers of small quantities of cores and the diversity of their conditions makes it difficult for the companies that receive them for remanufacturing to control their quality, as does the lack of closeness in the relationship between the remanufacturer and his supplier. Moreover, when selecting their suppliers, many remanufacturers choose the ones that offer the lowest prices [6]. This may lead to the purchase of cores in poor conditions to be remanufactured.

In some cases, the suppliers of cores are the clients that discard the product due to its end-of-service life or for other reasons, such as the launch of a more modern product. In this case, there are some problems concerning the lack of motivation to get these customers to return used

products to remanufacturing companies [30]. Below are some situations that may influence this lack of motivation:

- The remanufacturer's lack of contact with the client [28];
- The client's lack of knowledge and confidence about remanufacturing [31], e.g., the economic and environmental advantages of remanufacturing.

Thus, companies that want to succeed should think about remanufacturing strategies that encourage the client to make this return and that bring him close to the remanufacturing company [28].

Reverse logistics

1. Characteristics. According to the Council of Logistics Management [32], reverse logistics is the process of planning, implementing, and effectively controlling the flow of components, materials undergoing processing, end product, and related information from the consumption point to the source point [23].
2. Difficulties and practices. Reverse logistics is one of the great challenges of the remanufacturing system due to the difficulty in predicting product volumes, return times, and quality conditions, which makes planning of the remanufacturing operation difficult [24]. In recent years, RL has received more attention from managers due to its strategic implications [33].

Some of the issues that make it difficult for companies to implement RL are the lack of a system that integrates the activities of direct and reverse logistics [34], the difficulty of measuring the impact and of controlling the return of products and materials, and the fact that reverse flow is considered a cost for companies and therefore is given little or no priority as a business strategy [32]. The costs of reverse logistics activities can rely on different actors. The company that will remanufacture can be the one to handle with this cost by acquiring cores directly from clients. In a different scenario, these cores can be stored in deposits by an intermediate actor that will sell them to the one responsible for performing remanufacturing. In both cases, there is a great probability that the costs of reverse logistics activities are added to the remanufactured product final costs.

Hence, most companies are uninterested in implementing RL. In addition, there is a lack of studies by companies to assess the effects of the practice of RL on the success of organizations, the relationship between the actors involved in the activities of reverse logistics is poorly structured, there is little closeness between the plants and the suppliers of cores, and the cost of shipping is higher due to the lower volumes transported (Ballou 2006 in [35]).

Unlike distribution in direct logistics, which is designed to transport large volumes of the same product from the manufacturer to a few local clients, in reverse logistics, the product mix may vary considerably and the volume may be very low. This can make economic transportation difficult to achieve [29].

Another aspect to consider is the fact that, contrary to the situation in direct logistics, cores are not packaged and are therefore unprotected and susceptible to damage, which limits their recovering. Therefore, a packaging system is necessary to protect the product's residual value [29]. Pires [35] points out some characteristics specific to RL, which affect its efficiency:

- A convergent network structure, i.e., products from numerous sources and with few destinations;
- Geographically dispersed sources, usually not homogeneous in quantity, availability, and quality of product and/or parts with each other and over time, making planning activities difficult;
- Difficulty in achieving an economy of scale due to the small quantity of products collected from each source;
- Higher tendency for products to stay longer in reverse channels, resulting in higher inventory, transport, and storage costs, as well as reduced income due to the possibility of product obsolescence and degradation;
- Entrance of products into the flow that should not enter (e.g., products that cannot be reused), generating unnecessary costs;
- When the client is the core supplier, RL depends on his willingness to cooperate and reinsert the post-consumer material into the reverse logistics flow.

For Lacerda [36], the main factors that influence the efficiency of reverse logistics activities are good input and output controls, mapped and formalized processes, short cycle times, accurate information systems, planned logistics network, and collaborative relationships between clients and suppliers.

Element 3: information flow in the remanufacturing system
Characteristics
The main role of the information flow in the remanufacturing system is to deal with uncertainties concerning the return of products. An efficient information flow is an important tool to reduce these uncertainties and to help establish an effective system.

Due to the characteristics of the remanufacturing system, the information required for planning the system becomes accessibly very late, hampering its operation. If the remanufacturer does not reduce the lead time of information, the coordination of the remanufacturing operations and reverse logistics is impaired, thereby increasing the costs of the system [6]. Information about the product is important both in planning the reuse activities and in avoiding the transportation and reprocessing of products that have no potential to be recovered.

Difficulties and practices
To reduce these uncertainties and their consequences on the remanufacturing system, the remanufacturer must manage the following information [6]:

- Which products should be returned to the remanufacturer?
- When will these products arrive?
- Where are these products located?
- How many of these products can be remanufactured?

According to Thierry et al. [3], information related with product return management can be classified in four categories:

- Information about the composition of the product: types of material, their quantities, value, potential of harmfulness to nature, and how the different types of materials are combined.
- Information about the magnitude and uncertainty of the return flow: according to the type of commercialization chosen for the product, e.g., traditional sale, leasing, rental.
- Information about the market for remanufactured products: the perceived difference in the quality and cost of remanufactured and new products affects the acceptance of these products.
- Information about how product returns are currently done includes an analysis of the organizations involved, the obstacles, and the quantity of product that is remanufactured (for each returned product), the costs, and the overall environmental impact of the remanufacturing system.

Most companies experience difficulties in obtaining accurate information about aspects of product return management in their supply chains. However, companies can obtain such information by collaborating with their suppliers and others in the chain. To obtain this information, it is essential to deal with issues involving the characteristics of the product, the supply of cores, and the demand for remanufactured products, and to balance the supply and demand. Obtaining and processing this information require the development of an adequate information system [3].

Element 4: employees' knowledge and skills in remanufacturing

Characteristics

From the moment of contact with the supplier, acquisition, collection of cores (transport, storage), and the phases of the remanufacturing operation (inspection, disassembly, cleaning, etc.), to the sale of the remanufactured product, the employees should be perfectly familiar with the remanufacturing system in order to deal correctly with each stage of the system.

Difficulties and practices

According to Jacobsson [6], the success of the implementation of the remanufacturing operation often does not require a more highly qualified work force than the one in the manufacturing operation, but the qualifications required in each of these operations are different. In the remanufacturing operation, the employees should be trained and qualified to deal with variability and uncertainties, especially insofar as quality and quantity of cores are concerned.

Uncertainties lead to situations in which resources will sometimes be scarce and at other times abundant, thus requiring people with a good vision of the overall operation and with the necessary flexibility to deal with the different stages of the operation [37]. For the aforementioned authors, both qualified and non-qualified employees are necessary since non-qualified people are usually better able to think outside the box and contribute with new ideas.

Training for the remanufacturing operation should preferentially be given by the remanufacturing company itself which can train its employees according to the specificities of its operations and the characteristics and complexity of the product. For example, employees responsible for disassembling products should take them apart without damaging them. Cleaning, handling the product, and assembling it require less specialized qualifications, less precision, experience, and skills than inspection and testing [6].

Element 5: the remanufacturing operation

Characteristics

The remanufacturing operation begins with the arrival of the core at the remanufacturer's facilities, where it will go through several stages that include its complete disassembly, cleaning of its parts, inspection, reconditioning of the parts that will be reused, replacement of non-remanufacturable components, and assembly, resulting in a remanufactured product. This product is then tested to ensure that its quality is equivalent to that of a new product [16,31,38]. The order of these activities may differ according to the characteristics and type of the product [16]. It should be noted that, in this paper, the steps of reconditioning and replacement of used parts for new ones, with or without possible product updates, will be called reprocessing.

Next, a brief description of the stages of remanufacturing operation is given, according to Steinhilper [39]:

- Product disassembly: The purpose of this stage is the total disassembly of the product. This is one of the most time-consuming activities due to its degree of complexity.
- Cleaning of the components: Each of the components is cleaned with a different cleaning product, according to the material composing it. Four process variants can be cited that contribute to this cleaning: chemical effects (e.g., detergents), influence of temperature (e.g., heat), mechanical action (e.g., removal by high-pressure water jetting), and time (e.g., duration of the process).
- Inspection and storage of components: This stage proposes an identification to classify the components as well as an inspection to determine which should be replaced or allocated for other purposes, such as cannibalization, repair, reconditioning, or recycling. Storage refers to the site where the material will be deposited for subsequent assembly of the products.
- Reconditioning and replacement of components and parts (reprocessing): Components and parts are recovered. Therefore, some of them are replaced with new ones because they do not satisfy the minimal requirements to ensure the quality of the remanufactured product.
- Product reassembly: This consists of the assembly of the remanufactured product. A final test will ensure that the remanufactured product performs similarly to a new one, with the same characteristics, functionalities, and quality.

Steinhilper [39] states that the final test should not be considered a step but a certification that the product will have the same characteristics as a new one. The stages of the remanufacturing operation may follow a different order, depending on the type of product to be remanufactured.

Difficulties and practices

Remanufacturing operations require lower investments than manufacturing operations since the number of new parts produced is smaller, and a large part of the efforts and resources required has already been invested by the original manufacturer (Lund and Skeels 1983 in [4]).

Element 6: commercialization of the remanufactured product

Characteristics

This element encompasses sales, distribution, and relationship with the client who bought the remanufactured product. Additionally, it explores potential market segments and strategies to increase the attractiveness of remanufactured products.

Remanufactured products can be commercialized in different ways. Some companies may choose to sell them, while others will turn to leasing, using them as replacement products for warranties or selling their functionality to the client. The clients' preferences, the nature of the product, and its technical maturity are the main factors that influence the decision about which channel to use.

Atasu et al. [40] conducted a study involving a marketing approach to remanufacturing, related with the market demand for remanufactured products. These authors treat remanufacturing as a strategic marketing tool that strongly enhances the company's competitive advantage, which differs from the idea of remanufacturing as cost savings or an obligation with legal implications. The authors concluded that the main factors that influence the decision to remanufacture are competition, market growth, and increase in the 'green' range of the market.

An important decision about the commercialization of remanufactured products is whether or not they will be sold through the same distribution channels as new products. Because remanufactured products cost less than new products, they may cannibalize the sales of new ones. Therefore, many companies do not offer remanufactured products together with new ones. This is the case of Dell, Inc. (Round Rock, USA), which has a separate site for the sale of remanufactured products (www.delloutlet.com). On the other hand, selling remanufactured products together with new ones enables companies to better segment their market and to sell also to clients who cannot afford new products [41].

Difficulties and practices

Few studies explain how to place remanufactured products on the market [6]. In this regard, remanufacturers face major challenges because even though there is an increasing demand for environmentally attractive products, the remanufactured product contains parts, components, or materials that have been used previously. Therefore, marketing strategies must be adapted to market this product, especially with respect to issues such as below [6]:

- Market segment: Lower prices for products with the same performance as new ones expand the market range by making them accessible to clients who cannot afford new original products.
- Buying behavior: The remanufactured product offers the same functionality as a new product, at a lower price, but may not offer the client the same shopping experience.
- Client profile: Issues involving 'fashion' and 'currentness of the fashion' also affect remanufacturing. Clients may prefer the latest version of the product, regardless of the quality and cost of the remanufactured

alternative. Some clients demand novelty, and the selection and purchase of new products is a lifestyle.
- Perception about the product: Even if the price of remanufactured products is lower, some clients are not interested in them because of a perceived risk.
- Warranty: Remanufactured products should come with a guarantee that they meet the client's needs just like a new product would.
- Complementary services, e.g., warranties and maintenance.

From the standpoint of how the remanufactured product is commercialized, it has been demonstrated that the product-service system (PSS), through leasing or the offer its functionality, appears to be a promising approach [27]. Therefore, from the marketing point of view, products whose function is prioritized by the client are the most suitable ones for remanufacturing [6].

PSS can be defined as the result of a strategic innovation in the business model of companies, in which the focus shifts from the design and sale of physical products to the offer of a system of products and services that, together, can meet customer expectations. Moreover, the value is functionality rather than the physical properties of individual products [42].

The PSS is based on a fundamental change in the relationship between the manufacturers and consumers of a product and/or service. Instead of focusing on the traditional form of sale, consumption, and disposal of the product, the PSS focuses on the delivery of a function to the client, which means a combination of products and services that, together, meet the client's needs [43].

Another important point is the need to gain the client's confidence in the remanufactured product. In a case study with a manufacturer of copier machines, which remanufactures its products, it was observed that the sales department faces challenges to persuade customers of the equivalent quality of remanufactured and new copiers [3]. Thus, ensuring the performance, reliability, and quality of remanufactured products is essential to creating and sustaining the demand for them.

Results and discussion

Analyzing the literature and characterizing the elements of the system, one sees a considerable interdependence and interaction among the elements. Firstly, with regard to the RSC, several points are relevant.

The degree of structuring of a RSC can be defined based on the existence of structured organizational practices of core returns, the relationship and information exchanged among companies that belong to the reverse chain, and the level of resources these companies make available, e.g., employees' skills in remanufacturing [6,21,32,44]. The proper management of reverse chains also serves as an excellent source of information about clients' expectations and habits, contributing for the company to provide differentiated services and to increase the value perceived by its clients [21].

A relationship with the core supplier can augment information about the product's remanufacturing conditions, since the remanufacturer will have knowledge about the performance of the product during its service life. This knowledge is useful for the remanufacturing operation as well as for the improvement of the conception of the product as a whole and for remanufacturing (design for remanufacturing).

Still with regard to the RSC, the activities of reverse logistics require skills and information to correctly carry out the transportation, storage, and warehousing of cores. Information about the volume, condition, and time of return of products is also essential for planning the remanufacturing operation.

It is essential for the employees involved in the commercialization of the remanufactured product to be properly trained, to inform the client about the benefits of the product, about what remanufacturing is and how it works, and to dispel doubts regarding the quality of the product [41].

It is also clear that an efficient information flow is necessary to underpin the relationship of the core supplier with the remanufacturer, since information about the availability and quality of the core is essential for the remanufacturing system to work properly, particularly for the planning of remanufacturing operations. Information flow between the remanufacturer and the designers of the product is also essential in case of doubts emerging during the remanufacturing operation about how the product was designed, or even for suggestions to be made to the designers, aimed at improving the design for remanufacturing.

It is also clear that the skills, knowledge, and experience of employees involved in the stages and activities of the remanufacturing system are important. The literature places particular emphasis on the importance of employees qualified for the steps of the remanufacturing operation, since this is when the used product undergoes the transformations needed to turn it into a remanufactured product with the same quality as a new one [6,37,45].

In addition, the type of product and its complexity, which are issues of design for remanufacturing, influence the sequence and difficulty of the stages of the remanufacturing operation [38]. Jacobsson [6] also mentions the importance of the qualifications of the work force for the design for remanufacturing and to deal with the financial and legal aspects of remanufacturing.

Conclusions

This work involved the organization of the body of knowledge about remanufacturing by conceptualizing the remanufacturing system and its elements, contributing toward an integrated vision and expanding the theoretical knowledge about the theme. The elements of this system were consolidated and organized, enabling a better understanding of remanufacturing and facilitating the work of future studies as well as of companies that are restructuring their remanufacturing operations or intend to start them.

This paper clearly shows the interactions and interdependence among the elements of the remanufacturing system. Dividing the elements in this paper was not intended to omit or conceal these interactions, but instead, to make it easier to understand and organize the remanufacturing system.

In terms of its academic contribution, this paper aims to promote knowledge about remanufacturing and the development of studies on the theme, particularly in Brazil, since there are still very few companies that remanufacture and few studies about remanufacturing in the country. It is also hoped that companies will feel encouraged to implement remanufacturing, since this research describes characteristics and provides a better overall understanding about the remanufacturing system.

Competing interests
The authors declare that they have no competing interests.

Authors' contributions
APB has done the literature research on the main publications concerning the remanufacturing topic. HR and FAF supported on the development of this article. All the authors read and approved the final manuscript.

Authors' information
APB studied Pharmacy and Food Technology in the Federal University of Santa Catarina (2007). In 2010, she got her masters degree in Industrial Engineering at the Federal University of Santa Catarina and her dissertation was focused on the development of the first version of the remanufacturing system. Now, she is a Ph.D. candidate in the Industrial Engineering Department, School of Engineering of São Carlos, University of São Paulo. She has experience in industry in the area of product development and food technology, and researches on the following topics: product development, remanufacturing, product-service systems. HR studied Mechanical Engineering in the University of São Paulo (1980), has masters degree in Mechanical Engineering from the University of São Paulo, and has a dissertation on Production Planning and Control (1983). HR has a Ph.D. degree in Systematic Production in WZL RWTH Aachen, and his thesis was on Planning Process Computer CAPP (1988). He is a professor at the University of São Paulo, in the Faculty of Engineering School of São Carlos (EESC) USP since 1982. Nowadays, he is the deputy head of the Production Engineering, the coordinator of the Graduate Program in Production Engineering, and the coordinator of the Center for Advanced Manufacturing (NUMA), and Integrated Engineering Group Engineering and Integration (GEI2) of NUMA in University of São Paulo. He is currently researching on the development of products and services, and product lifecycle management and has a international project on BRAGECRIM Program which the main topic of research is remanufacturing. HR is supervisor of APB during her Ph. D. and supported the improvement of the first version of the remanufacturing system by means of case study execution in companies that do remanufacturing. FAF studied Mechanical Engineering in the Federal University of Santa Catarina, has masters degree in Mechanical Engineering from the Federal University of Santa Catarina, has Ph.D. in Mechanical

Engineering from the same university (1994), and has post-doctorate degree in Industrial Engineering from the University of São Paulo in 2008. FAF is currently the associate professor at the Federal University of Santa Catarina. He researches in the areas of development, modeling, improvement and management of processes, products and services. FAF was the supervisor of APB during her master degree and supported the development of the first version of the remanufacturing system, which was improved in this paper.

Acknowledgements
The authors would like to extend sincere thanks to Conselho Nacional de Desenvolvimento Científico e Tecnológico (CNPq) and Coordenação de Aperfeiçoamento de Pessoal de Nível Superior (CAPES) for supporting this research topic.

Author details
[1]São Carlos School of Engineering, University of São Paulo (USP), Av. Trabalhador São-carlense, 400, São Carlos, Sao Paulo 13566-590, Brazil. [2]Industrial Engineering Department, Federal University of Santa Catarina (UFSC), Campus Reitor João David Ferreira Lima, Florianópolis Santa Catarina 88040-970, Brazil.

References
1. Guide Jr, VDR, Harrison, TP, Van Wassenhove, LN: The challenge of closed-loop supply chains. Interfaces 33(6), 3–6 (2003)
2. Rose, C: Design for environment: a method for formulating product end-of-life strategies. Stanford University, Dissertation (2000)
3. Thierry, M, Salomon, M, Nunen, JAEE, Van Wassenhove, LN: Strategic issues in product recovery management. Calif. Manage. Rev. 37(2), 114–135 (1995)
4. Amezquita, T, Hammond, R, Salazar, M, Bras, B: Characterizing the remanufacturability of engineering systems. In: Proceedings of ASME Advances in Design Automation Conference. Boston, Massachusetts (1995)
5. Lindahl, M, Sundin, E, Östlin, J: Environmental issues within the remanufacturing industry. In: Proceedings of LCE: 13th CIRP International Conference on Life Cycle Engineering, Katholieke Universiteit Leuven. Belgium (2006)
6. Jacobsson, N: Emerging product strategies: selling services of remanufactured products. Lund University, Dissertation (2000)
7. Giuntini, R, Gaudette, K: Remanufacturing: the next great opportunity for boosting US productivity. Bus Horiz 46(6), 41–48 (2003)
8. Ferguson, ME, Toktay, LB: The effect of competition on recovery strategies. INSEAD, Fontainebleau (2004)
9. Johnson, RA, Kast, FE, Rosenweig, JE: The Theory and Management of Systems. McGraw-Hill, New York (1963)
10. Östlin, J: On remanufacturing systems: analyzing and managing material flows and remanufacturing processes. Linkoping University, Dissertation (2008)
11. Bertalanffy, VL: Teoria Geral Dos Sistemas. Vozes, Petrópolis (1975)
12. Uhlmann, GW: Teoria Geral Dos Sistemas: Do Atomismo Ao Sistemismo. Instituto Siegen, São Paulo (2002)
13. Barquet, APB, Rozenfeld, H, Forcellini, FA: Remanufacturing System: characterizing the reverse supply chain. In: Camarinha-Matos, LM, Pereira-Klen, A, Afsarmanesh, H (eds.) Adaptation and Value Creating Collaborative Networks, pp. 556–563. Springer, Heidelberg (2011)
14. Park, M: Sustainable consumption in the consumer electronics sector: design solutions and strategies to minimize product obsolescence. In: Proceeding of the 6th Asia Pacific Roundtable for Sustainable Consumption and Production., Melbourne (2005)
15. Nasr, N, Thurston, M: Remanufacturing: a key enabler to sustainable product systems. In: Proceedings of LCE. 13th CIRP International Conference on Life Cycle Engineering, Katholieke Universiteit Leuven., Belgium (2006)
16. Ijomah, WL, McMahon, CA, Hammond, GP, Newman, ST: Development of design for remanufacturing guidelines to support sustainable manufacturing. Robot Comput Integrated Manuf 23(6), 712–719 (2007)
17. Hatcher, GD, Ijomah, WL, Windmill, JFC: Design for remanufacture: a literature review and future research needs. J Clean Prod 19, 17–18 (2011)
18. Pigosso, DCA, Zanette, ET, Guelere Filho, A, Ometto, A, Rozenfeld, H: Ecodesign methods focused on remanufacturing. J Clean Prod 18, 21–31 (2010)
19. Guide Jr, VDR, Van Wassenhove, LN: The reverse supply chain. Harv Bus Rev 80(2), 25–26 (2002)
20. Corbett, C, Savaskan, C: Contracting and coordination in closed-loop supply chains. In: Dekker, R, Fleischmann, M, Inderfurth, K, van Wassenhove, LN (eds.) Quantitative Models for Closed Loop Supply Chain Management, pp. 1–23. Springer, New York (2002)
21. Kopicki, R, Berg, MJ, Legg, L: Reuse and recycling-reverse logistics opportunities. Oak Brook, United States (1993)
22. Guide Jr, VDR, Van Wassenhove, LN: The evolution of closed-loop supply chain research. Oper Res 57(1), 10–18 (2009)
23. Fioravanti, RD, Carvalho, MFH: Aplicações de modelos de cadeia reversa em uma operação de serviços: estudo de caso no setor de serviços de impressão. XI SIMPOI - Symposium on Production, Logistics and International Operations Management, FGV. São Paulo, Brazil (2008)
24. Guide Jr, VDR: Production planning and control for remanufacturing: industry practice and research needs. J Oper Manag 18, 467–483 (2000)
25. Guide Jr, VDR, Jayaraman, V, Linton, JD: Building contingency planning for closed-loop supply chains with product recovery. J Oper Manag 21(3), 259–279 (2002)
26. Blackburn, JD, Guide Jr, VDR, Souza, GC, Van Wassenhove, LN: Reverse supply chains for commercial returns. Calif Manage Rev 46(2), 6–22 (2004)
27. Sundin, E, Ostlin, J, Rönnbäck, AÖ, Lindahl, M, Sandström, GÖ: Remanufacturing of products used in product service system offerings. In: Proceeding of the 41st CIRP conference on manufacturing systems. Tokyo, Japan (2008)
28. Östlin, J, Sundin, E, Björkman, M: Importance of closed-loop supply chain relationships for product remanufacturing. Int J Prod Econ 115, 336–348 (2008)
29. Ferrer, G, Whybark, DC: From garbage to goods: successful remanufacturing systems and skills. Business Horizons 43(6), 55–64 (2000)
30. King, AM, Burguess, SC: The development of a remanufacturing platform design: a strategic response to the directive on waste electrical and electronic equipment. Proc IMechE Part B: J Eng Manufacture 219, 623–631 (2005)
31. Seitz, MS: A critical assessment of motives for product recovery: the case of engine remanufacturing. J Clean Prod 15, 1147–1157 (2006)
32. Rogers, DS, Tibben-Lembke, RS: Going Backwards: Reverse Logistics Practices and Trends. Reverse Logistics Executive Council, Reno, Nevada (1998)
33. Daugherty, PJ, Autry, CW, Ellinger, AE: Reverse logistics: the relationship between resource commitment and program performance. J Bus Logist 22(1), 107–123 (2001)
34. Daher, CE, Silva, EPS, Fonseca, AP: Logística reversa: oportunidade para redução de custos através do gerenciamento da cadeia integrada de valor. Brazilian Business Review 3(1), 58–73 (2006)
35. Pires, N: Modelo para a logística reversa dos bens de pós-consumo em um ambiente de cadeia de suprimentos. Universidade de Santa Catarina, Thesis (2007)
36. Lacerda, L: Logística Reversa, uma Visão sobre os Conceitos Básicos e as Práticas Operacionais. Center of Logistics Research, Rio de Janeiro (2004)
37. Hermansson, H, Sundin, E: Managing the remanufacturing organization for an optimal product life cycle. In: Yamamoto, R (ed.) Proceedings of the fourth international symposium on environmentally conscious design and inverse manufacturing, Tokyo, 12–14 December 2005, pp. 143–156. IEEE, New York (2005)
38. Sundin, E: Product and process design for successsful remanufacturing. Linkoping University, Thesis (2004)
39. Steinhilper, R: Remanufacturing: the ultimate form of recycling. The Remanufacturing Institute. (1998). http://www.reman.org/Publications_main.htm. Accessed May 23, 2012
40. Atasu, A, Sarvary, M, Van Wassenhove, LN: Remanufacturing as a marketing strategy. Manag Sci 54(10), 1731–1746 (2008)
41. Ovchinnikov, A: Revenue and cost management for remanufactured products. Prod Oper Manag 20(6), 1–17 (2011)

42. United Nations Environment Programme (UNEP): Product-service Systems and Sustainability. Opportunities for Sustainable Solutions. UNEP, Paris (2002)

43. Goedkoop, MJ, Van Halen, CJG, Riele, HRMT, Rommens, PJM: Product Service Systems: Ecological and Economic Basics. VROM and Economic Affairs, Netherlands (1999)

44. Leite, PR: Logística Reversa: Meio Ambiente e Competitividade. Prentice Hall, São Paulo (2003)

45. Ferrer, G: Yield information and supplier responsiveness in remanufacturing operations. Eur J Oper Res 149, 540–556 (2003)

15

A facility location model for socio-environmentally responsible decision-making

Dominic Ansbro and Qing Wang[*]

Abstract

The consideration of external costs is becoming more important in supply network design, as companies are under increasing pressure to reduce the environmental and social impacts of their operations. This paper presents a single time period, single-product mixed integer linear programming formulation, which considers such external costs, as well as the impact of waste disposal. The model presented considers a network of suppliers, manufacturing facilities, customers, scrap recyclers, general recycling facilities and landfill sites and makes facility location and allocation decisions so as to minimise both the economic and external costs of all network operations. The model was formulated using the What's Best Excel add-in and tested on a commercial case study concerning the supply network operations of Hydram, a leading sheet metal fabrication company, considering three different scenarios. Details of how the external and economic costs were determined are included, with reference to the literature. By analysis of the experimental results, commercial recommendations for facility location are made, and the managerial uses of the model for socio-environmentally responsible decision-making are discussed. The benefits and limitations of the proposed model are also discussed.

Keywords: External costs; Mixed integer linear program; Supply network; Facility location-allocation problem

Background

Introduction

A supply chain, or supply network, may be defined as an integrated process whereby raw materials are acquired, converted into products and delivered to customers [1]. Research attention on the design and analysis of supply networks has increased, and it has become apparent that companies that wish to remain competitive must increasingly pay attention to their supply networks and aim to increase the efficiency of their logistics operations [2]. Furthermore, companies are under increasing pressure to behave in an environmentally and socially responsible manner. By considering sustainability issues, companies can reduce costs whilst enhancing their reputations among customers and investors [3], leading to increased profits and sales revenues. In other words, the consideration of sustainability issues can be an important contributor to successful business performance.

The increasing attention to supply network design and sustainability issues as separate disciplines has naturally led to sustainability considerations becoming more important within the field of supply network design. A 2010 survey [4] in accordance with the UN Global Compact and Accenture found that company CEOs increasingly believe that sustainability issues should be fully integrated into the company supply network, strategy and operations, with 88% of the CEOs interviewed citing that they should be integrating sustainability through their supply chains. The study also finds a 'significant performance gap' between the companies who embed sustainability throughout their supply network and those who do not, indicating the importance of such considerations. Whilst these findings highlight the current importance of considering sustainability issues, various factors such as tightening environmental regulation, increasing environmental concern of customers and increasing demand for 'green' goods and services mean that sustainability considerations will only become more significant in the future [5].

One method of embedding sustainability into the supply network is through the consideration of the external costs of supply network operations during decision-making. The ExternE Project defines an external cost (also known as an externality) as a cost that 'arises when

* Correspondence: qing.wang@durham.ac.uk
School of Engineering and Computing Sciences, Durham University, Durham DH1 3LE, UK

the social or economic activities of one group of persons have an impact on another group and when that impact is not fully accounted or compensated for, by the first group' [6]. Broadly, external costs quantify the effects of a process or action on the environment and society. The most efficient solution to externalities is to require them to be included in the costings of the engaged activity [7]. This method of taking account of external costs is known as 'internalisation', whereby policies such as taxation or environmental regulations are employed to incentivise their minimisation. Internalisation of external costs is becoming a common strategy for ensuring sustainable development [6], evidenced by stricter environmental regulations, and rising tariffs for activities that result in external costs such as the landfill tax in the UK.

Supply network operations result in various external costs. Transportation activities cause environmental impacts, accidents and congestion [8], whilst the treatment of waste causes pollution and disamenity effects [9]. Disamenity effects are localised impacts that generate negative local reactions, reflected in reduced house prices in the area surrounding a waste treatment facility.

Although external costs often do not affect companies directly, it is suggested that their consideration during supply network design, in conjunction with traditional economic considerations, would allow the designer to make decisions that are not only cost-effective, but also socially and environmentally responsible. It is argued that this will equip them for the likelihood of stricter regulations and higher environmental taxes in the future and also enhance their customer reputation by projecting an ethos of corporate responsibility.

The rest of the paper is structured as follows: the rest of the 'Background' section presents a literature review to outline past research in the field of sustainable supply chain design, and the quantification of external costs. In the 'Methods' section, the methods employed are presented, including the presentation of a generic model for determining an optimal supply network structure, taking socio-environmental externalities into account, and the application of the model to a case study from the sheet metal industry. The 'Results and discussion' section presents and discusses the results of the case study, and the 'Conclusions and further work' section presents the conclusions of the report and discusses possible extensions to the model.

Literature review

A great deal of research has already been conducted on the mathematical location modelling for supply network design, and most of the literature focuses on the key questions of location and allocation [10]. Deterministic analytical models are the most common type for supply chain design and analysis, cost minimisation is usually the objective function [1], and most of the past literature

fails to consider external costs. However, environmental considerations are incorporated within the objective function in [11], whilst [12] and [13] suggest the use of objective functions in future research that consider factors other than just cost minimisation, such as environmental costs and responsiveness. It is argued that the incorporation of social and environmental impacts in the objective function would give a more comprehensive picture of the total supply chain cost and would be useful to help managers to make more responsible supply chain decisions.

Organizations are facing growing pressure to increase environmental awareness and act in a socially responsible manner. These philosophies are being incorporated throughout all business operations, including their supply chains. As such, interest in reverse logistics is interesting, and the number of publications has been growing steadily. The field of reverse logistics is clearly significant: terms such as recycling, reuse and remanufacture are now known to the general public. As shown by [14], the inclusion of reverse logistics processes in the supply chain can have environmental, economic and social benefits. All of these factors justify reverse logistics as an important and relevant area of research.

One way in which sustainability considerations have been considered within the supply chain design is through the consideration of reverse logistics processes, fitting with the trend that manufacturers are becoming more responsible for the recovery of their products [15]. Mixed integer linear programming (MILP) reverse logistics network models are proposed in [5], [16] and [17], based on traditional warehouse location models. Although reverse flows have been considered in a range of publications for reuse and remanufacture, the consideration of waste disposal appears to be an area that has been neglected in much of the reviewed supply network design literature.

Much of the literature neglects the costs of landfilling, despite the fact that sending waste to a landfill has both economic costs (in the form of tipping fees) and external costs [14]. However, [13] and [15] consider the external costs of landfilling operations, which include impacts to human health, crops, materials and buildings [9]. Landfill sites also cause disamenity costs associated with odour, dust, litter, noise, vermin and visual intrusion [18].

Transport activities also result in significant costs to the environment and society, caused by accidents, congestion, noise and pollution. The contribution of transport to greenhouse gas emissions is widely realised, and transport accounted for 23% of global CO_2 emissions in 2007 [19]. The external costs associated with freight transportation are particularly relevant: the total costs caused by goods vehicles in 2000 for the EU15 amounted to EUR 135.85 billion [20]. If 'real prices', which incentivise the best choice of mode of transport for sustainable mobility, are to be used in transport, then internalisation must be pursued, and not just at the minimum level for political acceptance [21]. Despite the obvious significance of transportation externalities, at the time of research, their inclusion into supply network models also appears to have been neglected in past research.

Methods
Problem definition and mathematical modelling

In this section, the problem to be solved and the mathematical formulation of the sustainable supply network design model will be presented in detail. The model is inclusive of waste disposal considerations and considers external costs of supply chain operations within the objective function as an attempt to contribute some innovative research to the field of supply network design and management, as these points seem to have been neglected in past research.

The logistics network discussed in this paper is a multi-stage forward logistics network, including customers, potential facilities, raw material suppliers, landfill facilities, recycling facilities for general waste and scrap recyclers who pay for scrap material. As shown in Figure 1, the facilities receive raw material from an allocated supplier. The raw material is converted into a product to satisfy the demand of customers. In producing a unit of product, the facilities also produce a given amount of scrap which is sold to an allocated scrap facility. It is assumed that all of the raw material is either converted into either product or scrap. During operations, the factory also produces a given amount of domestic waste for every amount of produced product.

The domestic waste is either recyclable general waste which is sent to an allocated recycling facility or nonrecyclable general waste which is sent to an allocated landfill site.

Opening a facility in the network incurs a designated cost. Additionally, all transportation operations have economic and external costs, and the disposal of general waste has differing environmental and economic implications depending upon the method of disposal. The model assumes negligible production costs.

This structure can be translated into a MILP facility location model, as below. The objective of the model is to determine which facilities to open, how much product to produce at the open facilities and how to allocate the product to customers so as to minimise the sum of economic and external costs. The use of a weighting factor on the external cost allows the user to decide the extent to which these economic costs are included.

Index sets

$G = \{1,...N_G\}$ Set of material suppliers, $\forall\, g \in G$
$I = \{1,...N_I\}$ Set of potential facilities, $\forall\, i \in I$
$J = \{1,...N_J\}$ Set of customers, $\forall\, j \in J$
$K = \{1,...N_K\}$ Set of recycling sites, $\forall\, k \in K$
$L = \{1,...N_L\}$ Set of landfill sites, $\forall\, l \in L$
$M = \{1,...N_M\}$ Set of scrap purchasing centres, $\forall\, m \in M$

Costs
Economic costs

f_i — Per time-period operating cost of facility i

CT_{giv} — Transportation cost per unit distance per unit of raw material v from supplier g to facility i

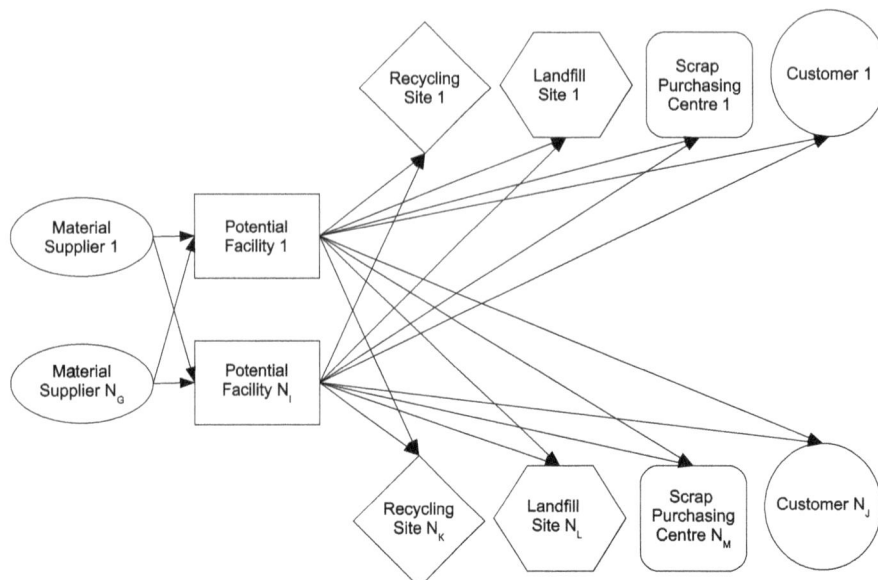

Figure 1 A network diagram illustrating the logistics network considered.

CT_{ijp} Transportation cost per unit distance per unit of product p from facility i to customer j

CT_{ikr} Transportation cost per unit distance per unit of recyclable general waste r from facility i to recycling site k

CT_{ilq} Transportation cost per unit distance per unit of non-recyclable general waste q from facility i to landfill site l

CT_{imz} Transportation cost per unit distance per unit of scrap waste z from facility i to scrap recycling centre m

CD_{lq} Unit disposal cost (landfill tax) at landfill site l of non-recyclable general waste q

CD_{kr} *Unit revenue at scrap recycling centre m of scrap waste z*

External costs

XT_{giv} External cost of transportation per mile per unit of raw material v from supplier g to facility i

XT_{ijp} External cost of transportation per mile per unit of product p from facility i to customer j

XT_{ikr} External cost of transportation per mile per unit of recyclable general waste r from facility i to recycling site k

XT_{ilq} External cost of transportation per mile per unit of non-recyclable general waste q from facility i to landfill site l

XT_{imz} External cost of transportation per mile per unit of scrap waste z from facility i to scrap recycling centre m

XD_{lq} External cost of disposing one unit of non-recyclable general waste q at landfill site l

XD_{kr} External benefit of recycling one unit of recyclable general waste r at recycling site k

XD_{mz} External benefit of disposing of one unit of scrap z at scrap purchasing centre m

Parameters

T_{gi} Distance between supplier g and facility i
T_{ij} Distance between facility i and customer j
T_{ik} Distance between facility i and recycling centre k
T_{il} Distance between facility i and landfill site l
Z_{gv} Supply capacity of supplier g for raw material v
Z_{ip} Capacity of potential facility i for product p
Z_{kr} Capacity of recycling site k for recyclable general waste r
Z_{lq} Capacity of landfill site l for non-recyclable general waste q
Z_{mz} Capacity of scrap recycling centre m for scrap waste z
$Y_{\min}$ Minimum number of facilities to open
$Y_{\max}$ Maximum number of facilities to open
d_{jp} Demand of customer j for product p
α Weighting factor for the inclusion of external costs

b_{pv} Amount of raw material v required per unit of produced product p

b_{pr} Amount of recyclable general waste r produced per unit of produced product p

b_{pq} Amount of non-recyclable general waste q produced per unit of produced product p

b_{pz} Amount of scrap waste z produced per unit of produced product p

Decision variables

$$Y_i = \begin{cases} 1 \text{ if a potential facility is opened at location } i \\ 0 \text{ otherwise} \end{cases}$$

Flow variables

s_{ijp} Amount of product p transported from facility i to customer j

$s_{giv} = b_{pv}\, s_{ijp}$ Amount of raw material v transported from supplier g to facility i

$s_{ikr} = b_{pr}\, s_{ijp}$ Amount of recyclable general waste r transported from facility i to recycling site k

$s_{ilq} = b_{pq}\, s_{ijp}$ Amount of non-recyclable general waste q transported from facility i to landfill site l

$s_{imz} = b_{pz}\, s_{ijp}$ Amount of scrap waste z transported from facility i to scrap recycling centre m

Objective function

The objective function minimises the sum of the economic costs (A_1) and external costs (A_2) by setting the decision variable and flow variables. The weighting factor α allows the user to determine the extent to which external costs are included:

$$\text{Min } A = A_1 + \alpha A_2 \tag{1}$$

$$
\begin{aligned}
A_1 = &\sum_{i\in I} f_i Y_i + \sum_{g\in G}\sum_{i\in I} CT_{giv}s_{giv}T_{gi} + \sum_{i\in I}\sum_{j\in J} CT_{ijp}s_{ijp}T_{ij} \\
&+ \sum_{i\in I}\sum_{k\in K} CT_{ikr}s_{ikr}T_{ik} + \sum_{i\in I}\sum_{l\in L} CT_{ilq}s_{ilq}T_{il} \\
&+ \sum_{i\in I}\sum_{m\in M} CT_{imz}s_{imz}T_{im} + \sum_{i\in I}\sum_{l\in L} CD_{lq}s_{ilq} - \sum_{i\in I}\sum_{k\in K} CD_{kr}s_{ikr}
\end{aligned} \tag{2}
$$

$$
\begin{aligned}
A_2 = &\sum_{g\in G}\sum_{i\in I} XT_{giv}s_{giv}T_{gi} + \sum_{i\in I}\sum_{j\in J} XT_{ijp}s_{ijp}T_{ij} \\
&+ \sum_{i\in I}\sum_{k\in K} XT_{ikr}s_{ikr}T_{ik} + \sum_{i\in I}\sum_{l\in L} XT_{ilq}s_{ilq}T_{il} \\
&+ \sum_{i\in I}\sum_{m\in M} XT_{imz}s_{imz}T_{im} + \sum_{i\in I}\sum_{l\in L} XD_{lq}s_{ilq} \\
&- \sum_{i\in I}\sum_{k\in K} XD_{kr}s_{ikr} - \sum_{i\in I}\sum_{m\in M} XD_{mz}s_{imz}
\end{aligned} \tag{3}
$$

The economic cost formulation (2) includes the cost of opening facilities, the economic cost of all transportation operations, the cost of landfilling non-recyclable waste and the revenue achieved from the sale of scrap (represented as a negative cost). The external cost formulation (3) includes the external cost of all transportation operations and the environmental cost of landfilling waste.

Constraints

$$\sum_{g\in G} s_{giv} = \sum_{j\in J} s_{ijp} + \sum_{m\in M} s_{imz} \quad \forall i\in I \tag{4}$$

$$\sum_{j\in J} s_{ijp} \leq Z_{ip} Y_i \quad \forall i\in I \tag{5}$$

$$\sum_{i\in I} s_{giv} \leq Z_{gv} Y_i \quad \forall g\in G \tag{6}$$

$$\sum_{i\in I} s_{ikr} \leq Z_{kr} Y_i \quad \forall k\in K \tag{7}$$

$$\sum_{i\in I} s_{ilq} \leq Z_{lq} Y_i \quad \forall l\in L \tag{8}$$

$$\sum_{i\in I} s_{imz} \leq Z_{mz} Y_i \quad \forall m\in M \tag{9}$$

$$\sum_{i\in I} s_{ijp} \geq d_{jp} \quad \forall j\in J \tag{10}$$

$$Y_i \in \{0, 1\} \quad \forall i\in I \tag{11}$$

$$s_{giv} \geq 0 \quad \forall g\in G, \forall i\in I, \forall v\in V \tag{12}$$

$$s_{ijp} \geq 0 \quad \forall i\in I, \forall j\in J, \forall p\in P \tag{13}$$

$$s_{ikr} \geq 0 \quad \forall i\in I, \forall k\in K, \forall r\in R \tag{14}$$

$$s_{ilq} \geq 0 \quad \forall i\in I, \forall l\in L, \forall q\in Q \tag{15}$$

$$s_{imz} \geq 0 \quad \forall i\in I, \forall m\in M, \forall z\in Z \tag{16}$$

$$Y_{min} \leq \sum_i Y_i \leq Y_{max} \quad \forall i\in I \tag{17}$$

Balance of material in the potential facilities is guaranteed by constraint (4), which ensures that all of the raw material going into a potential facility is either converted into a product or scrap. Constraints (5) to (9) are capacity constraints for facilities, suppliers, recycling sites, landfill sites and scrap purchasing centres. Constraint (10) ensures that all customer demand is satisfied. Constraint (11) defines the plant opening decision variable as binary, and constraints (12) to (16) ensure that there are no negative flows of raw material, product, recyclable general waste, non-recyclable general waste or scrap so that all flows follow the arrow directions indicated in

Figure 1. Constraint (17) dictates that the number of facilities to open must be between specified bounds.

It was decided that the model should be tested on a case study for validation purposes and to assess its functionalities. The model proposed is highly generic and theoretically widely applicable, as the logistics problem considered is familiar to many companies. However, there were still some challenges to find an appropriate case study to test the model. Firstly, it was decided that the model should be tested on a company which is genuinely looking to solve a facility location problem, to ensure that the results of this paper are useful from a commercial perspective as well as an academic perspective. Secondly, it was realised that in order for the model to be applied properly to a case study, a company should be found with a sufficient amount of data readily available regarding the amount of raw material used, the amount of waste, scrap and product produced, as well as demand data and customer locations for a given time period.

Case study: Hydram
Brief
After analysing various companies according to the aforementioned criteria, Hydram was chosen as a suitable case study upon which the MILP model could be evaluated. Hydram is a subcontract sheet metalwork and fabrication company, based in County Durham, with customers throughout the UK. The sheet metalwork products produced at Hydram are suitable for modelling as a single product, using weight as the demand quantity. The model presented in this paper is a single time period model, and a time period of 1 year was selected as appropriate for analysis of the Hydram problem.

Hydram is looking at the possibility of opening a smaller 'satellite' facility, either in Yorkshire or in London. Using commercial property websites, three potential facilities were located for evaluation, in Halifax, Bradford and London.

Data provided
Firstly, sales data were provided for the 12-month period between 1 March 2010 and 1 March 2011. This detailed the total spend during the period for 64 customers. Data were also provided for the amount of metal raw material purchased during this period and the amount of metal that was scrapped. Environmental key performance indicators' data were also provided, detailing all general waste disposed of by landfill and recycling during the same time period, as well as the total fleet fuel consumption.

The sales data for the period closely follow the Pareto principle, with the top 20.3% of customers accounting for 81.20% of sales. It was decided that it would be sufficient to consider only the top 31 customers in the model, as they account for 88.5% of all demand. As such, all other data were scaled to 88.5% of the original figures to accommodate the neglect of the bottom 33 customers.

Using the data provided, standard units were determined for use in the model. Material flows are in tonnes (t), distances in kilometres (km) and costs in pounds sterling (£). The time period considered is 1 year. The rest of this paper will use these units as standard.

Fitting the data to the model

Assuming that all purchased material is either turned into product or scrapped, the total amount of product produced was calculated as the difference between the amount of raw material purchased and the amount of metal scrapped. Using this calculation, a value was calculated for the amount of product produced per customer £ spent. The results of these calculations, along with some other key input data, are shown in Table 1.

Using the total product produced per customer £ spent value, the customer sales data were next converted to sales in terms of weight for each customer, to be used as d_{jp} inputs for the model. The general waste volumes were also converted to weights, assuming that average general waste has a density of 170 kg/m^3 [22]. Finally, using the data in Table 1, the ratios b_{pv}, b_{pr}, b_{pq} and b_{pz} were calculated. These ratios are shown in Table 2.

Using the customer postcodes, the distances between every potential facility and every customer (T_{ij} distances) were calculated. Additionally, each potential facility was allocated a local material supplier, scrap purchasing centre, landfill site and recycling facility. Although it is not necessary in the model to allocate a material supplier, scrap purchasing centre, landfill site and recycling facility to each potential facility, after correspondence with Hydram, it was deemed that this would be the most likely option. Therefore, for the purposes of the case study, $G = I = K = L = M$.

Using online map tools, the distances between these allocated facilities and the potential facilities were

Table 1 Key input data

Parameter	Value
Total customer sales	£5,078,964.85
Total metal purchased (t)	1,336.32
Total metal scrapped (t)	553.58
Total product produced (t)	782.73
Total fleet fuel consumption (L)	27,827.36
Total general waste recycled (m^3)	476.80
Total general waste landfilled (m^3)	100.55
Total product produced per customer £ spent (t/£)	0.000154113

Table 2 Material ratios

Ratio	Calculation	Value
b_{pv}	Total metal purchased/Total product produced	1.707
b_{pr}	Total general waste recycled/Total product shipped	0.104
b_{pq}	Total general waste landfilled/Total product shipped	0.022
b_{pz}	Total metal scrapped/Total product shipped	0.707

calculated. A network map of the 4 potential facilities, 31 customer locations, chosen material suppliers, recycling sites, landfill sites and scrap purchasing centres as well as all possible product flow routes is shown in Figure 2.

The currently open facility in County Durham has a production capacity of 1,000 tonnes per year, and the potential facilities have initial production capacities of 400 tonnes. For the Hydram problem, it is assumed that the suppliers, recycling sites, landfill sites and scrap recycling centres have capacities that are significantly larger than the flows that they will be required to handle, which means that the constraints (6) to (9) can be ignored.

Allocation of economic costs

The cost of setting up a new satellite facility was estimated to equal £100,000 per year to cover the purchasing of equipment and managerial expenses, assuming a 10-year facility lifetime. Property expenses were also evaluated for each of the potential facilities, using either rent prices or mortgage calculations for the same 10-year period. The sum of the setup cost and property expenses gives a complete yearly operational cost for each facility. As the County Durham facility already owns all equipment and property outright, the operating cost is assumed to be zero which essentially models the facility as an existing facility rather than a potential facility. The yearly operational costs of all facilities are shown in Table 3.

For the Hydram problem, it was established that the suppliers pay the economic cost of transportation. Furthermore, it is assumed that non-recyclable waste, waste and scrap are collected by third parties and transported to their respective facilities. Therefore, Hydram incurs no direct economic cost from any of these transportation activities. As such, for the purposes of the case study,

$$\mathrm{CT}_{giv}, \mathrm{CT}_{ilq}, \mathrm{CT}_{ikr}, \mathrm{CT}_{imz} = 0 \qquad (18)$$

However, the costs of transportation to customers are directly incurred by Hydram. In order to calculate CT_{ijp}, firstly, the T_{ij} distances were calculated using online mapping tools, and they were used with the scaled sales data to calculate the total number of tonne-kilometres (tkm) travelled during the 12-month period as 74,367.41 tkm.

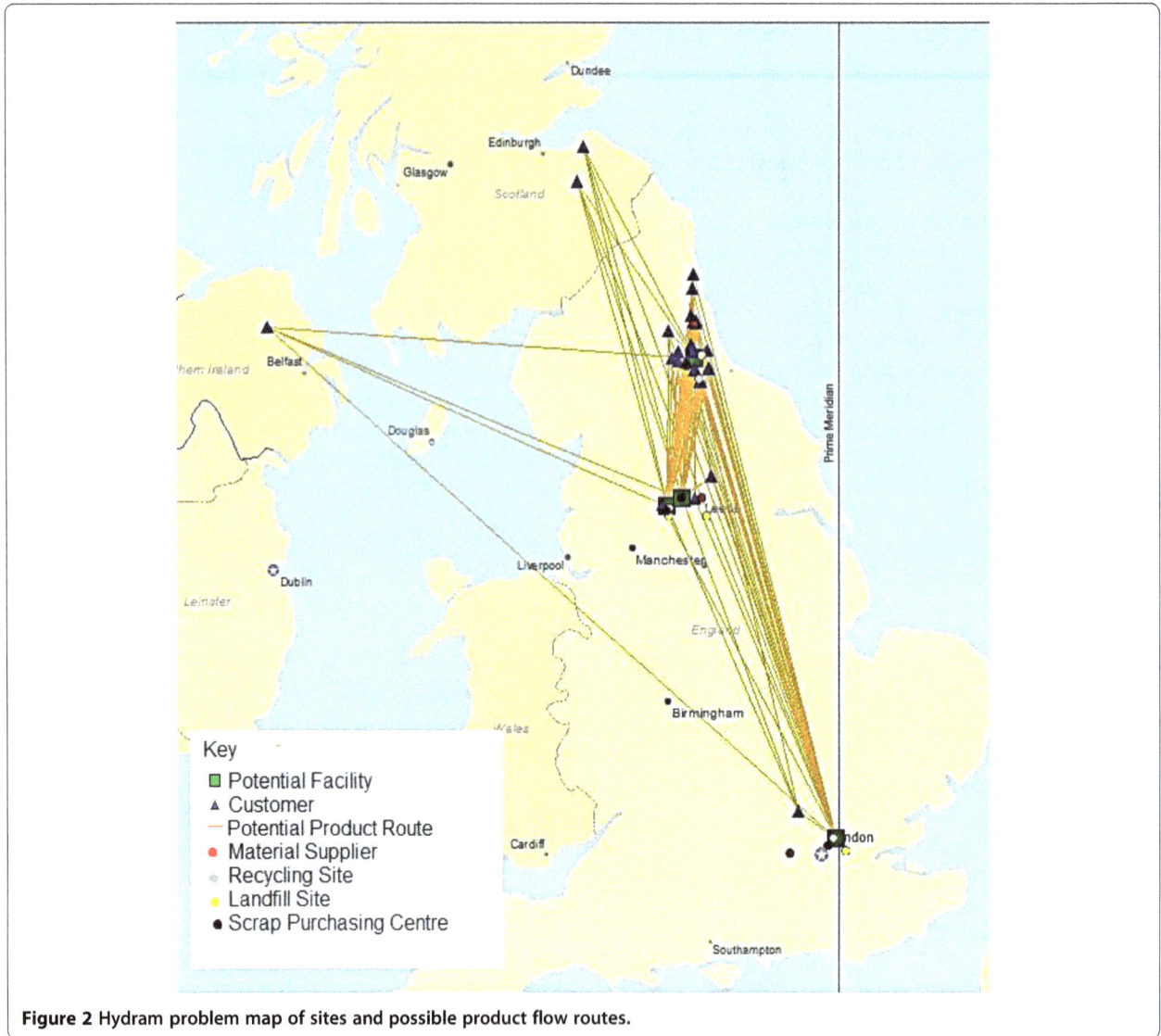

Figure 2 Hydram problem map of sites and possible product flow routes.

Using the total fleet fuel consumption value from Table 1 and assuming a diesel cost of £1.4305 per litre, the total expenditure on fuel was calculated as £39,807.03.

Hydram uses 13.5-t trucks, for which fuel costs only amount to 24% of total vehicle operating costs [23]. As such, the total economic cost of transportation for Hydram during the 2010-2011 sales period was calculated as £165,862.63, which gives a CT_{ijp} value of £2.23 per tkm by dividing the total cost by the total number of tkm travelled.

Finally, economic costs and revenues arise out of Hydram's waste operations. It is assumed that only inactive waste is landfilled, so a rate £2.50 per tonne is used for CD_{lq} [24]. A revenue value of £180 per tonne was allocated for CD_{kn}, using current ferrous metal scrap prices as a guideline.

Allocation of external costs

In order to include external costs into the supply network design model, the external costs must first be quantified.

Table 3 Yearly operational costs and capacities of the Hydram potential facilities

Potential facility	Property cost per year	Setup cost per year	Total cost per year	Production capacity (t)
Durham	£0.00	£0.00	£0.00	1,000
Halifax	£148,024.43	£100,000.00	£248,024.43	400
Bradford	£120,000.00	£100,000.00	£220,000.00	400
London	£180,057.30	£100,000.00	£280,057.30	400

However, the quantification of external costs is a difficult process, and different studies rely on different assumptions to estimate externality costs [25]. Although the estimation of external costs comes with inherent uncertainties, there is a wide consensus on which methods are most appropriate for identifying different types of external cost [8]. This section reviews some of the literature on the external costing of transport operations and landfilling to determine quantitative external costs for use in (3).

Although no economic costs of transportation are incurred between the potential facilities and suppliers, recycling sites, landfill sites or scrap recycling centres, such operations are still within the supply network. As such, their external costs are still included in the model, as the external costs are directly related to decisions made by Hydram. For the purpose of this analysis, it is assumed that all transportation operations within the network use the same type of vehicle and resultantly produce the same external costs per tonne of material per kilometre travelled. As such, for the analysis of the case study problem,

$$XT_{giv} = XT_{ijp} = XT_{ikr} = XT_{ilq} = XT_{imz} \qquad (19)$$

By quantifying the effects of noise, accidents, pollution (and its effects on health, nature and buildings), climate change and effects on the natural landscape, a marginal external cost of 0.3 to 1.2 EUR/tkm for interurban road freight transport and 1.1 to 4.4 EUR/tkm for urban road freight transport is calculated in [26]. Using an intermediate value, but assuming that most transport is interurban, it was decided that an external cost of 1.5 EUR/tkm would be appropriate for application to the model, which translates to £1.25/tkm. The costs in [26] assume heavy goods vehicles with an average load factor of 15 tonnes per vehicle, so are suitable for application to the case study problem, which assumes 13.5-tonne trucks.

For the case study, there are significant revenues to be made from recycling scrap metal. It was decided that the economic benefit achieved is essentially internalisation of the external cost of landfilling metal scrap, as there is a significant direct incentive not to landfill the scrap material. As such, it is argued that the external benefit of recycling the metal scrap is already reflected in the

revenue obtained by scrapping the metal and should therefore be ignored to avoid counting it twice.

Similarly, recycling general waste presents an avoided cost of £2.50 per tonne compared to landfilling it, which essentially internalises the external benefit of recycling. As a result of this, as well as a lack of consensus on the literature regarding the environmental benefits of recycling general waste, it was decided that the external benefit of recycling general waste should also be ignored.

Resultantly, for analysis of the case study,

$$XD_{rk} = XD_{mz} = 0 \qquad (20)$$

The external costs of landfilling were quantified by [9] and found to range between 10 and 13 EUR/t waste, dominated by the emission of un-captured methane. Using the mid-range value and converting the currency at current rates, this figure translates to £9.74 per tonne of landfilled waste. The external costs associated with transporting waste to a landfill are neglected in [9], so the inclusion of XT_{ilq} is still appropriate in the model.

Disamenity costs were also neglected in [9], but [18] quantified a fixed disamenity cost of between £1.52 and £2.18 per tonne of landfill in 2000. The Nationwide House Price Index (HPI) calculator indicates an increase of 112.08% on the average house price since 2000, so an appropriate current disamenity cost, using the results of [18] adjusted in line with the HPI, is therefore somewhere between £3.22 and £4.62 per tonne of landfill generated. It was decided that the mid-range value of £3.92 per tonne would be appropriate for use in the model. As such, the total cost of XD_{lq} is allocated as £13.66 per tonne, using [9] and [18]. A summary of the economic and external cost inputs is provided in Table 4.

Experimentation

Three key scenarios were analysed using three discrete experiments.

Experiment 1 Experiment 1 was performed to establish the optimum choice of which facilities to open, using the operational costs and capacities as per Table 3. In order to model this open scenario, constraint (17) was effectively neglected by setting Y_{min} to 0 and Y_{max} to 4.

Table 4 Summary of economic and external cost inputs

Economic cost	Value	External cost	Value
CT_{giv}, CT_{ilq}, CT_{ikr}, CT_{imz}	0	XT_{giv}, XT_{ijp}, XT_{ilq}, XT_{ikr}, XT_{imz}	£1.25/tkm
CT_{ijp}	£2.23/tkm	XD_{rk}, XD_{mz}	0
CD_{lq}	£2.50/t	XD_{lq}	£13.66/t
CD_{kr}	£180.00/t (revenue)		

Experiment 2 Experiment 2 was performed to establish which facilities should be open if it is specified that the network should contain at least two facilities, so Y_{min} was set to 2, and Y_{max} was set to 4.

Experiment 3 Experiment 3 was performed to evaluate a different scenario whereby 400 t worth of production capacity from the County Durham facility is moved to a new satellite facility. Through consultation, it was assumed that moving the equipment rather than purchasing new equipment at the new facility would reduce the cost of setting up the new facility to £30,000 per year. However, in order to provide 400 t per year of capacity for a satellite facility, the capacity at the Durham facility would have to reduce by 400 t per year. The new operational costs and capacities due to this proposal are shown in Table 5. Under these conditions, it would only be possible to have a maximum of two facilities in the network, as there is currently only enough equipment to equip two facilities. Resultantly, Y_{min} was set to 0, and Y_{max} was set to 2.

Each of the experiments was modelled using α values between 0 and 1, in increments of 0.2, to evaluate the effects of including external costs during the decision-making process. To solve the problems, the model was formulated using What'sBest, an add-in for Excel, produced by Lindo Systems Incorporated (Chicago, IL, USA), which uses a branch-and-bound algorithm. More information about the allocation of operational costs in Tables 3 and 5 is given in Additional file 1.

Results and discussion

Table 6 shows the Y_i outputs of the model, indicating which facilities the model nominates to open, and the total flows from each facility. Figure 3 shows the total cost outputs (the objective value produced by the model) for each of the experiments, using different alpha weightings. A breakdown of the total economic and external costs calculated for each experiment is shown in Figure 4. Figure 5 shows the capacity utilizations of each of the open plants for each of the experiments. The allocation of customers to plants for experiments 2 and 3 is shown in Figure 6.

The major result of the experiments is that external costs are significant in the field of supply network logistics. By analysis of Figure 4, it can be seen for all cases that the external costs form a significant portion of the

total cost. In the worst case, experiment 1, the external costs of the Hydram logistics network are more than twice the magnitude of the economic cost.

Initially, it was only intended to model scenario 1 to determine the feasibility of opening a satellite facility for alpha values between 0 and 1. As the alpha value represents the extent to which external costs are factored in, it is suggested that the maximum useful value of alpha to be considered should be 1, as no rational manager would wish to over compensate for the potential of future rising costs brought about by governmental policy which aims to internalise externalities (and no rational government would over-internalise externalities). By analysis of Figure 3 and Table 6, it is clear that the model indicates that it is currently not feasible for Hydram to open up a new facility for all alpha values in this range. However, although the decision not to open a new facility currently has the lowest total cost, it has the greatest external cost and is therefore the most damaging to the environment and society, mainly due to the impact of transport operations. By analysis of the projected flow from the County Durham facility in experiment 1 and the available production capacity, it was calculated that the County Durham facility will run at 78.3% of capacity under this scenario, as shown in Figure 5.

As it was deemed a somewhat trivial result that the model allocated not to open any new facilities when the alpha value is set between 0 and 1, values greater than 1 were considered. The alpha value inputs were varied to determine the extent to which external costs must be considered to make the opening of a facility feasible for the Hydram case, when there is no constraint imposed on the minimum number of facilities to be opened. It can be seen by analysis of the results that the model indicates that it is currently not feasible for Hydram to open up a new facility for all alpha values less than 3.6. At this point, the model indicated that the County Durham and Bradford facilities should be opened. The fact that the facility location decision changes according to the alpha weighting as shown in Table 6 validates the model.

The somewhat trivial initial result in experiment 1 also led to the formulation of experiment 2, which was conducted to determine which facilities should be open when it is specified that there should be at least two facilities in the network, i.e., $Y_{min} = 2$. The model indicated that the County Durham and Bradford facilities should be open under these circumstances. However, it was

Table 5 Yearly operational costs and capacities of the Hydram potential facilities for experiment 3

Potential facility	Property cost per year	Setup cost per year	Total cost per year	Production capacity (tonnes)
Durham	£0.00	£0.00	£0.00	600
Halifax	£148,024.43	£30,000.00	£178,024.43	400
Bradford	£120,000.00	£30,000.00	£150,000.00	400
London	£180,057.30	£30,000.00	£210,057.30	400

Table 6 Choice of facilities to open and product flow from each facility to the nearest tonne

		Experiment					
		Y_i values			Product flows (t)		
		1	2	3	1	2	3
$0 < a < 3.6$	County Durham	1	1	1	783	524	524
	Halifax	0	0	0	0	0	0
	Bradford	0	1	1	0	258	258
	London	0	0	0	0	0	0
$a > 3.6$	County Durham	1	1	1	524	524	524
	Halifax	0	0	0	0	0	0
	Bradford	1	1	1	258	258	258
	London	0	0	0	0	0	0

noticed that both plants would be running significantly under capacity based on the allocated product flows in Table 6. In experiment 2, the County Durham facility is only operating at 52.4% of capacity, and the Bradford facility also only operates at 64.5% of capacity. Overall, the facilities operate at 55.8% of capacity.

It was this under use of capacity that led to the formulation of experiment 3, which moves 400 t of the Durham production capacity to a new facility, in order to reduce the setup costs of the satellite facility and improve the overall capacity utilisation. Under experiment 3, the model nominates to open the same facilities, Bradford and Durham, and allocate the same production to each facility. Resultantly, the external costs for experiments 2 and 3 are the same, as shown by Figure 4. However, due to reduced operating costs, the economic costs of the proposal are significantly less than those of experiment 2. Furthermore, the projected capacity

utilisation figures are better for scenario 3 than for scenario 2: there is an overall 78.3% capacity utilisation as per example 1, as the same amount of product is produced using the same overall capacity.

Therefore, if it is decided by Hydram that a satellite facility is to be opened, it is suggested that the Bradford location should be chosen and that scenario 2 should be employed. By moving 400 t of production capacity in terms of tools and equipment from the County Durham facility to the Bradford facility, Hydram can minimise setup costs and maintain acceptable capacity utilisation, whilst reducing the external costs of the logistics network.

Finally, it is proposed that customer demand is partly affected by the geographical location of facilities. Figure 6 shows that current customers are somewhat clustered geographically around the current County Durham facility. Therefore, it is suggested that if the Bradford satellite facility were to be built, then total demand would likely increase due to the addition of extra customers local to the new satellite facility. If demand in the Bradford area does increase in this manner, then new tools and equipment can be purchased to increase production capacity if necessary.

Conclusions and further work

The concept of green supply chain management is gaining increasing interest among both researchers and practitioners of operations and supply chain management [27], and the consideration of external costs has been steadily growing. The main conclusion from this report is that external costs can form a significant portion of the total cost when considered in a logistics network.

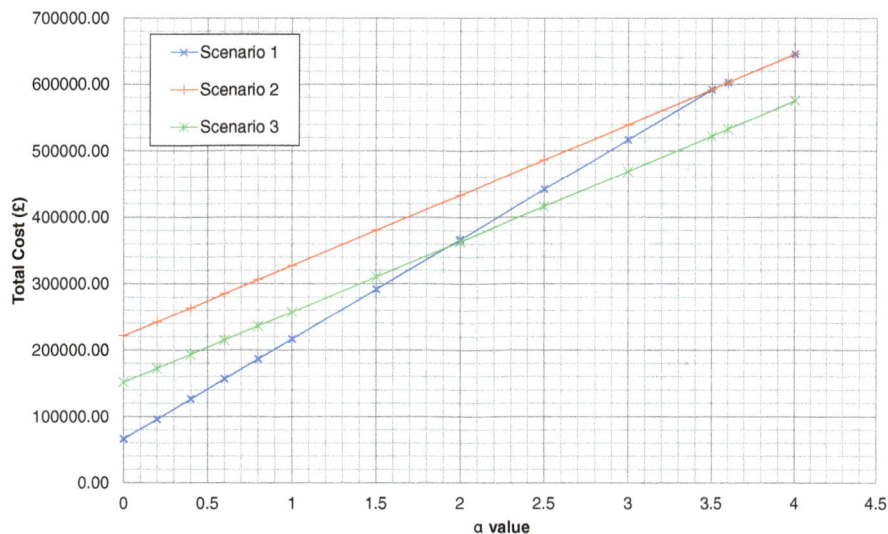

Figure 3 Total cost (objective value) using different alpha weightings.

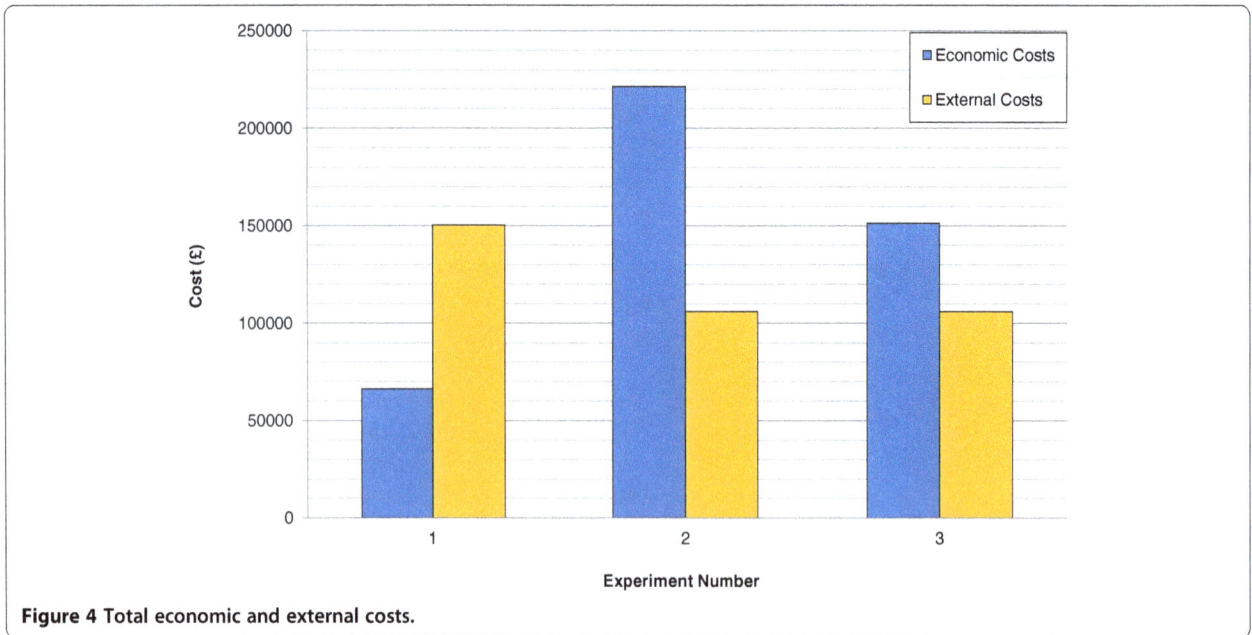

Figure 4 Total economic and external costs.

The model presented in this paper incorporates social and environmental impacts of operations through the inclusion of external costs, but allows user discretion on the extent to which external costs are included when making location and allocation decisions. Even if the user chooses to fully ignore the external cost (and set alpha to 0) within their objective function when making their logistics decisions, the spread sheets generated for this paper still give an indication of the external cost that their decisions will make. However, it is suggested that there is a critical need to consider external costs in the supply network design to cope with increasing economic and regulatory pressure and increasing consumer awareness. As legal restrictions tighten and taxes increase to attempt to internalize external costs, it is argued that consideration of the total external cost by the use of a non-zero alpha value would be prudent so as to prepare for the external costs becoming internalized by future governmental policy.

Although there is a wide consensus on which methods are most appropriate for identifying different types of external cost [7], the quantification of external

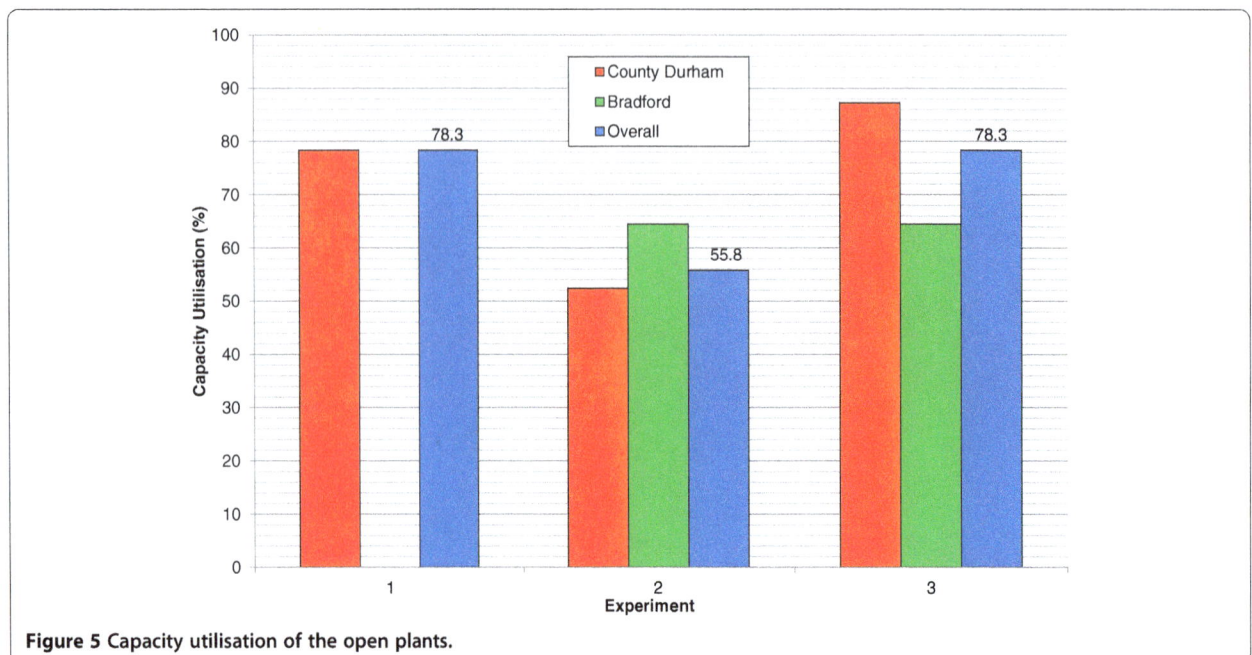

Figure 5 Capacity utilisation of the open plants.

Figure 6 Allocation of customers to plants for experiments 2 and 3.

costs is still a difficult process, with inherent uncertainties. For example, even when accredited techniques such as the willingness-to-pay method are used, it is still difficult to put an economic value on something as qualitative as a human life. Therefore, the external costs generated by the model are somewhat uncertain due to the uncertainties in the external costs used as model inputs. It is suggested that more work must be done by the academic community on estimating external costs of logistics operations so that they can be incorporated into future supply network models with more credibility and certainty.

With regard to the Hydram case study, it is concluded that it is currently not feasible to open up a satellite facility, based on the output of the developed model. However, if the managers do wish to open a new facility, it is suggested that the facility should be in

the Bradford location suggested and that 400 t of production capacity should be transferred from the County Durham facility, as per experiment 3, in order to maximize capacity utilization and minimise operating costs. Such a decision may also improve customer sales for Hydram, as demand tends to come from customers within a close proximity to the facilities.

This paper presented a single product, single time period model for minimizing both economic and external costs in a logistics network brought about by facility opening, transportation and waste disposal considerations. Real-world supply chains are often more complicated than the one considered in this paper. As such, some extensions to the model have been envisioned, in order to extend the current MILP formulation to more realistic real-world supply network structures, including the following:

- The consideration of multiple time periods to analyse dynamic situations
- The incorporation of production costs within the model (both economic and external), such as the costs associated with running facilities
- Inclusion of reverse logistic options in the model such as reuse and remanufacture to make the model more widely applicable

However, it is recognised that such extensions may prove very difficult to implement. The modelling of more complex supply chains across multiple time periods would lead to much more complicated mathematical models, and it is suggested that the formulation of an MILP that is applicable to a wide range of real-world supply chains would take a great deal of effort to formulate and solve.

Additional file

Additional file 1: Operational cost breakdown for experiments 1, 2 and 3.

Competing interests
The authors declare that they have no competing interests.

Authors' contributions
QW defined the research aim and identified the problem. DA carried out the literature review on external costs, and QW carried out the literature review on logistic network structures. DA carried out mathematical modelling and conceived the case study. QW participated in the experimental design and discussion of the results and corrected the manuscript. Both authors read and approved the final manuscript.

Acknowledgements
Special thanks to Andrew Jordan at Hydram for providing the case study data and to Tom Davies for providing technical support.

References
1. Beamon, BM: Supply chain design and analysis: models and methods. Int. J. Prod. Econ. 2(3), 281–294 (1998)
2. Ambrosino, D, Scutella, M: Distribution network design: new problems and related models. Eur. J. Oper. Res. 165, 610–624 (2005)
3. Copps, A: Emission maps put heat on supply chain. The Times (2012)
4. Lacy, P, Cooper, T, Hayward, R, Neuberger, L: A New Era of Sustainability: UN Global Compact-Accenture CEO Study, 2010. Accenture, Ilinois (2011)
5. Fleischmann, M, Beullens, P, Bloemhof-Ruwaard, JM, Van Wassenhove, L: The impact of product recovery on logistics network design. Prod. Oper. Manage. 10, 156–173 (2001)
6. European Commission: External costs: research results on socio-environmental damages due to electricity and transport. http://bit.ly/JjYvoL. Accessed 5 Apr 2011
7. The Economist: Economics A-Z terms beginning with E. http://www.externe.info/externe_2006/externpr.pdf]. Accessed 5 Apr 2011
8. Maibach, M, Schreyer, C, Sutter, D, van Essen, HP, Boon, BH, Smokers, R, Schroten, A, Doll, C, Pawlowska, B, Bak, M: Handbook on Estimation of External Costs in the Transport Sector. Internalisation Measures and Policies for All External Cost of Transport (IMPACT). CE Delft, the Netherlands (2008)
9. Rabl, A, Spadaro, S, Zoughaib, A: Environmental impacts and costs of solid waste: a comparison of landfill and incineration. Waste Manage. Res. 26, 147–162 (2008)
10. Domschke, W, Drexl, A: Location and Layout Planning. Lecture notes in Economics and Mathematical Systems, Springer, Berlin (1985)
11. Pourmohammadi, H, Dessouky, M, Rahimi, M: A reverse logistics model for the distribution of waste/by-products. J. Cleaner Prod (2013). in press
12. Pishvaee, M, Kianfar, K, Karimi, B: Reverse logistics network design using simulated annealing. Int. J. Adv. Manufact. Technol. 47, 269–281 (2009)
13. Benaissa, M, Benabdelhafid, A: A multi-product and multi-period facility location model for reverse logistics. Polish J. Manage. Stud. 2, 7–19 (2010)
14. Yongsheng, Z, Shouyang, W: Generic model of reverse logistics network design. Transpn Sys Eng & IT 8(3), 71–78 (2008)
15. Krikke, H, Kooi, E, Schuur, P: Network design in reverse logistics: a quantitative model. Lect. Notes Econ. Math. Syst. 480, 45–62 (1999)
16. Salema, M, Pvoa, A, Novais, A: A warehouse-based design model for reverse logistics. J. Oper. Res. Soc. 57(6), 615–629 (2006)
17. Locklear, EC: Product take-back using geographic information systems. University of South Carolina, Thesis (2000)
18. Cambridge Econometrics in association with EFTEC and WRC: A Study to Estimate the Disamenity Costs of Landfill in Great Britain: Final report. Defra, Cambridge (2003)
19. Bräuninger, M, Schulze, S, Leschus, L, Perschon, J, Hertel, C, Field, S, Foletta, N: Achieving sustainability in urban transport in developing and transition countries (EURIST/HWWI study, 2012). Hamburg Institute of International Economics, Hamburg (2011)
20. Persson, J, Song, D: The Land Transport Sector: Policy and Performance. OECD Economics. Department Working Papers, No. 817. OECD Publishing, Paris (2010)
21. International Union of Railways, Greening Transport: reduce external costs. http://www.allianz-pro-schiene.de/presse/pressemitteilungen/2012/013-studie-lkw-maut-pkw-maut/kurzfassung-studie-externe-kosten-verkehr-cer-uic.pdf]. Accessed 5 Apr 2011
22. The Environment Protection Agency, Vicoria, Austrailia, Waste materials: density data. http://bit.ly/I6iWp5. Accessed 5 Apr 2011
23. The Freight Transport Association, Fuel as a percentage of HGV operating costs. http://www.fta.co.uk/policy_and_compliance/fuel_prices_and_economy/fuel_prices/fuel_fractions.html. Accessed 6 Apr 2011
24. HMRC Reference: Notice LFT1 (May 2012): A general guide to landfill tax. http://customs.hmrc.gov.uk/channelsPortalWebApp/channelsPortalWebApp.portal?_nfpb=true&_pageLabel=pageExcise_ShowContent&propertyType=document&id=HMCE_CL_000509#P311_23550. Accessed 6 Apr 2011
25. Koomey, K, Krause, F: Introduction to environmental externality costs. CRC Press, Boca Raton, In CRC Handbook on Energy Efficiency (1997)
26. Essen, HP, Boon, BH, Maibach, M, Schreyer, C: Methodologies for external cost estimates and internalisation scenarios. Discussion paper, Delft (2007). http://www.ce.nl/4288_Inputpaper.pdf. Accessed 6 Apr 2011
27. Srivastava, S: Network design for reverse logistics. Omega 36, 535–548 (2008)

16

A comparison of repaired, remanufactured and new compressors used in Western Australian small- and medium-sized enterprises in terms of global warming

Wahidul K Biswas[1*], Victor Duong[2], Peter Frey[3] and Mohammad Nazrul Islam[4]

Abstract

Repaired compressors are compared with remanufactured and new compressors in terms of economic and environmental benefits. A detailed life cycle assessment has been carried out for compressors under three manufacturing strategies: repaired, remanufactured and new equipment. The life cycle assessment of the global warming potential of repaired compressors varies from 4.38 to 119 kg carbon dioxide equivalent (CO_2-e), depending on the type of components replaced. While greenhouse gas emissions from the remanufactured compressors (110 to 168 kg CO_2-e) are relatively higher than those from the repaired ones (4.4 to 119 kg CO_2-e), a new compressor has been found to produce a larger amount of greenhouse gas emissions (1,590 kg CO_2-e) compared to both repaired and remanufactured compressors. Repairing failed compressors has been found to offer end users both dollar and carbon savings in contrast to remanufactured and new compressors. The research also found that extended lifetime is more important than the manufacturing processes in terms of greenhouse gas emissions. Since a remanufactured compressor offers a longer life than a repaired compressor, the replacement of the latter with the former can avoid 33% to 66% of the greenhouse gas emissions associated with a new compressor production with a lifetime of 15 to 25 years.

Keywords: End-of-life product, Life cycle assessment, Global warming

Introduction

This paper compares the economic and environmental implications of repaired, remanufactured and new compressors. It focuses on refrigeration and air-conditioning compressors.

With the increase in the world's consumption of household and industrial products, there is a need to reduce the consumption of mineral resources and the amount of waste generated and end-of-life products sent to scrap yards. To implement this resource efficiency objective, recoverable manufacturing systems could be used. These include repair and remanufacturing, which differ in key control aspects [1]. From ACDelco's experience in Australia, repairing

time-consuming and expensive task than completely stripping the engine and then rebuilding or remanufacturing it [2]. For customers, the availability of remanufactured goods means more uptime for their products, which translates into significant production and financial benefits. For some existing manufacturers, the economic efficiency of remanufacturing is clear, and it has become a widely held assumption that such systems would also be more eco-efficient [3].

By utilising recovered end-of-use products and parts, remanufacturing could reduce the manufacturing and disposal costs of heavy and material-intensive industrial machinery and electronic equipment [3-6]. In a carbon-constrained economy, it would be useful to work out the carbon-saving benefits of the replacement of an original equipment manufacturer (OEM) product with a remanufactured one. OEM refers to the company that makes a product using original parts and virgin materials.

* Correspondence: w.biswas@curtin.edu.au
[1]Sustainable Engineering Group, School of Civil and Mechanical Engineering, Curtin University, Bentley, Perth, Australia
Full list of author information is available at the end of the article

Life cycle assessment (LCA) has been widely used to analyse the environmental benefits of the replacement of a new product with an end-of-life product for internal combustion engines, electrical appliances, gear boxes and compressors [7-10]. The most detailed LCA remanufacturing study in Australia to date was on the remanufacturing of photocopiers at Fuji Xerox and of a compressor at Recom Engineering [2,11].

The environmental impact of options following a machinery failure needs to be assessed in order to reduce emissions from the manufacturing sectors. These options can either be repairing or replacement with an OEM or remanufactured machinery. In addition to the energy and materials required for repairing, remanufacturing and the production of new machinery, the end-of-life situation could affect the life cycle environmental performance of these options. However, so far, the literature reviewed did not estimate the environmental advantages of remanufacturing over repairing in Australia and elsewhere. The environmental performance of repair and remanufacturing was thus carried out by a detailed LCA.

This paper provides options for choosing repairing, remanufacturing or purchasing new compressors to reduce the carbon price. It demonstrates that a repaired compressor can perform as well and as long as remanufactured and new compressors within the first 3 years of the compressor's life. Performance depends on the type of fault in the air-conditioning compressors and on the repairs being of the highest quality.

This research considers a 'cradle to gate' assessment of compressors for a Western Australian small- and medium-sized enterprise. This means that the LCA does not take into account product use and disposal, including factors such as recycling and recovery, which would offset carbon emissions and hence the carbon price. The LCA includes only global warming impacts which can be directly attributed to a repaired compressor and the production of a remanufactured compressor and a new compressor.

Methodology

LCA has been carried out to assess the carbon-saving benefits of a repaired compressor over remanufactured and new compressors.

The LCA follows the ISO14040-43 guidelines [12] in four steps:

1. Goal and scope definition
2. Inventory analysis
3. Impact assessment
4. Interpretation

Goal and scope

The goal of this LCA is to determine and compare the economic and environmental implications of repaired compressors with remanufactured and new compressors. The compressor used in this case study is a 20 HP Bitzer compressor (Bitzer Kühlmaschinenbau GmbH, Sindelfingen, Germany) for refrigeration and/or air conditioning. This paper determines the difference in the carbon footprint (greenhouse gas (GHG) emissions) associated with repairing compressors and remanufacturing them. These results will also be compared against the carbon footprint of new (OEM) compressors to demonstrate the carbon and cost savings.

There are some limitations that determine the scope of this LCA. First, the analysis only took into account repairs that are of A+ quality (as good as new with 2 years of product starts). This quality is not always achieved, with some repairs being of lower quality. Second, refrigeration and air-conditioning compressors are inherently reliable and can survive in service for decades. In Recom Engineering's experience, the mean time to failure for a compressor is 15 years. Third, this analysis is only based on the GHG emissions associated with repairing, remanufacturing and producing new refrigeration and air-conditioning compressors. Other than global warming or GHG emissions, no other environmental impacts have been estimated for this LCA.

Life cycle inventory

A life cycle inventory (LCI) considers the amount of each input and output of different stages of a product's life cycle and is a necessary initial step in an LCA. An LCI was constructed that includes all inputs involved in the repairing of a compressor in a specific scenario, where the scenarios vary with the number and type of parts replaced and the type of repairing and machining operations.

The LCA of a repaired compressor is better described as preventative maintenance and/or when the compressor has had a minor failure. After the failed compressor has been cleaned and prepared for repair, the repair process involves replacing the damaged components that were removed during the disassembly stage. These were remanufactured parts (valve plate, terminal block) or new components (oil pump, shaft seal). Table 1 shows the differences between repairing and remanufacturing operations in order to clarify the environmental implications of these recovery or reuse operations.

The benefit of repair is that it is on-site. There is no need to disassemble the compressor and transport it to another location, and this saves time, money and, most importantly, carbon emissions. Repairing a compressor involves the following:

- Inspection and diagnostics: once a failed compressor has been inspected and the unit is deemed repairable, the repair can commence. Inspection and diagnostics are done by hand which can take 30 min to 1 h.

Table 1 Differences between the repairing and remanufacturing processes

Stages	Repair	Remanufacture
Stage 1: disassembly	Disassembly of the faulty component/part is manually done	The whole compressor is stripped down, using an electric- or air-powered rattle gun. The energy used during cleaning and machining is approximately 13.5 MJ
Stage 2: cleaning and washing	Minimal cleaning and washing with repairs done on-site, using a universal cleaning agent (200 mL) to wipe down the surface that is being repaired	Most components, as they are reused, are thoroughly cleaned and washed with a variety of chemicals, using approximately 10.5 MJ
Stage 3: machining	There is usually no machining during a repair, depending on the part; in some cases, valve plates need to be remanufactured, requiring only minor cleaning, washing and surface grinding (4.23 MJ)	Parts that can be remanufactured and reused are often machined to ensure that they are useable; machining includes polishing, surface grinding and rewinding, using a total energy of 114.8 MJ
Stage 4: part replacement	Components are usually replaced as necessary with new parts during repairs	Most components are reused, and components that are worn out are replaced with new components
Stage 5: assembly	Reassembly is on-site using hand tools or manual labour	All components are reassembled using an air gun and hand tools, using 13.5 MJ

The data on energy consumption were obtained from Recom Engineering, Perth, Western Australia.

- Part disassembling: after inspection, the faulty part is detached from the compressor using hand tools which can take 1 to 2 h.
- Minor cleaning and washing: the faulty part or component is removed and cleaned to remove any dirt and debris. Universal flushing agents are used for the cleaning of the compressor and for the remanufacturing of valve plates and terminal blocks. Chemicals such as alkalis, phosphoric acid and decarboniser are mixed with hot water for cleaning and washing operations.

Repairing involves machining a part or component to be remanufactured. For example, in 80% of cases, valve plates are remanufactured which requires cleaning and then removing the broken reeds. The valve plate is then machined using a surface-grinding machine.

Part reassembling involves simply reassembling the compressor with a new part or a remanufactured valve plate or terminal block. This is done with hand tools or an electric impact gun. On the other hand, remanufacturing a compressor involves disassembling, cleaning and washing (C&W), machining operations, replacements, assembly and testing [11].

Figure 1 shows a detailed version of the LCI of a repaired compressor. It shows the inputs and outputs associated with the processes and the steps that a failed compressor goes through to be returned to service.

Life cycle impact assessment

The carbon footprint of a repaired compressor has been measured in terms of GHG emissions. From the energy and material data in the LCI, the GHG emissions have been calculated and converted to carbon dioxide equivalent (CO_2-e). Simapro 7.2 software [13] has been used to calculate these GHG emissions from paired compressors. The result is also compared with the carbon footprint of remanufactured and new compressors [11]. The input and output data of different stages from the LCI were used in the Simapro 7.2 software. This allowed the GHG emissions to be calculated for the repair of refrigeration and air-conditioning compressors. These inputs and outputs are linked to relevant Simapro 7.2 libraries which are databases of energy consumption, emission and material data for the production of one unit of a product. The units of input and output data from the LCI depend on the units of the relevant emission database in the software or its libraries [11].

Since local chemicals were used for cleaning and washing, the libraries from the Australian LCA database [14] for these chemicals were used for the analysis. The main chemicals used were phosphoric acid, thinner, decarboniser, alkali and penetrene. Phosphoric acid was available on the Simapro 7.2 databases, whereas sodium carbonate (alkali), thinner (methyl ethyl ketone) and decarboniser (dimethlamine) were obtained from the ecoinvent database [15] because they were unavailable in the local database [14]. Penetrene is used in both repairing and remanufacturing, and its data were available neither on the Simapro 7.2 database nor in the literature. Penetrene is comprised mainly of petroleum distillate and tetrachloroethylene, so the energy consumption values of these chemicals were used to estimate the approximate emission factor of penetrene.

Recom Engineering is a Western Australian company which disassembles, cleans, washes, machines, reassembles and tests parts and equipment. Therefore, the Western Australian electricity mix has been considered for calculating GHG emissions. Repairing requires the replacement with new components, and so, the emissions for the production of these new parts were estimated by summing all GHG emissions from mining, processing, foundry and

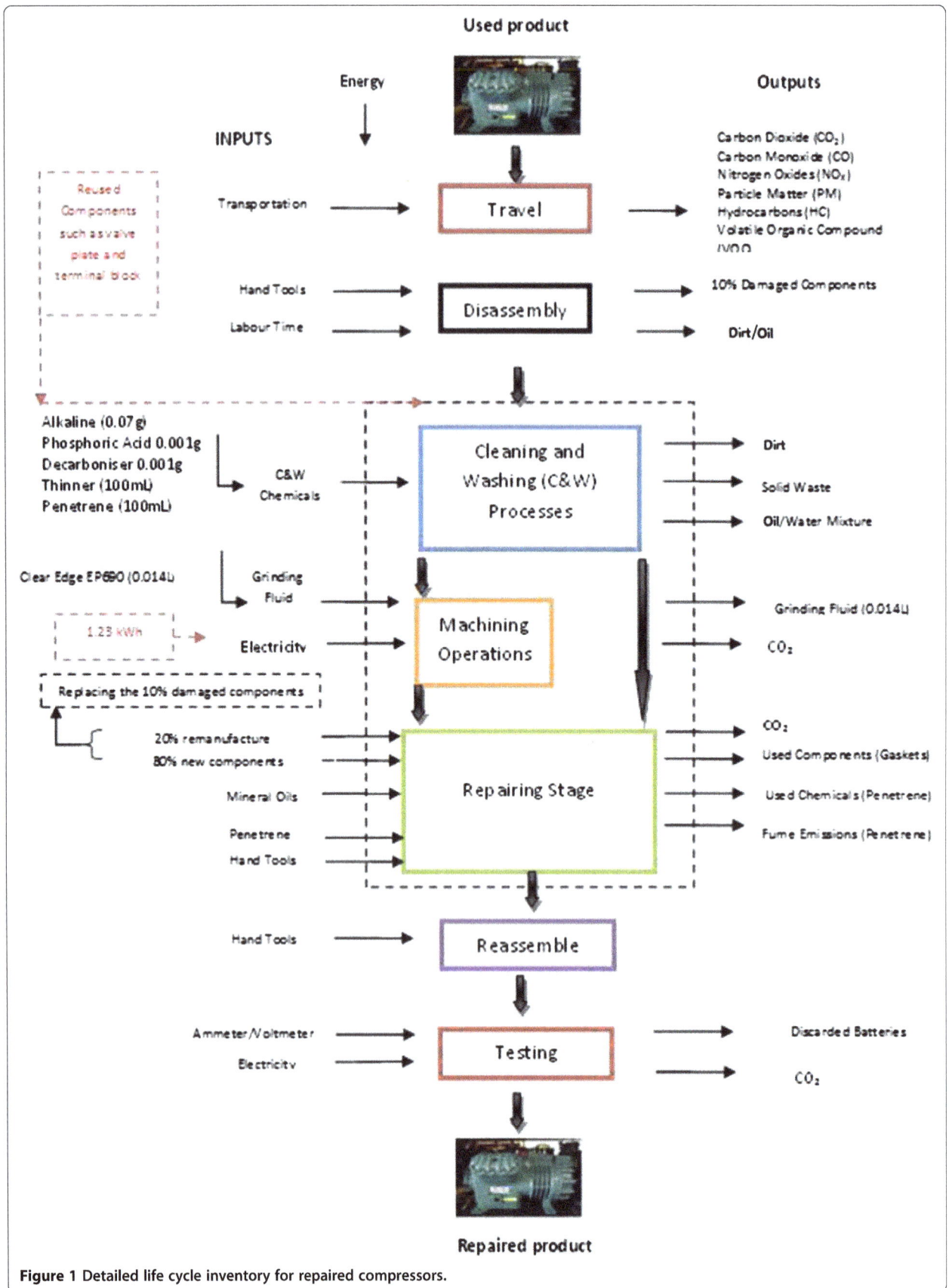

Figure 1 Detailed life cycle inventory for repaired compressors.

assembling processes. Transport is done using a truck and van with an average distance of 100 km. The unit for transport is tonne-kilometre travelled.

Results and discussion

Carbon saving benefits of the replacement of remanufactured and new compressors with repaired compressors

Firstly, the carbon-saving benefits have been calculated for the following scenarios:

- Scenario 1: repairing with valve plate replaced
- Scenario 2: repairing with oil pump replaced
- Scenario 3: repairing with terminal block replaced

The carbon footprints of repaired compressors are 4.4, 6.7 and 119.5 kg CO_2-e for the three scenarios, respectively. Scenario 3 produces 12 and 27 times more GHG emissions than scenarios 1 and 2, respectively. This is because the valve plate that was replaced with the new one in scenario 1 was comprised of stainless steel, which is a very energy-intensive material consuming 108 times more energy than normal steel (Figure 2).

The carbon footprint of the same-size (20 HP Bitzer) repaired semi-hermetic reciprocating compressor was compared with the carbon footprint of a remanufactured compressor and a new compressor [11]. The production of a new compressor would result in a total of 1,590 kg CO_2-e, while the remanufactured one would produce 110 kg CO_2-e in remanufacturing scenario 1 and 168 kg CO_2-e in remanufacturing scenario 2 [11].

Table 2 shows that GHG emissions could be mitigated by the replacement of remanufactured and new compressors with a repaired compressor. The repaired compressor in scenario 1 produced 29% and 92.5% less GHG emissions than the remanufactured compressor in remanufacturing scenario 2 (with 96.5% of total parts reused) and a new compressor, respectively, but produces 8% more GHG emissions than remanufacturing scenario 1 (with 99% of the total parts reused). As explained before, the replacement of a valve plate with the new one, which is made of energy-intensive stainless steel, increased the GHG emissions significantly. The repaired compressor in scenario 2 (with oil pump) has produced 94%, 96% and 99.6% less GHG emissions than those in remanufacturing scenarios 1 and 2 and new compressor production, respectively. The repaired compressor in scenario 3 can mitigate 96%, 97.4% and 99.7% of the total GHG emissions by replacing remanufacturing scenarios 1 and 2 and the production of a new compressor, respectively. Thus, repairing is less energy intensive and produces lower GHG emissions compared to remanufacturing or the production of new compressors.

In remanufacturing scenario 1 (minimum part replacement), 99% of the parts (on the basis of weight) were reused by cleaning, washing and machining, and less than 1% were replaced with new parts.

In remanufacturing scenario 2 (maximum part replacement), 96.5% of the total parts were reused, and the rest were replaced with new parts.

When the use (or operation) stage is included, the additional GHG emissions would be 189,000 kg CO_2-e, which is significantly higher than the emissions from the manufacturing stage. In most of the repairing of the compressor, replacement accounts for a significant portion of the total emissions, followed by repairing and machining. For scenario 1, 95%, 3% and 2% of the GHG emissions result from replacement, machining and repair, respectively. These values for scenario 2 are 62%, 20% and 18%, respectively. For scenario 3, 49%, 30% and 22% of the total GHG emissions result from replacement, machining and repair, respectively. Therefore, replacement needs to be avoided, especially for high-energy-intensive materials - those associated with the processing and manufacturing of the valve plate. In all

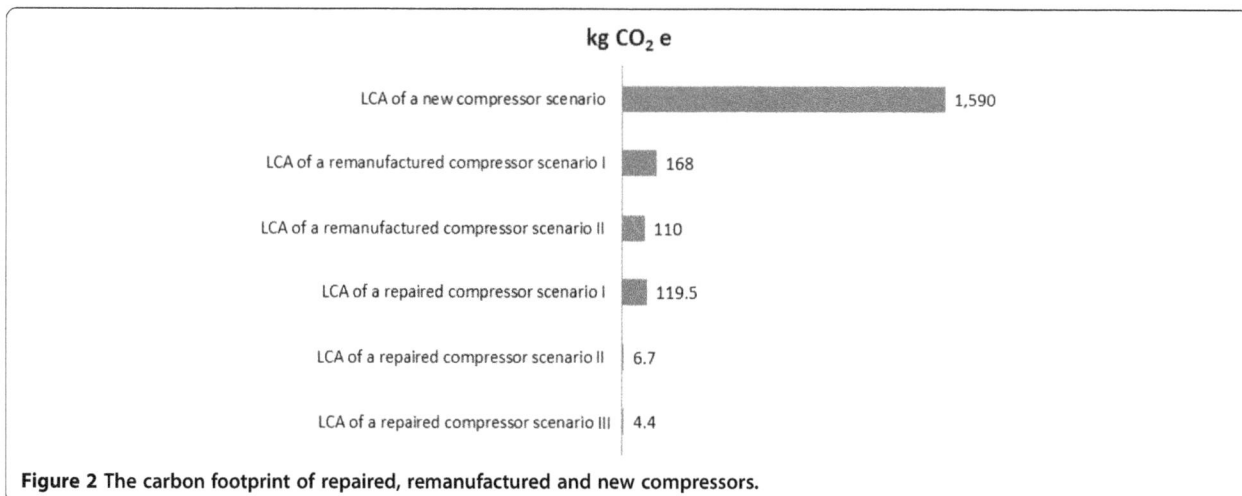

Figure 2 The carbon footprint of repaired, remanufactured and new compressors.

Table 2 GHG emission mitigation due to the replacement of remanufactured and new compressors with repaired compressors

Repairing scenario	GHG emission mitigation by the three repairing scenarios (%)		
	Remanufacturing scenario 1	Remanufacturing scenario 2	New
1	−8	29	92.5
2	94	96	99.6
3	96	97.4	99.7

Remanufacturing scenario 1 (99% of the parts were reused and 1% were replaced with new parts) and scenario 2 (96.5% of the parts were reused and the rest were replaced with new parts (3.5%)) were extracted from the work of Biswas and Rosano [11].

cases, machining and repairing produce a very small portion of GHG emissions compared to replacement.

Economic benefits

Since repairing emits less CO_2-e, the carbon tax will be reduced in the current Australian carbon pricing scheme. If the price of carbon were set at Australian $25 per tonne of CO_2-e, a new compressor which emits 1.590 tonnes CO_2-e equates to an additional cost of Australian $39.75. Repairing scenarios 1, 2 and 3 would have an additional cost of Australian $2.98, $0.20 and $0.10, respectively. Remanufacturing scenarios 1 and 2 have an additional cost of Australian $2.75 and $4.20, respectively. In the case of a 20 HP Bitzer semi-hermetic reciprocating compressor, the estimated cost of repaired compressors (Australian $1,000) is 73% less than that of a remanufactured compressor (Australian $3,752) and 82.4% less than purchasing a new compressor (Australian $5,686) (PF, personal communication).

Implication of the lifetime of compressors in the life cycle assessment

Since the lifetime of a remanufactured compressor is more than that of a repaired compressor (Figure 3), a different conclusion is obtained. Figure 3 was derived by interviewing local users, repairers and manufacturers. It can be seen that the replacement of a repaired compressor with a remanufactured compressor can avoid two thirds of an OEM compressor production over a lifetime of 25 years or one third for a lifetime of 15 years. This shows that the replacement of a repaired compressor with a remanufactured compressor can avoid 33% to 66% of the GHG emissions associated with the OEM compressor production between 15 and 25 years. Therefore, repairing appears to be a carbon emission reduction option, and remanufacture is more like a carbon sequestration option from a long-term perspective.

Conclusions

Repairing as the main option for the management of end-of-life products can help reduce the stress on natural resources and can also be economically and environmentally beneficial. Repairing compressors can potentially reduce the GHG emissions associated with a remanufactured or new compressor. About 29% to 97% of the total GHG emissions can be mitigated by replacing a remanufactured compressor with a repaired compressor, and a maximum of 99% of the total GHG emissions can be mitigated by replacing with a new compressor. Repaired compressors are cheaper than both remanufactured and new compressors

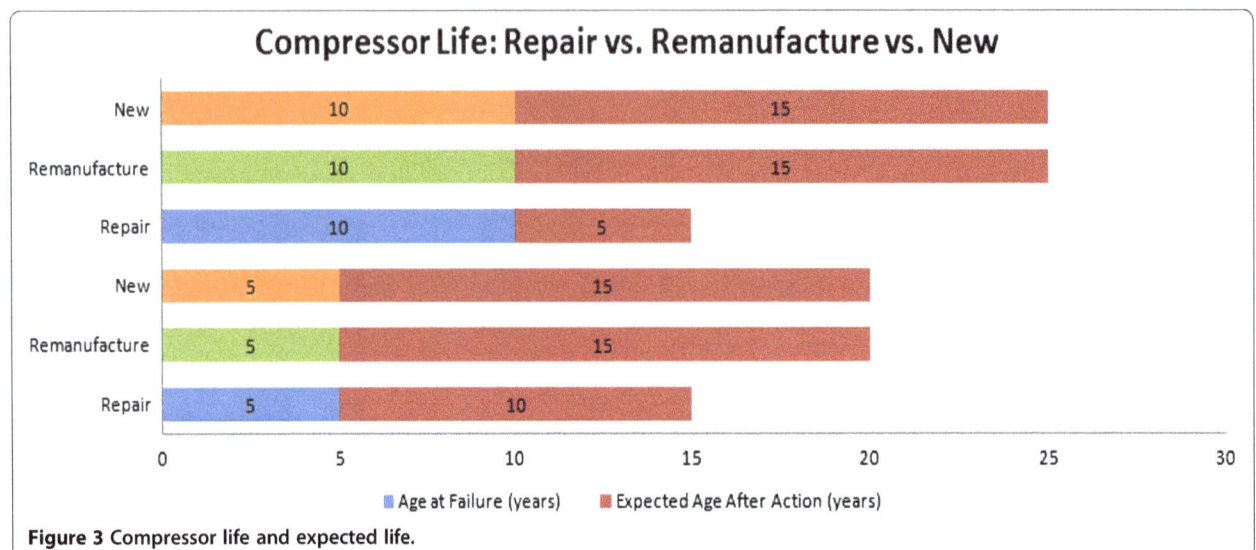

Figure 3 Compressor life and expected life.

and conserve non-renewable mineral resources for future generations.

From a short-term perspective, repairing compressors can potentially reduce the GHG emissions associated with a remanufactured or new compressor. The research also found that lifetime durability matters more than the manufacturing processes. Remanufacturing can offer significant carbon-saving benefits rather than repairing from a long-term perspective. Finally, in the case of either repairing or remanufacturing, it is crucial that the replacement with energy- and carbon-intensive new parts be avoided to increase the carbon-saving benefits.

Competing interests
The authors declare that they have no competing interests.

Authors' contributions
WKB carried out the life cycle assessment analysis and completed the manuscript. VD collected data, developed the life cycle inventory and carried out a literature review and analysis. PF provided industry data, technical support and guidance. MNI participated in the review and consultation process. All authors read and approved the final manuscript.

Acknowledgements
The authors acknowledge Recom Engineering, Osborne Park, Perth, Australia, for allowing VD to collect data on compressors in their workshop.

Author details
[1]Sustainable Engineering Group, School of Civil and Mechanical Engineering, Curtin University, Bentley, Perth, Australia. [2]Otraco International Pty Ltd, Technology Park, Bentley, Perth, Australia. [3]Recom Engineering, Carbon Court, Osborne Park, Perth, Australia. [4]Department of Mechanical Engineering, Curtin University, Bentley, Perth, Australia.

References
1. Guide, VDR, Srivastava, R: Recoverable manufacturing systems: a framework for analysis. In: Innovation in Technology Management - The Key to Global Leadership, pp. 675–678. Portland International Conference on Management and Technology, Portland (1997)
2. ACDelco: Remanufactured engines & cylinder heads. Australian Catalogue Issue 3. http://www.acdelco.com.au/PDFs/Catalogue_ACDelco_RemanEngines.pdf (2006). Accessed 30 September 2011
3. Kerr, W, Ryan, C: Eco-efficiency gains from remanufacturing a case study of photocopier remanufacturing at Fuji Xerox. Australia. J. Clean. Prod. **9**, 75–81 (2001)
4. Seliger, G, Kernbaum, S, Zettl, M.: Remanufacturing approaches contributing to sustainable engineering. Gestão & Produção **13**(3), 367–384 (2006)
5. Skerlos, SJ, Morrow, WR, Chan, K, Zhao, F, Hula, A, Seliger, G, Basdere, B, Prasitnarit, A: Economic and environmental characteristics of global cellular telephone remanufacturing. In: Proceedings of the IEEE International Symposium on Electronics and the Environment, pp. 99–104. Boston (2003)
6. Kumar, V, Sutherland, JW: Sustainability of the automotive recycling infrastructure: review of current research and identification of future challenges. Int. J. Sustainable Manufacturing **1**(1/2), 145–167 (2008)
7. Kondo, Y, Nakamura, S: Evaluating alternative life-cycle strategies for electrical appliances by the waste input–output model. Int. J. Life Cycle Assessment **9**(4), 236–246 (2004)
8. Shi-can, L, Pei-jing, S: Benefit analysis and contribution prediction of engine remanufacturing to cycle economy. J. Central South University Tech. **12**(2), 25–29 (2004)
9. Smith, VM, Keoleian, GA: The value of remanufactured engines life-cycle environmental and economic perspectives. J. Indus Ecology **8**(1–2), 193–221 (2004)
10. Kara, H: Carbon Impacts of Remanufactured Products, Gear Box (6 Speed Automatic). Centre for Remanufacturing and Reuse, Aylesbury (2009)
11. Biswas, WK, Rosano, M: A life cycle greenhouse gas assessment of remanufactured refrigeration and air conditioning compressors. Int. J. Sustainable Manufacturing **2**(2–3), 222–236 (2011)
12. ISO (International Standard Organization): Environmental management – life cycle assessment – principles and framework, ISO 14040. International Organization for Standardization (ISO), Geneva (1997)
13. PRé Consultants: Simapro Version 7.2. PRé Consultants, Amersfoort (2010)
14. RMIT (Royal Melbourne Institute of Technology): Australian LCA database. Centre for Design, Royal Melbourne Institute of Technology, Melbourne (2005)
15. SCLCI: The ecoinvent database. Swiss Federal Laboratories for Materials Testing and Research, Switch Centre for Life Cycle Inventory, Zurich (2007)

Permissions

List of Contributors

Pedro Piñeyro
Departamento de Investigación Operativa, Instituto de Computación, Facultad de Ingeniería, Universidad de la República, Julio Herrera y Reissig 565, Montevideo, CP 11300, Uruguay

Omar Viera
Departamento de Investigación Operativa, Instituto de Computación, Facultad de Ingeniería, Universidad de la República, Julio Herrera y Reissig 565, Montevideo, CP 11300, Uruguay

Erwin M. Schau
Department of Environmental Technology, Chair of Sustainable Engineering, Technische Universitaet Berlin, Office Z1, Strasse des 17. Juni 135, BerlinD-10623, Germany

Marzia Traverso
Department of Environmental Technology, Chair of Sustainable Engineering, Technische Universitaet Berlin, Office Z1, Strasse des 17. Juni 135, BerlinD-10623, Germany

Matthias Finkbeiner
Department of Environmental Technology, Chair of Sustainable Engineering, Technische Universitaet Berlin, Office Z1, Strasse des 17. Juni 135, BerlinD-10623, Germany

Wahidul K Biswas
Sustainable Engineering Group, School of Civil and Mechanical Engineering, Curtin University, Bentley, Perth, Australia

Victor Duong
Otraco International Pty Ltd, Technology Park, Bentley, Perth, Australia

Peter Frey
Recom Engineering, Carbon Court, Osborne Park, Perth, Australia

Mohammad Nazrul Islam
Department of Mechanical Engineering, Curtin University, Bentley, Perth, Australia

Farazee MA Asif
Department of Production Engineering, KTH Royal Institute of Technology, Stockholm, Sweden

Carmine Bianchi
Department of Political Sciences, University of Palermo, Palermo, Italy

Amir Rashid
Department of Production Engineering, KTH Royal Institute of Technology, Stockholm, Sweden

Cornel Mihai Nicolescu
Department of Production Engineering, KTH Royal Institute of Technology, Stockholm, Sweden

Michael Mutingi
Mechanical Engineering Department, University of Botswana, P/Bag 0061 UB Post, Gaborone, Botswana
Faculty of Engineering and the Built Environment, University of Johannesburg, Johannesburg, South Africa

Herbert Mapfaira
Mechanical Engineering Department, University of Botswana, P/Bag 0061 UB Post, Gaborone, Botswana

Robert Monageng
Mechanical Engineering Department, University of Botswana, P/Bag 0061 UB Post, Gaborone, Botswana

Akram El Korchi
ENSA AGADIR, Université Ibn ZOHR, Agadir 80000, Morocco

Dominique Millet
SUPMECA, Toulon 83957, France

Mark Errington
College of Engineering, Mathematics and Physical Sciences, University of Exeter, EX4 4QF, Exeter, UK

Stephen J Childe
College of Engineering, Mathematics and Physical Sciences, University of Exeter, EX4 4QF, Exeter, UK

Dominic Ansbro
School of Engineering and Computing Sciences, Durham University, Durham DH1 3LE, UK

Qing Wang
School of Engineering and Computing Sciences, Durham University, Durham DH1 3LE, UK

Xavier Liang
Department of Mechanical Engineering, University of Michigan,1031 H.H. Dow, 2300 Hayward Street, Ann Arbor, MI 48109, USA

Xiaoning Jin
Department of Mechanical Engineering, University of Michigan,1031 H.H. Dow, 2300 Hayward Street, Ann Arbor, MI 48109, USA

Jun Ni
Department of Mechanical Engineering, University of Michigan,1031 H.H. Dow, 2300 Hayward Street, Ann Arbor, MI 48109, USA

Yuchun Xu
School of Applied Sciences, Cranfield University, Cranfield, Bedford MK43 0AL, UK

Wei Feng
School of Applied Sciences, Cranfield University, Cranfield, Bedford MK43 0AL, UK

Gillian D Hatcher
Design Manufacture and Engineering Management, 4th Floor Architecture Building, University of Strathclyde, Glasgow G4 0NG, UK

Winifred L Ijomah
Design Manufacture and Engineering Management, 4th Floor Architecture Building, University of Strathclyde, Glasgow G4 0NG, UK

James F C Windmill
Electronic and Electrical Engineering, R3.40a, Royal College Building, University of Strathclyde, Glasgow G1 1XQ, UK

Wei-Fen Hsieh
Computer Science and Information Engineering, National Chi Nan University, Puli, Taiwan

Lieu-Hen Chen
Computer Science and Information Engineering, National Chi Nan University, Puli, Taiwan

Hao-Ming Hung
Computer Science and Information Engineering, National Chi Nan University, Puli, Taiwan

Eri Sato-Shimokawara
Faculty of System Design Tokyo Metropolitan University, Hino, Tokyo, Japan

Yasufumi Takama
Faculty of System Design Tokyo Metropolitan University, Hino, Tokyo, Japan

Toru Yamaguchi
Faculty of System Design Tokyo Metropolitan University, Hino, Tokyo, Japan

Eric Hsiao-Kuang Wu
Computer Science and information Engineering, National Central University, Taoyuan, Taiwan

Yu-Wei Chen
Department of Internal Medicine, Landseed Hospital, Taoyuan, Taiwan

Mitsutaka Matsumoto
Center for Service Research, National Institute of Advanced Industrial Science and Technology (AIST), 1-1-1, Umezono, Tsukuba, Ibaraki, 305-8568, Japan

Akira Ikeda
Shin-Etsu Denso Co., Ltd, 2656-210, Taira, Omachi, Nagano 398-0001, Japan

www.ingramcontent.com/pod-product-compliance
Lightning Source LLC
Chambersburg PA
CBHW050453200326
41458CB00014B/5165